P9-DXD-499

THE WORLD IN 2050

THE WORLD IN 2050

*Four Forces Shaping Civilization's
Northern Future*

LAURENCE C. SMITH

DUTTON

DUTTON

Published by Penguin Group (USA) Inc.

375 Hudson Street, New York, New York 10014, U.S.A.

Penguin Group (Canada), 90 Eglinton Avenue East, Suite 700, Toronto, Ontario M4P 2Y3, Canada (a division of Pearson Penguin Canada Inc.); Penguin Books Ltd, 80 Strand, London WC2R 0RL, England; Penguin Ireland, 25 St Stephen's Green, Dublin 2, Ireland (a division of Penguin Books Ltd); Penguin Group (Australia), 250 Camberwell Road, Camberwell, Victoria 3124, Australia (a division of Pearson Australia Group Pty Ltd); Penguin Books India Pvt Ltd, 11 Community Centre, Panchsheel Park, New Delhi–110 017, India; Penguin Group (NZ), 67 Apollo Drive, Rosedale, North Shore 0632, New Zealand (a division of Pearson New Zealand Ltd); Penguin Books (South Africa) (Pty) Ltd, 24 Sturdee Avenue, Rosebank, Johannesburg 2196, South Africa

Penguin Books Ltd, Registered Offices: 80 Strand, London WC2R 0RL, England

Published by Dutton, a member of Penguin Group (USA) Inc.

First printing, September 2010

10 9 8 7 6 5 4 3 2 1

Library of Congress Cataloging-in-Publication Data has been applied for.

ISBN 978-0-525-95181-0

Printed in the United States of America

While the author has made every effort to provide accurate telephone numbers and Internet addresses at the time of publication, neither the publisher nor the author assumes any responsibility for errors, or for changes that occur after publication. Further, the publisher does not have any control over and does not assume any responsibility for author or third-party Web sites or their content.

For my brilliant, beautiful Abbie,
who is a part of this story

CONTENTS

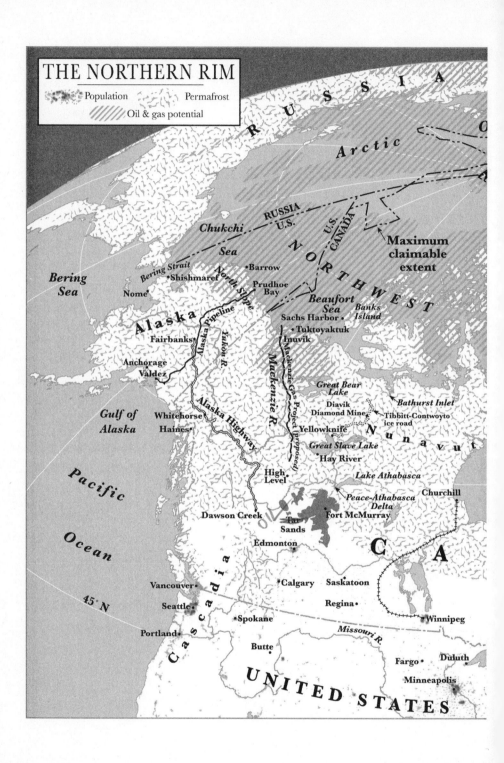

THE NORTHERN RIM

Population   Permafrost
Oil & gas potential

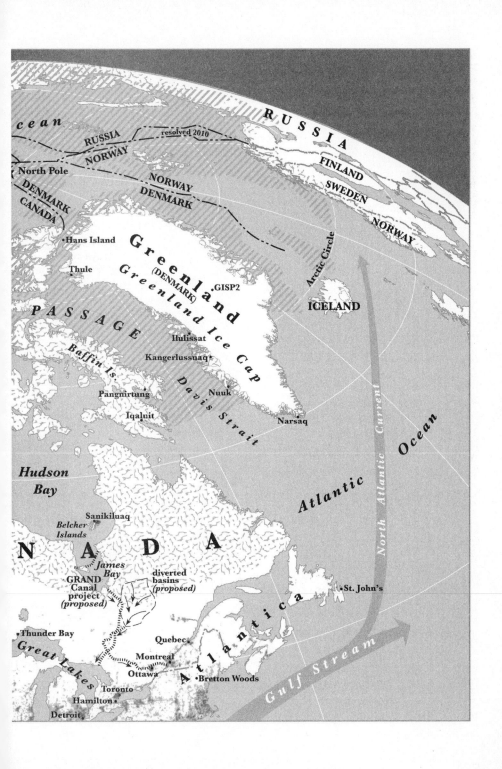

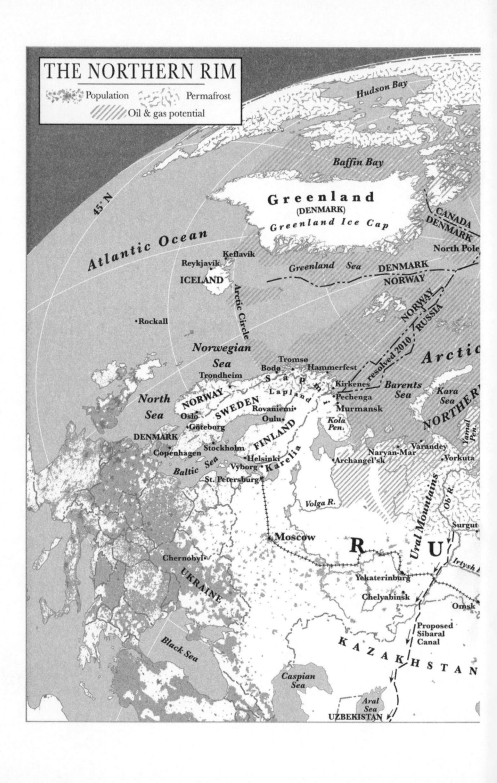

THE NORTHERN RIM

Population   Permafrost
Oil & gas potential

Hudson Bay

Baffin Bay

Greenland
(DENMARK)
Greenland Ice Cap

CANADA
DENMARK

North Pole

45° N

Atlantic Ocean

Keflavík

Reykjavík

ICELAND

Greenland Sea   DENMARK
NORWAY

NORWAY
RUSSIA
resolved 2010

Arctic

•Rockall

Arctic Circle

Norwegian
Sea

Tromsø

Bodø   Hammerfest

Trondheim   Sápmi

Kirkenes   Barents
Sea

Kara
Sea

Lapland   •Pechenga

North
Sea

NORWAY

SWEDEN

Rovaniemi•

Oulu•

Murmansk

Kola
Pen.

NORTHERN

Yamal
Pen.

Oslo

•Göteborg

FINLAND

Naryan-Mar   •Varandey

DENMARK

Stockholm

Karelia

Archangel'sk   •Vorkuta

Copenhagen

Baltic Sea

•Helsinki

Vyborg•

St. Petersburg•

Ob R.

Volga R.

Ural Mountains

Surgut

Moscow   R   U

Chernobyl•

Yekaterinburg

Irtysh R.

UKRAINE

Chelyabinsk

Omsk

Proposed
Sibaral
Canal

KAZAKHSTAN

Black Sea

Caspian
Sea

Aral
Sea

UZBEKISTAN

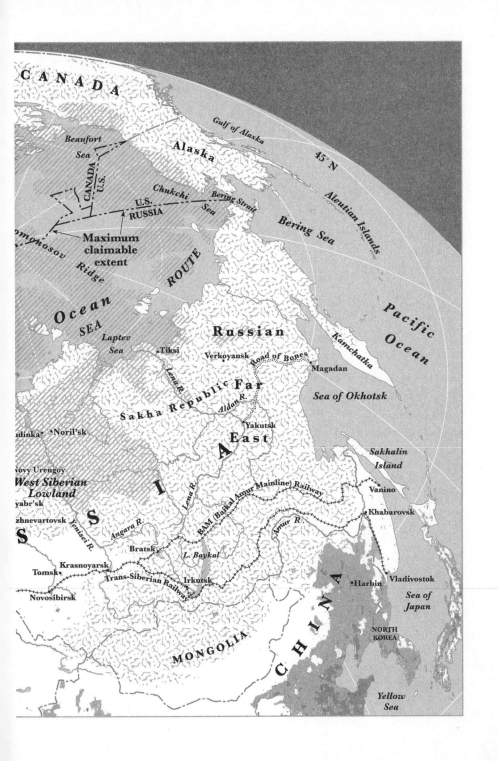

Flying into Fort McMurray

My nose was pressed against the rear window glass of a Boeing 747. It was a direct flight from Edmonton to the booming new oil city of Fort McMurray, Alberta, in the broad belt of boreal forest that girdles the globe through Alaska, Canada, Scandinavia, and Russia. The scene below morphed from urban concrete to canary-yellow canola fields, then gradually from fields to a deep shag carpet of evergreen forest jeweled with bogs. The forest was crisscrossed here and there by roads, and patched with clearings, but grew more desolate by the minute. In under an hour, the transformation from urban metropolis to farmland to wilderness was complete.

Then, suddenly, the woods dissolved into gleaming homes, the newest residential subdivision of Fort McMurray. Freshly cut survey lines radiated outward in all directions through the woods. Bulldozers and work crews ground away at roadbeds and building pads, engraving a sort of master blueprint into the landscape for hundreds more homes in waiting. Small wonder. The median price of a home in Fort McMurray had just surged to $442,000, more than $100,000 higher than in my home city of Los Angeles.[1] The aggressive transformation taking place beneath my window was just one of many I was about to see over the next fifteen months.

This was not my first trip to the North. I'd already been studying cold, remote places for fourteen years, beginning with a doctoral dissertation studying the Iskut River, a tree-tossing torrent that rips through a remote corner of British Columbia. Something about the rawness of the place, the sense of danger and frontier, hooked me hard. The sight of fresh grizzly-bear footprints, smashed just minutes before over my own, was a shivery thrill.

I finished school, became a geography professor at UCLA, and started a long series of research projects in Alaska, Canada, Iceland, and Russia.

My specialty was the geophysical impacts of climate change. In the field I would measure stream flows, survey glacier snouts, sample soil, and the like. Back home in Los Angeles I would continue the research from my desk, extracting numbers from satellite data like little digital polyps. But all this would change in 2006. The flight to Fort McMurray was the beginning of my attempt to gain deeper understanding of other phenomena now unfolding around the northern quarter of our planet, and how they fit in with even bigger global forces reverberating throughout the world as a whole.

From my scientific research, I knew that amplified climate warming had begun in the North, but what might that mean for the region's people and ecosystems? What about its ongoing political and demographic trends, or the vast fossil fuel deposits thought to exist beneath its ocean floors? How would it be transformed by even bigger pressures building around the world? And if, as many climate models suggest, our planet becomes one of killer heat waves, fickle rain, and baked croplands, might new human societies emerge in places currently unappealing for settlement? Could the twenty-first century see the decline of the southwestern United States and European Mediterranean, but the ascent of the northern United States, Canada, Scandinavia, and Russia? The more I looked, the more it seemed this northern geographic region was highly relevant to the future of us all.

I was about to burn through almost two years of my life going to places you've heard of, like Toronto, Helsinki, and Cedar Rapids, and others you maybe haven't, like High Level, Tromsø, and the Belcher Islands. I was about to fly on helicopters and airplanes, rent cars, ride buses and trains, and live on a ship. My goal was to see with my own eyes what is happening with these places, and to ask the scientists, business owners, politicians, and ordinary residents who live and work in them what they saw happening and where they thought things might be heading. After studying it for years, I was about to discover the North—and its broader importance to our future—for the very first time.

CHAPTER 1

Martell's Hairy Prize

"Prediction is very difficult. Especially about the future."
—Niels Bohr (1885–1962)

"The future is here. It's just not evenly distributed yet."
—William Gibson (1948–)

On a cold April day in 2006, Jim Martell, a sixty-five-year-old businessman from Glenns Ferry, Idaho, shot and killed a strange beast. Cradling his rifle, he ran with his guide, Roger Kuptana, to where it lay slumped on the snow. Both men wore thick parkas as protection against the icy wind. They were on Banks Island, high in the Canadian Arctic, some 2,500 miles north of the U.S. border.

Martell was an avid big-game hunter and had paid about forty-five thousand dollars for the right to bag *Ursus maritimus,* a polar bear, one of the most coveted trophies of his sport. Kuptana was an Inuit tracker and guide who lived in the nearby village of Sachs Harbor. Polar bear hunting is strictly regulated but legal in Canada, and the hefty license and guide fees provide important revenue to Sachs Harbor and other Inuit towns like it. Martell had permission to take down a polar bear and only a polar bear. But that was not what lay bleeding in the snow.

At first glance the creature looked well enough like a polar bear, albeit a small one. It was seven feet long and covered with creamy white fur. However, its back, paws, and nose were mottled with patches of brown. There were dark rings, like a panda's, around its eyes. The creature's face was flattened, with a humped back and long claws. In fact, it had many of the features of *Ursus arctos horribilis,* the North American grizzly bear.

Martell's bear triggered an international sensation. Canadian wildlife officials seized the body and submitted DNA samples to a genetics lab to find out what it was. Tests confirmed the animal was indeed a half-breed, the product of a grizzly bear father and a polar bear mother.[2] It was the first evidence of grizzly/polar bear copulation ever reported in the wild. News outlets announced the arrival of a "Hairy Hybrid"[3] and the blogosphere erupted with either wonderment and proposed names—"pizzly?" "grizzlar?" "grolar bear?"—or outrage that the only known specimen had been shot dead. A "Save the Pizzly" Web site hawked T-shirts, coffee mugs, and stuffed dolls. Martell was subjected to angry criticism; in response he pointed out that the world would never have learned of the thing—whatever it was called—if not for his fine shot.

For this bizarre tryst even to have happened required that a grizzly wander far north into polar bear territory, a formerly rare phenomenon that biologists are now beginning to see more often. Journalists were quick to make a climate change connection: Was this, they wondered, a preview of nature's response to global warming? But scientists like Ian Stirling, a leading polar bear biologist, were rightly skeptical of drawing strong conclusions from what was, after all, an isolated event. That changed in 2010, when a second specimen was shot. Tests confirmed it was the offspring of a hybrid mother; in other words, *they are breeding.*[4] The coming decades will show whether Martell's bear, now stuffed and snarling in his living room, is just the latest biological indicator among many that something big is going down on our planet.

If you enjoy watching wildlife in your backyard, perhaps you've noticed something. All around the world there are animals, plants, fish, and insects creeping to higher latitudes and elevations. From spittlebugs in California to butterflies in Spain and trees in New Zealand, it is a broad pattern that biologists are discovering. By 2003 a global inventory of this phenomenon found that on average, plants and animals are shifting their ranges about six kilometers toward the poles, and six meters higher in elevation, every decade. Over the past thirty years phenological cycles—the annual rhythm of plant flowering, bird migrations, birthing babies, and so on—have shifted earlier in the spring by more than four days per decade.[5]

If these numbers don't sound large to you, they should. Imagine your lawn crawling north, away from your house, at a speed of five and one-half feet each day. Or that your birthday arrived ten hours sooner each year. That's how fast these biological shifts are happening. Life-forms are migrating—and it's going on right outside your window.

The 2006 pizzly story—like the record-shattering Atlantic hurricane season in 2005, or the strange weather patterns that rained out the Winter Olympic Games in Vancouver while burying the eastern U.S. seaboard with "Snow-pocalypse" in 2010[6]—is yet another example of something that might have been triggered by climate change or might not. Such events are eye-catching in the news of the day but not, taken in isolation, conclusive of anything. In contrast, while the painstaking statistical analyses of decades of field research on spittlebugs and trees may not rouse the daily news cycle, it does me. It is deeply important, a compelling discovery that provides real understanding about the future. It is a megatrend, and megatrends are what this book is all about.

The Thought Experiment

This is a book about our future. Climate change is but one component of it. We will explore other big trends as well, in things like human population, economic integration, and international law. We will study geography and history to show how their preexisting conditions will leave lasting marks on the future. We will look to sophisticated new computer models to project the futures of gross domestic product, greenhouse gases, and supplies of natural resources. By examining these trends collectively, and identifying convergences and parallels between them, it becomes possible to imagine, with reasonable scientific credibility, what our world might look like in forty years' time, should things continue on as they are now. This is a thought experiment about our world in 2050.

It can be fun imagining what our world might look like by then. Robots and flying cars? Custom-grown body parts? A hydrogen economy? As any disappointed sci-fi buff will tell you, the pace of reality is usually slower than human imagination. Fans of George Orwell's book *1984,* the television series *Lost in Space* and *Space 1999,* the films *2001: A Space Odyssey,* and (it's looking like) *Blade Runner*—set in a perpetually raining 2019 Los

Angeles—see their landmark years come and go. But outside of the ongoing technical explosions in information and biotechnologies, our lives are considerably less different than the writers of these fictional works imagined they would be.

We've discovered quarks and flung people into space, yet still depend on the internal combustion engine. We've cracked DNA and grown a human ear on a mouse's back, yet are still dying of cancer. We've created fluorescent green pigs by inserting jellyfish genes into them (*Green Eggs and Ham,* anyone?), yet still catch wild fish from the sea and use dirt and water to grow our food. Nuclear power is but a pale shadow of what was hoped for it in the 1950s. We still use boats, trucks, and trains to move goods. And even in this unprecedented era of globalization, the fundamental principles of our markets and economies differ surprisingly little from the days of Adam Smith, more than two hundred years ago.

But in other, sneakier ways, things have changed profoundly. Imagine describing to a 1950 California tomato farmer how in the next fifty years he would grow genetically programmed seeds, see the water in his state tilted from one end to the other, and experience a tripling of the state's population. Imagine explaining he would one day compete with Chinese farmers to sell tomatoes to Italians who would blend them with beans from Mexico to make canned goods for British supermarkets.[7]

Any of these would blow our yesteryear farmer's mind. But to us, they are familiar, even boring. They fly below our radar because they have crept upon us, hiding in plain sight over the course of decades. But that doesn't mean transformations like these aren't huge, fast, and profound. Big changes often just sort of ease their way in. And quietly change the world.

What will our world look like in 2050? Our distribution of people and power? The state of the natural world? Which countries will lead, and which ones suffer? Where do you think *you'll* be in 2050?

The answers to these questions, at least in this book, derive from a core argument: The northern quarter of our planet's latitudes will undergo tremendous transformation over the course of this century, making them a place of increased human activity, higher strategic value, and greater economic importance than today. I loosely define this "New North" as all land and oceans lying 45° N latitude or higher currently held by the United

States, Canada, Iceland, Greenland (Denmark), Norway, Sweden, Finland, and Russia.

These eight countries, which control vast territories and seas extending as far north as the Arctic Ocean, comprise a new "Northern Rim" roughly encircling that ocean. Developments in these Northern Rim countries, here coined the NORC countries, or "NORCs," are explored in Parts II and III (Chapters 5 through 10). Part I (Chapters 2 through 4) presents powerful worldwide trends in human population, economics, energy and resource demand, climate change, and other factors keenly important to our global civilization and ecosystem. Besides imagining what life might be like for most of us by 2050, these first chapters identify some critical world pressures that are stimulating the New North into existence.

Before beginning our travels around this 2050 world, there are some rules to establish.

The Rules

Fortunately, we have the tools, the models, and the knowledge to construct an informed thought experiment of what we might expect to see unfold over the next forty years. However, as in any experiment, we must first define the assumptions and ground rules upon which its outcomes are contingent.

1. **No Silver Bullets.** Incremental advances in technology for the next forty years are assumed. No cold nuclear fusion or diesel-growing fungus[8] will suddenly solve all our energy problems; no God-like genetic engineering to grow wheat without water. This is not to say a radical technology breakthrough can't or won't happen; only that the possibility will not be entertained here.

2. **No World War III.** The two "world" wars in the first half of the twentieth century recast the map and forged economic, political, and infrastructural changes that reverberate to this day. A nuclear or major, multicountry, conventional war like World War II would be a game changer and is not imagined here (indeed, empirical evidence suggests that over the long run we may be becoming somewhat less violent[9]). However, the possibility of lesser armed conflicts, like the

ones ongoing today in the Middle East and Africa, is evaluated. Major laws and treaties, once made, are assumed to stay in place.

3. **No Hidden Genies.** A decades-long global depression, an unstoppable killer disease pandemic, a meteorite impact, or other low-probability, high-impact event is not imagined here. However, this rule is relaxed in Chapter 9 to explore six plausible, if unlikely, outcomes, like an abrupt climate change or the collapse of global trade—both of which have happened before and could happen again.

4. **The Models Are Good Enough.** Some of the conclusions reached in this book stem from experiments using computer models of complex phenomena, like climate and economies. Models are tools, not oracles. All have their flaws and limitations.[10] But for the broad-scale purposes of this book, they are excellent. I will focus on the robust, uncontroversial messages of these models rather than push the limits of their capabilities. As before, this rule is relaxed in Chapter 9 to explore some plausible outcomes lying outside our current modeling capacity.

The purpose of these rules is to introduce conservatism to the thought experiment. By favoring likely, forseeable trajectories over unlikely, exciting ones, we avoid sacrificing a more probable outcome to a good story. By pursuing multiple lines of argument rather than one grand idea, we avoid the so-called "foxes and hedgehogs" trap, by lessening the likelihood that an important actor will be overlooked.[11] By concentrating on the most robust simulations of computer models, we steer the conversation toward the science that is best understood, rather than poorly understood.

Why even try to project forty years into the future anyway? To imagine the world in 2050, we must closely study what is happening today, and why. By forcing our minds to take the long view, we can identify factors that might seem beneficial in the near term, but lead to undesired consequences in the long term, and vice versa. After all, doing good things (or at least, less bad things) for the long term is a worthy goal. I certainly don't believe the future is predetermined: Much of what does or does not happen forty years from now rests on actions or inactions taken between now and then.

Some of the changes I will present will be perceived as good or bad, depending on the reader's own perspective. To be sure, some of them, like species extinctions, no one wishes to see. But others, like military spending and energy development, evoke valid, strongly opposed reactions. My goal is not to argue one side or another, but to pull together trends and evidence into a bigger picture as well and objectively as I can. The reader can take it from there.

But before we can intelligently discuss the future, we must first understand the past. In roughly historical order of their rise in significance, here are four global forces that have been busily shaping our 2050 world for tens to hundreds of years.

FOUR GLOBAL FORCES

The first global force is demography, which essentially means the ups, downs, and movements of different population groups within the human race. Demographic measures include things like birth rates, income, age structure, ethnicity, and migration flows. We shall examine all of these in due course but for now, let us start with the most basic yet profound measure of all: the total number of people living on Earth.

Before the invention of agriculture some twelve thousand years ago, there were perhaps one million persons in the world.[12] That is roughly the present-day population of San Jose, California. People foraged and hunted the land, living in small mobile clans. It took *twelve thousand* years (until about 1800 A.D.) for our numbers to grow to one billion. But then, oh boy, liftoff.

Our second billion arrived in 1930, a mere 130 years later. The global Great Depression was under way. Adolf Hitler led his Nazi Party to stunning victory in Germany's Reichstag elections. My Italian immigrant grandfather, then living in Philadelphia, was thirty-three years old.

Our third billion came just thirty years later in 1960. John Kennedy beat Richard Nixon in the U.S. presidential race, the first satellites were orbiting the Earth, and I was a scant seven years from being born.

Our fourth billion took just fifteen more years. It was 1975 and I was eight. The U.S. president Gerald Ford escaped two assassination attempts (one by Charles Manson's murderous henchwoman Lynette "Squeaky" Fromme), the Khmer Rouge had taken over Cambodia, and the movie *Godfather II* ran away with six Academy Awards, including one to the Italian-American actor Robert De Niro.

Our fifth billion came in 1987, now just twelve years after the fourth. The Dow Jones Industrial Average closed above 2,000 for the first time in history and the Irish rock band U2 released their fifth album, *The Joshua Tree*. Standing outside Berlin's Brandenburg Gate, U.S. president Ronald Reagan exhorted Soviet leader Mikhail Gorbachev to "tear down this wall." The world's last dusky seaside sparrow died of old age on a tiny island preserve in Florida's Walt Disney World Resort. A self-absorbed college sophomore at the time, I only noticed *The Joshua Tree*.

Our sixth billion arrived in 1999. This is now very recent history. The United Nations declared 1999 the International Year of Older Persons. The Dow Jones climbed above 11,000 for the first time in history. Internet hookups ballooned and millions of songs, to the dismay of U2 and the rest of the music industry, were swapped for free on Napster. Hugo Chávez became president of Venezuela, and a huge chunk of northern Canada quietly assumed self-rule as the new territory of Nunavut. By then, I was a young professor at UCLA, working toward tenure and starting to notice things. The world vacillated between nervous fretting about Y2K and excitement over the dawn of a new millennium.

11,800 years . . . 130 years . . . 30 years . . . . 15 years . . . 12 years. . . . The length of time we need to add another billion has petered down to nearly nothing. One billion is more than triple the 2010 population of the United States, the third most populous country on Earth. Imagine a world in which we added one-plus USA, or two Pakistans, or three Mexicos, *every four years*. . . . Actually, this requires no imagination at all. It is reality. We will add our seventh billion some time in 2011.

This extraordinary acceleration, foreseen over two centuries ago by Thomas Malthus,[13] burst into popular culture again in 1968 when Paul Ehrlich, then a young biology professor at Stanford, jolted the world with *The Population Bomb,* a terrifying book forecasting global famines, "smog deaths," and massive human die-offs if we didn't somehow control our numbers.[14] He became a frequent guest on *The Tonight Show Starring Johnny Carson* and his ideas almost certainly helped nudge China toward its "One-Child" population control policy implemented in 1979.

Arguments against Ehrlich's ecological approach to human beings charged that it underestimated the limits of our technology and ingenuity. So far, these arguments appear to have been correct. Our numbers have surged on and Ehrlich's scariest predictions have, as yet, failed to materialize. But even so, generations from now, our descendants will marvel at the twentieth century, a time when our numbers shot from 1.6 to 6.1 billion in a mere blink of time.

What triggered this enormous twentieth-century population spurt? Why did it not happen before, and is it likely to continue into the future?

Fast population growth behaves a lot like a personal savings account. Just as its account balance depends on the spread between the rates of deposit versus spending, the balance of people on Earth depends on the rates at which new people are created (the fertility rate) versus how fast existing people disappear (the death rate).[15] When the two rates are equal, population holds steady. When they diverge or converge, population rises or falls accordingly. It doesn't really matter whether birth rates rise or death rates fall; what matters is the spread and whether rate adjustments are staggered in time or happen simultaneously. Most importantly, once a run-up (or decline) has happened, we are stuck with the new population level, even if the gap between fertility and death rates is then closed and population stability is returned.

From our earliest beginnings until the late nineteenth century, our fertility and death rates, on average, were both high. Mothers had more babies than today, but few of them survived to old age. In the preindustrial era, famine, warfare, and poor health kept death rates high, largely offsetting high fertility. The global population of humans trickled higher, but only very slowly.

However, by the late nineteenth century, industrialization had changed everything in Western Europe, North America, and Japan. Mechanized food production and distribution reduced famine deaths. Local warfare disappeared under the rising control of central governments. Death rates dropped as doctors discovered modern medical procedures and drugs. But fertility rates fell more slowly—cultural expectations are slower to change—so the human population took off. By 1950, New York was the first city in the world to break the ten million mark.

Not only did the Industrial Age bring machines and medicine, it also spurred migration from farms to cities. People increasingly bought what they needed rather than growing or making things themselves. The cost of housing rose; the economy grew. More women entered college and the workplace, squeezing down the number of children families wanted or could afford. Fertility rates began to drop and families became smaller. When fertility rates at last fell to match the death rates, population growth halted, and the industrialized societies that had participated in all this were transformed. Instead of being small, poor, prolific, and death-prone they were now large, rich, and long-lived with few children.

This chain of events, in which a population run-up is at first initiated, then later stabilized, by the forces of modernization is called the Demographic Transition and is a bedrock concept in demography.[16] The Demographic Transition supposes that modernization tends to reduce death and fertility rates, but not simultaneously. Because people tend to readily adopt technological advances in medicine and food production, death rates fall first and quickly. But fertility reductions—which tend to be driven by increased education and empowerment of women, an urban lifestyle, access to contraception, downsized family expectations, and other cultural changes—take more time. And just like a bank account, when the death (spending) rate falls faster than the birth (savings) rate, the result is a rapid run-up in the sum total. Even if fertility rates later fall to match death rates—thus completing the Demographic Transition and halting further growth—a new, much larger population balance is then carried forward.

In the twentieth century, one Demographic Transition concluded and another began. In Europe and North America it took from about 1750 to 1950 to complete, making these places the fastest-growing in the world while most of Asia and Africa grew slowly. This growth then slowed or stopped as industrialized countries completed the Demographic Transition, their fertility rates falling to near or even below the death rate.

But in the developing world, a new Demographic Transition that began in the early twentieth century with the arrival of modern medicine has still not finished. Thanks to the inventions of antibiotics and vaccines, along with insecticides to control diseases like malaria, death rates have plummeted[17] but fertility rates, while dropping, have fallen less quickly. In some countries they haven't fallen at all, defying the classic Demographic Transition notion that all modernized women prefer fewer babies. Such discrepancies underline a known weakness of the Demographic Transition model: Not every culture will necessarily adopt the western ideal of a small nuclear family, even after women's rights, health, and security conditions improve.

So somewhere around 1950, our fastest population growth rates left the OECD countries[18] and went to the developing world. Because the base population levels in the latter are so much larger, the resulting surge in world population has been nothing short of phenomenal. In most developing countries the spread between fertility and death rates, while narrowing, remains substantial. This second Demographic Transition is not yet finished, and unlike before, it involves the vast majority of the human race. Until a few decades after it ends—if it ends—world population will continue to grow.

The second global force, only partly related to the first, is the growing demand that human desires place upon the natural resources, services, and gene pool of our planet. *Natural resources* means both finite assets like hydrocarbons, minerals, and fossil groundwater; and renewable assets like rivers, arable land, wildlife, and wood. *Natural services* include life essentials like photosynthesis, absorption of carbon dioxide by oceans, and the labors of bees to pollinate our crops. And by *gene pool* I mean exactly that—the diversity of genes being carried around by all living organisms still existing on Earth.

It's difficult to comprehend how fully dependent we are upon these things. Steel machines burn oil to grow and harvest our grains, with fertilizers made from natural gas, generating many times over what a farmer and mules could produce on the same land. From the genetic code of organisms we take the building blocks for our food, biotech, and pharmaceutical industries. We frame our buildings with timber, steel, and cement. We take water from the ground or trap it behind dams to grow alfalfa and cotton in the desert. We need trucks and diesel and giant metal-hulled ships to move ores and fish and manufactured goods from the places that have them to places that want them. The resulting trade flows have grown entire economies and glittering cities, with their music and culture and technology. Coal-fired electricity zaps through billions of miles of metal cable to power buildings, electric cars, cell phones, and the Internet. Airplanes and cars burn the sludge of long-dead things, granting us personal freedom and the chance to see the world.

It's no secret that our twentieth-century expansions in population, modernization, trade, and technology have escalated demand for all of these. Public concern—both for the stability of raw commodity supplies and for the health of the natural world—has been high since the 1970s, especially after the OPEC oil embargo crisis of '73–'74 and NASA's launch of ERTS-1 (later renamed Landsat), the first civilian satellite to disseminate graphic images of clear-cuts gnawing away the vast rain forests of the Amazon basin. Today, news feeds crackle with stories about dwindling oil, fights over water, and soaring food prices. Many plants and animals are disappearing as their habitats are converted to plantations and parking lots. Still others have been harvested into oblivion. Fully *four-fifths* of the world's land surface (excluding Antarctica) is now directly influenced by human activities.[19] The lingering exceptions to this are those places that are truly remote: the northern forests and tundra, the shrinking rain-forest cores of the Congo and Amazon basins, and certain deserts of Africa and Australia and Tibet.

Perhaps no resource pressure has grown faster than our demand for fossil hydrocarbon fuels. This began in Europe, North America, Australia, and Japan and has now spread to China, India, and other modernizing nations. Because the United States has been (and still is) the largest consumer of these fuels, let's illustrate the rapacity of this phenomenon as it has unfolded there.

In 1776, when the United States of America declared independence from Great Britain after a little over a year of war, most of the fledgling country's energy came from wood and muscles. Yes, there were sawmills turning waterwheels to cut logs, and coal was used to make coke for casting iron cannons and tools, but the vast majority of America's energy came from fuelwood, horses, mules, oxen, and human backs.

By the late 1800s, the Industrial Revolution, steam locomotive, and westward expansion had changed all that. Dirty black coal was the shining new prince—fueling factories, coke ovens, foundries, and trains all across the young nation. Coal consumption grew from 10 million short tons per year in 1850, to 330 million short tons just fifty years later.[20] Little mining towns sprang up all over Appalachia, like now-defunct Ramseytown in western Pennsylvania, where my grandmother was later born. Nearby Rossiter produced my grandfather, who worked in the coal mines as a teenager.

But in the twentieth century, coal was surpassed. Oil, first drilled out of a quiet Pennsylvania farm in 1859 to make lamp kerosene, caught on slowly at first. Gasoline was originally a junk by-product that some people dumped into rivers to get rid of. But then someone thought of pouring it into a combustion engine, and gasoline became the fuel of Hercules.

Packed inside a single barrel of oil is about the same amount of energy as would be produced from eight years of day labor by an average-sized man. Seizing oil fields became a prime strategic objective in both world wars. The Baku fields of Azerbaijan were a prime reason that Hitler invaded Russia, and it was their oil supply, gushing north to the Russian army, that stopped him.

By the end of World War II, cars and trucks had outgrown the rail system, locomotives had switched to diesel, and the liquid-fuels market was really taking off. Oil consumption surpassed coal in 1951, though sales of both—along with natural gas—continued to rise strongly. In just one hundred years (1900–2000) Americans ramped up their coal consumption from about 330 million to 1.1 billion short tons per year,[21, 22] a 230% increase. Oil-burning grew from 39 million to 6.6 *billion* barrels per year,[23] a 16,700% increase. In comparison, that old stalwart fuelwood rose a measly 12%, from 101 million to just 113 million cords per year.[24]

Although the U.S. population also rose quickly over this same time period (from 76 to 281 million, or +270%), oil consumption rose far faster on a per capita basis. By the beginning of the twenty-first century the average American was burning through more than twenty-four steel drums of oil every year. In 1900, had my Italian grandfather already emigrated to the United States, he would have used just twenty-two gallons, about one-half of one steel drum.

The twentieth century saw similar extraordinary growth in American consumption of iron, nickel, diamonds, water, softwood, salmon, you name it. To varying degrees, this rapid escalation of resource consumption has either happened or is now happening in the rest of the world.

So we see that resource consumption, much like our global population, grew ridiculously fast in a single century. But while the two certainly feed off one another, rising resource demand has less to do with population growth per se than with modernization. My UCLA colleague Jared Diamond illustrates this by considering an individual's "consumption factor."[25] For the average person living in North America, Western Europe, Japan, or Australia, his or her consumption factor is 32.

If your consumption factor, like mine, is 32, that means you and I each consume thirty-two times more resources and produce thirty-two times more waste than the average citizen of Kenya, for example, with a consumption factor of 1. Put another way, in under two years we plow through more stuff than the average Kenyan does in his entire life. Of the 6.8+ billion of us alive on Earth now, only about a billion—15%—enjoy this lavish lifestyle. The vast majority of the human race lives in developing countries with consumption factors much lower than 32, mostly down toward 1.

Places with a consumption factor of 1 are among the most impoverished, dangerous, and depressing on Earth. Regardless of what country we live in, we all want to see these conditions improve—for security as well as humanitarian reasons. Many charitable people and organizations are working toward this goal, from central governments and NGOs to the United Nations to local churches and individual donors. Most developing countries, too, are striving mightily to industrialize and improve their lot. Organizations large and small, from the World Bank and International Monetary Fund (IMF), to the Grameen Bank and other microlenders, are providing loans to help. Who among us does not want to see such efforts succeed? Who does not want the world's lingering poverty, hunger, and disease brought to an end?

But therein lies a dilemma. What if you could play God and do the noble, ethically fair thing by converting the entire developing world's level of material consumption to that now carried out by North Americans, Western Europeans, Japanese, and Australians today. By merely snapping your fingers you could eliminate this misery. Would you?

I sure hope not. The world you just created would be frightening. Global consumption would rise *elevenfold*. It would be as if the world's population suddenly went from under 7 billion today to *72 billion*. Where would all that meat, fish, water, energy, plastic, metal, and wood come from?

Now let us suppose that this transformation were to happen not instantly but gradually, over the next forty years. Demographers estimate that total world population might level off at around 9.2 billion by 2050. Therefore, if the end goal is for everyone on Earth to live as Americans, Western Europeans, Japanese, and Australians do today, then the natural world must step up to provide enough stuff to support the equivalent of 105 billion people today.

Viewed in this light, lifestyle is an even more potent multiplier of human pressure on the world resource base than is total population itself. Global modernization and prosperity—an eminently laudable and desirable goal—are thus raising our demands upon the natural world now more than ever.

The third global force is globalization. A big word spanning many things, it most commonly refers to increasingly international trade and capital flows but also has political, cultural, and ideological dimensions.[26] Frankly, there are about as many definitions for globalization as there are experts who study it. For our purposes here let us simply think of "globalization" very broadly as a set of economic, social, and technological processes that are making the world more interconnected and interdependent.

Most people were aware of how interconnected the world economy had become long before the 2008–09 global financial crisis laid it bare. In his 2006 book *The World Is Flat*, the *New York Times* columnist and author Thomas Friedman famously asked, "Where were you when the world went flat?"[27] *Flat* is Friedman's simple metaphor for the opening and leveling of a global playing field for trade and commerce, one that in principle maximizes efficiency and profitability for all because the cheapest ore or cheapest labor can be hunted down to the last corners of Earth.

No doubt everyone has a different answer to Friedman's question. For me, it was in Burbank in 1998, while waiting in a queue at an IKEA home-furnishings store. It struck me that my arms were filled with products designed in Sweden, built in China, shipped to my store in California, and sold to me by a Mexican cashier. From a single store selling pens and seed packets in tiny Älmhult in 1958, IKEA had grown to three hundred franchises in thirty-seven countries by 2010. At €22 billion (USD $33 billion) annually its economy was bigger than that of the country of Jordan and adding twenty-plus new stores worldwide each year.[28] Not only is this single company now a planetary economic force, it is globalizing Swedish culture by cultivating a taste for juicy meatballs and clean Scandinavian furniture design from the United States to China to Saudi Arabia.

Globalization kills economies too. After years of slow bleeding, my wife's hometown in Michigan crashed when Delphi, a major supplier of automotive parts to General Motors Corporation, went bankrupt. Also, globalization's spread is very uneven: The world is not so much "flat" as it is lumpy. Some countries, like Singapore and Canada, are integrating broadly and rapidly whereas others, like Myanmar and North Korea, are isolated backwaters.

Taking the long view, the world appears to be in the early phase of an economic transformation to something bigger and more integrated than anything ever seen before. It is more far-reaching and sophisticated than any previous alliance in human history. We will all be potential rivals, but also all potential friends. Alongside the demise of entire sectors will be new markets, new trade, and new partnerships. Gone are the days when General Motors could import rubber and steel and export automobiles. The design, raw materials, components, assembly, and marketing of today's cars might come from fifty different countries around the world.

But what unleashed this new era of global integration upon us? Was it the blazing speed and easy reach of the Internet—or something deeper? I only noticed it in 1998, but might this phenomenon be older than we think?

Like rising world population and natural resource demand, the present global integration lifted off in the middle of the twentieth century. But unlike the first two, it happened deliberately. It all began with a big conference in the Mount Washington Resort near Bretton Woods, New Hampshire, in July 1944. Over seven hundred delegates from forty-four countries—including Britain's John Maynard Keynes (whose ideas later found new life in the wake of the 2008 global credit meltdown)—were in attendance.

World War II was drawing to a close. Governments were turning their attention to their shattered economies and how to rebuild them after two catastrophic wars, a global depression, a long escalation of protectionist tariffs, and some crazy currency devaluations. Everyone at the conference wanted to figure out how to stabilize currencies, get loans to war-ravaged countries for rebuilding, and get international trade moving again.

The outcome of this conference was something called the Bretton Woods Agreement. Among other things, it stabilized international currencies by pegging them to the value of gold (which lasted until 1971, when President Richard Nixon dropped the U.S. dollar from the gold standard). But its most persistent legacy was the birth of three new international institutions: the International Monetary Fund (IMF) to administer a new monetary system; the International Bank for Reconstruction and Development (IBRD) to provide loans—today, the World Bank; and the General Agreement on Tariffs and Trade (GATT) to fashion and enforce trade agreements—today, the World Trade Organization (WTO). These three institutions guided much of the global reconstruction effort after the war; and during the 1950s their purpose expanded to giving loans to developing countries to help them industrialize. Today these three powerful institutions—the IMF, World Bank, and WTO—are the prime actors making and enforcing the rules of our global economy.

Up until the demise of the Bretton Woods monetary regulatory system in the early 1970s, it presided for three decades over what some have called the "golden age of controlled capitalism."[29] But by the 1980s, "controlled capitalism" had fallen to a revolution of "neoliberalism"—the deregulation and elimination of tariffs and other controls on international trade and capital flows. The neoliberalism movement was championed by British prime minister Margaret Thatcher and U.S. president Ronald Reagan, but was rooted in the ideas of Adam Smith.

Throughout the 1980s and 1990s the IMF, WTO, and World Bank aggressively pursued agendas of liberalizing (deregulating) trade markets around the world, vigorously urged on by the United States.[30] A common tactic was to require developing countries to accept neoliberal reforms to qualify for IMF or World Bank loans. This practice was exemplified by the "Washington Consensus," a controversial list of hard-nosed reforms including trade deregulation, opening to direct foreign investment, and privatization of state enterprises.[31]

In the United States, presidents from both political parties also worked to dismantle international trade barriers. Of particular importance to this book was the North American Free Trade Agreement (NAFTA), proposed in 1991 by President George Herbert Walker Bush to remove trade barriers between the United States, Mexico, and Canada. Two years later, President Bill Clinton made NAFTA the cornerstone of his legacy. In his speech at the signing ceremony Clinton pressed the need "to create a new world economy," with former presidents Bush, Jimmy Carter, and Gerald Ford nodding in attendance. Clinton's successor also agreed: Fifteen years later, citing a near-quintupling of U.S. free-trade agreements under his watch, outgoing president George W. Bush stated that global trade expansion had been one of the "highest priorities of his administration."[32]

Notice that the origins of today's great global integration are at odds with one of its most widely promulgated myths: that globalization has emerged organically, born from fast Internet technology and the "invisible hand" of free markets. In truth, this global force owes its existence to a long history of entirely purposeful policy decisions, championed especially by the United States and Britain, dating to the waning days of World War II. Many who write about globalization see it as exploding suddenly in the 1970s or 1980s, thus missing the institutional groundwork laid first under Bretton Woods, pressed upon the developing world by its daughter institutions the IMF, WTO, and World Bank, and subsequently advanced by U.S. presidential administrations of both political parties ever since. Its foundations are now codified into decades of historical precedent and a plethora of free-trade treaties. They are engrained in generations of politicians and business

CEOs, and were reaffirmed even during the turmoil of the 2008–09 global financial crisis.[33] This megatrend has roots going back more than sixty years and is now a deep, powerful global force already shaping the twenty-first century economy.

The fourth global force is climate change. Quite simply, it is observed fact that human industrial activity is changing the chemical composition of the atmosphere such that its overall temperature must, on average, heat up.

The power of greenhouse gases is simply beyond dispute. Their existence was deduced in the 1820s by the French mathematician Joseph Fourier, who noticed that the Earth is far warmer than it ought to be, given its distance from the Sun. Without greenhouse gases our planet would be an icebox, like the Moon and Mars, with temperatures some 60° Fahrenheit colder than today.[34] Their magic comes from letting solar radiation easily in but not easily out, roughly analogous to how a closed-up car becomes hotter inside than out from sunlight passing through the window glass.[35]

The basic physics of this was worked out in the 1890s by the Swedish chemist Svante Arrhenius.[36] Like glass, greenhouse gases are transparent to short-wavelength sunlight, allowing it to pass unimpeded through the atmosphere to warm the Earth's surface (unless blocked by a cloud). But they are opaque to the (invisible) long-wavelength infrared radiation returned from the warmed Earth back to space, instead absorbing it and thus becoming infrared radiators themselves.

Arrhenius was trying to solve the puzzle of ice ages, so was initially interested in global cooling, not warming, but his calculations worked easily well in either direction. He later wondered if humans, by adding carbon dioxide to the air through fossil-fuel burning, could also influence the planet's climate. He ran the numbers and found that they certainly could, and substantially, too, if the gas's concentration was raised high enough. His initial estimate of +5°C warming for a doubling of atmospheric CO_2, calculated by hand, was remarkably close to the ones generated by far more sophisticated computer models running today. But Arrhenius didn't think much of this at the time, because he couldn't imagine humans ever releasing that much carbon dioxide. For humans to double the atmosphere's CO_2, he reasoned, would take at least three thousand years.[37]

Apparently, the physics of greenhouse gas warming is a lot easier to comprehend than the pace of human industrialization. We've already raised the concentration of CO_2 in the atmosphere nearly 40%, up from ~280 parts per million by volume (ppmv) in preindustrial times to ~387 ppmv as of 2009. Two-thirds of that rise has been carefully documented since 1958, when the first continuous air sample measurement program was begun by Charles Keeling at Hawaii's Mauna Loa Observatory as part of the International Geophysical Year. Atmospheric measurements of two other powerful greenhouse gases also released by human activity, methane and nitrous oxide levels, have followed a similar rising pattern. Depending on the choices we make about carbon emissions, CO_2 projections for century's end range anywhere from 450 to 1,550 ppmv, corresponding to a +0.6 to +4.0°C increase in average global temperature on top of the +0.7°C increase already experienced in the twentieth century.[38] Many policy pragmatists now feel a +2°C increase is all but assured, after the 2009 Copenhagen Climate Conference failed to produce anything resembling a binding international agreement to curb carbon emissions.

These numbers may sound small but they're not. At the height of the last ice age, when Chicago was buried under a mile-deep sheet of ice, global temperatures averaged just 5°C (9°F) cooler than today. From historical weather-station data the global average temperature is already +0.8°C warmer than in Arrhenius' time, with most of that rise since the 1970s. An increase of that magnitude is already much larger than the difference between any one year and the next. As expected, this warming trend varies strongly with geography, with even some local cooling in some places (the details and reasons for this are known and discussed further in Chapter 5). But the global average is trending upward, along with the steady measured growth of greenhouse gas concentrations in the atmosphere.

Not only are average temperatures rising, the *way* they are rising is consistent with the greenhouse effect but inconsistent with other natural cycles and processes also known to influence climate. Temperatures are warming more by night than by day; more in winter than in summer; more over oceans than over land; more at high latitudes than in the tropics; and in the troposphere but not the stratosphere. All of these are consistent with greenhouse gas forcing but inconsistent with other known causes, like urban heat islands, changing solar brightness, volcanic eruptions, and astronomical cycles. Those, too, influence climate, but none can fully explain what we are seeing today.

In addition to number-crunching weather data, there is plenty of anecdotal evidence that our climate is beginning to act strangely. A staggering thirty-five thousand people were killed in 2003 when a massive heat wave spilled across Europe. Lesser waves killed hundreds more in Japan, China, India, and the United States in the following summers, when the world suffered eleven of the top twelve hottest years ever recorded. That's dating all the way back to the first weather stations of 1850, when Zachary Taylor was president of the United States, and Italy wasn't even a country yet. Hurricane Katrina drowned New Orleans in 2005, a record year for tropical storms. Ironically, many of the displaced moved to Houston, where they got pounded again by Hurricane Ike in 2008. That one killed about two hundred people and put a tree through the roof of my best man, then proceeded to black out nearly a million homes across Ohio, Indiana, and Kentucky.

Like the pizzly bear, no one of these events is conclusive of anything. But after enough of them happen, the private sector gets moving. Goldman Sachs and the *Harvard Business Review* started writing reports on how to contain risk and maximize profits from climate change.[39] Multinational corporations like General Electric, Duke Energy, and Dupont began stumping green technology and formed the U.S. Climate Action Partnership, calling on the U.S. federal government to "quickly enact strong national legislation to require significant reductions of greenhouse gas emissions."[40] By 2008 its membership included American International Group, Inc. (AIG), Boston Scientific Corporation, Chrysler LLC, ConocoPhillips, Deere & Company, the Dow Chemical Company, Exelon Corporation, Ford Motor Company, General Motors Corporation, Johnson & Johnson, Marsh, Inc., the National Wildlife Federation, the Nature Conservancy, NRG Energy, Inc., PepsiCo, Rio Tinto, Shell, Siemens Corporation, and Xerox Corporation.[41] However, by late 2009 the rush of corporations to join the

U.S. Climate Action Partnership had slowed, following the failed climate treaty conference at Copenhagen, some dumb e-mails circulated among a clique of climate scientists (the so-called Climategate scandal, a scientifically minor but politically devastating public-relations fiasco), and a moribund cap-and-trade bill in the U.S. Senate. By 2010 ConocoPhillips, BP America, Caterpillar, and Xerox had pulled out.

Gas molecules are impervious to politics, so all of this is really just the beginning. To underscore just how dramatic our run-up of CO_2, methane, and nitrous oxide in the atmosphere is, let's place it within the much longer context of geological time. Greenhouse gases follow both natural cycles—which fall and rise with ice ages and warm interglacial periods, respectively—and human activity, which proceeds much faster. These two actors operate over totally different time scales, with the ice age variations happening over tens of thousands of years but our human excursion unfolding over tens of years. The natural processes that drive greenhouse-gas shifts—rock weathering, astronomical cycles, the spread of forests or wetlands, ocean turnover, and others—take thousands of years, whereas human excavation and burning of old buried carbon—as illustrated earlier from U.S. history—is an action both massive and brief. And because our human-generated carbon burst is perched atop an already large, slow-moving natural interglacial peak, we are taking the atmosphere to a place the Earth has not seen for hundreds of thousands, perhaps millions, of years.[42]

We know this from the ancient memories of glaciers, deep ocean sediments, tree rings, cave speleothems, and other natural archives. Most spectacular are tiny air bubbles trapped within Greenland and Antarctic ice, each a hermetically sealed air sample from the past. Loose air inside a glacier's surface snowpack gets closed off into bubbles as the weight of still more snowfall fuses it into ice. Annual layers of these bubbles have been quietly laid down for hundreds of thousands of years, before being drilled from the guts of Greenland and Antarctica by a rare breed of scientist. The gas levels inside them prove we have now elevated the concentrations of greenhouse gases in the Earth's atmosphere higher than they've been for at least eight hundred thousand years.

Eight hundred thousand years. Jesus Christ walked barely two thousand years ago, Egypt's pharaohs four. Our first agricultural civilizations began ten thousand years ago; twenty thousand years before that, there still were Neanderthals alive. But the world has not seen atmospheric CO_2 levels like today's for eight hundred thousand years—and they are now approaching those of *fifteen million* years ago in the Miocene Epoch, when the world's temperatures were 3° to 6°C warmer, its oceans acidic, polar ice caps diminished, and sea levels twenty-five to forty meters higher than today.[43]

This, too, is a global force to be reckoned with.

———

These four global forces—demographics, resource demand, globalization, and climate change—will shape our future and are recurring themes throughout this book. As each force comes up, the corresponding icon from the set that headed the four preceding sections will head the discussion. While I have described these forces separately they are, of course, intimately intertwined. Greenhouse gas comes from the exploitation of natural resources, which in turn tracks the global economy, which in turn relates partly to population dynamics, and so on.

A fifth force twining through the first four is technology. Fast global communications facilitate global financial markets and trade. Modern health care and pharmacology are shifting population age structures in the developing world. Advances in biotech, nanotech, and materials science affect demand for different resource stocks. Smart grids, solar panels, and geoengineering might combat climate change, and so on. Under our "No Silver Bullets" rule, technological advances like these are evaluated as enablers or brake pads on the four global forces, rather than as an independent force of its own.

The thought experiment is begun. Its assumptions and ground rules are stated, its four overarching themes defined. Let us turn now to the first subject of scrutiny for the year 2050—ourselves.

PART ONE

THE PUSH

CHAPTER 2

A Tale of Teeming Cities

"Tomorrow morning we will release our sales numbers for the month of November. This event is overshadowed by the tragic death of Jdimytai Damour at our Valley Stream, New York, store on November 28. . . ."

—Statement from the president of the Northeast Division, Walmart USA (December 3, 2008)

Italy, France, United Kingdom, Germany, Japan, and the United States.

—Economies projected by Goldman Sachs to be overtaken by one or more of China, India, Russia, or Brazil before 2050

"From here on out, it's an urban world."

—Joel E. Cohen, Professor of Populations, Rockefeller University and Columbia University

It was one o'clock in the morning when Leana Lockley, twenty-eight years old and five months pregnant, lined up with her husband and two family members outside the Green Acres Mall Walmart store in Valley Stream, New York. Whining engines and lights pierced the night as jets came and went from nearby JFK Airport. It was November 28, 2008, the day following Thanksgiving called "Black Friday," the busiest American shopping day of the year. The global economy was crashing, everyone was looking for bargains, and Walmart was cutting its prices for six hours only. By the time the store opened at five o'clock a.m. there were two thousand people crowded restlessly against the glass storefront, waiting to get in.

The doors unlocked and people surged forward. Lockley was literally picked up off her feet and carried through the door opening. There were loud cracking noises as hinges broke off, and the sounds of crashing glass. An older woman fell. Lockley tried to pick her up but was knocked to her knees. A large man saw her and tried to help. "He was facing the crowd, and he had his hands up, trying to push them back in order for me to escape," she later recounted on Fox News. "He was trying to block the people from pushing me to the ground and trampling me . . . he was on his knees, I could look at him eye to eye, and he was trying to push them back and then the crowd pushed him down, and he fell on top of me."[44] His body covering hers, hundreds of deal-hungry shoppers stomped and shoved over them and streamed into the store.

Lockley and her unborn daughter survived, as did three other injured shoppers sent to area hospitals. But the man who saved her life, thirty-four-year-old Jdimytai Damour, was killed. As paramedics attempted to resuscitate him, shoppers continued to jostle past; then became irate when officials announced the store was being closed.

The only son of Haitian immigrants, Jdimytai Damour was a very large but gentle man who enjoyed watching football and talked about being a teacher one day. The reason he was there that morning was not to buy, but to work. Because of his size—he was six feet five inches tall and weighed 270 pounds—he had been assigned to the front door. But he wasn't a trained security officer. He was a temporary worker, a subcontractor, hired by Walmart to help them help us to consume more stuff during the busiest retail season of the year.

Far less tragic—and certainly less attention-grabbing—was a second, very profound event that also happened in 2008. Its exact timing will never be known, but at some instant during the year, the number of people living in urban areas grew to briefly match, for a few seconds, the number of people living in rural areas. Then, somewhere, a city baby was born. From that child forward, for the first time in our history, the human race became urban in its majority.

For the first time ever, we have more people living in cities than out on the land. For the first time, most of us have no substantive ability to feed or water ourselves. We have become reliant upon technology, trade, and commerce to carry out these most primitive of functions. Sometime in 2008, the human species crossed the threshold toward becoming a different animal: an urban creature, geographically divorced from the natural world that still continues to feed and fuel us.

What does the awful death of Jdimytai Damour have to do with our transformation to an urban race? Aside from occurring in the same year, what connection may be drawn between these two events?

From a macroeconomic perspective, the frenzied horde that killed poor Mr. Damour was, in its own mindless way, helping to build cities all over the world. Most of the items for sale in the Valley Stream Walmart were made overseas, by urban workers in the hundreds of Asian towns and cities churning out the cell phones, flat-screen televisions, netbooks, and other essentials of twenty-first-century life. Cities around the world have participated in the process of getting those products onto Walmart's shelves.

A global supply web was needed to transfer the raw materials and components to manufacturing hubs like Shenzhen, Dongguan, Guangzhou, and Bangalore. Then, the finished goods were sent to the United States, very likely in freighters and steel shipping containers built in places like Geoje (South Korea), Nagasaki (Japan), and Ningbo (China). These vessels were unloaded in the American ports of Long Beach or Los Angeles before being trucked east to Gloucester City, New Jersey, for redistribution. From there, they were trucked once again to Valley Stream, New York. Financial transactions between New York City and Hong Kong, as well as Chicago, Tokyo, London, Paris, Frankfurt, Singapore, and Seoul were taking place. So every time a new flat-screen TV is sold by Walmart, urban agglomerations around the globe all win another little economic boost.

These invisible ties across the globe hum, the economic wheels turn. Consumption fuels commerce, thus growing cities, further enlarging the overall consumer base.[45] The urban economy grows—as it *must*, to support its growing number of residents and the many services they now require. Salaries rise such that even menial entry-level jobs pay better than a farmworker's wages.

The reason that the world's rural people are moving into cities is that they can make more money in town. This is partly because of the described growth of urban economies, and partly because demand for farm labor falls as agriculture commercializes, mechanizes, and becomes export-oriented. Worldwide employment in agriculture is falling fast and in 2006, for the first time ever, it was surpassed by employment in the services sector.[46]

And because every new urban resident is also a new urban consumer, the cycle is self-reinforcing. More urbanites buy more electronics, services, and imported processed food, prepared and served to them by others. More entry-level jobs for new migrants are created. More managerial posts are needed. Ladders rise and the urban economy grows.[47]

This urban shift is driving major demographic changes around the globe. City dwellers are projected to roughly double in number by 2050, rising from 3.3 billion in 2007 to 6.4 billion in 2050.[48] However, the geography of this is not uniform. Urban majorities came to Europe and America decades ago, in the 1960s, 1950s, or even sooner. These places are already more than 70% urban today. This new trend is most dramatic in the developing world, especially Asia and Africa, the most populous places on Earth.

For the last two decades, cities in the developing world have been growing by about three million people per week.[49] That is equivalent to adding one more Seattle to the planet every day. Asia is only about 40% urban today, but by 2050 that number will top 70% in China, with over one billion new city slickers in that country alone. Already, places like Chongqing, Xiamen, and Shenzhen are growing more than 10% annually.

About 38% of Africans live in cities today, but by 2050 more than half will. While Africa will still be less urbanized than Europe or North America today, this is nonetheless a profound transformation. When combined with its fast population growth rate, this means that Africa will *triple* the size of its cities over the next forty years.[50] At 1.2 billion people, Africa will hold nearly a quarter of the world's urban population.[51]

Tucked away in the back of a 2008 report by the United Nations Population Division are some stunning data tables.[52] They rank our past, present, and future "megacities"—urban agglomerations with ten million inhabitants or more—for the years 1950, 1975, 2007, and 2025. The projections may surprise you:

World Megacities of Ten Million or More
(population in millions)

1950
  New York–Newark, USA (12.3)
  Tokyo, Japan (11.3)

1975
   Tokyo, Japan (26.6)
   New York–Newark, USA (15.9)
   Mexico City, Mexico (10.7)

2007
   Tokyo, Japan (35.7)
   New York–Newark, USA (19.0)
   Mexico City, Mexico (19.0)
   Mumbai, India (19.0)
   São Paulo, Brazil (18.8)
   Delhi, India (15.9)
   Shanghai, China (15.0)
   Kolkata (Calcutta), India (14.8)
   Dhaka, Bangladesh (13.5)
   Buenos Aires, Argentina (12.8)
   Los Angeles–Long Beach–Santa Ana, USA (12.5)
   Karachi, Pakistan (12.1)
   Al-Qahirah (Cairo), Egypt (11.9)
   Rio de Janeiro, Brazil (11.7)
   Osaka–Kobe, Japan (11.3)
   Beijing, China (11.1)
   Manila, Philippines (11.1)
   Moskva (Moscow), Russia (10.5)
   Istanbul, Turkey (10.1)

2025
   Tokyo, Japan (36.4)
   Mumbai, India (26.4)
   Delhi, India (22.5)
   Dhaka, Bangladesh (22.0)
   São Paulo, Brazil (21.4)
   Mexico City, Mexico (21.0)
   New York–Newark, USA (20.6)
   Kolkata (Calcutta), India (20.6)

Shanghai, China (19.4)

Karachi, Pakistan (19.1)

Kinshasa, Democratic Republic of the Congo (16.8)

Lagos, Nigeria (15.8)

Al-Qahirah (Cairo), Egypt (15.6)

Manila, Philippines (14.8)

Beijing, China (14.5)

Buenos Aires, Argentina (13.8)

Los Angeles–Long Beach–Santa Ana, USA (13.7)

Rio de Janeiro, Brazil (13.4)

Jakarta, Indonesia (12.4)

Istanbul, Turkey (12.1)

Guangzhou, Guangdong, China (11.8)

Osaka–Kobe, Japan (11.4)

Moskva (Moscow), Russia (10.5)

Lahore, Pakistan (10.5)

Shenzhen, China (10.2)

Chennai, India (10.1)

Paris, France (10.0)

The century of megacities has already begun. From just two in 1950 and three in 1975, we grew to nineteen by 2007 and expect to have twenty-seven by 2025. Furthermore, in sheer size alone our global urban culture is shifting east. Of the eight new megacities anticipated over the next fifteen years, six are in Asia, two in Africa, and just one in Europe. Zero new megacities are anticipated for the Americas. Instead, this massive urbanization is happening in some of our most populous countries: Bangladesh, China, India, Indonesia, Nigeria, and Pakistan. New York City was the world's second-largest metropolis in 1977, when Liza Minnelli first sang the hit song "New York, New York" (later popularized by Frank Sinatra) to Robert De Niro in a Martin Scorsese movie. By 2050, the "City That Never Sleeps" will be struggling just to stay in the top ten.

The story doesn't end with megacities. People are flocking to towns of all sizes, large and small. Indeed, some of the fastest growth is happening in urban centers with less than five hundred thousand people. According to the United Nations model, the number of "large" cities—those with populations between five and ten million—will increase from thirty in 2007 to forty-eight by 2025. Three-quarters of these will be in developing countries. By 2050 Asia—the world's most populous continent and still dominated by farmers today—will be nearly as urbanized as Europe.[53]

What does all this mean for life in the countryside? The world's rural population is projected to peak somewhere around 3.5 billion in 2018 or 2019, then gradually fall to around 2.8 billion by 2050. Most of this rural depopulation will happen in the developing world, because OECD countries have now largely completed this shift. Take a drive through rural America. You'll find it littered with ghostly relics of formerly bustling farm towns. The developing world is repeating now—on a much grander scale—the same emptying out of rural regions that began in developed countries in the 1920s.

If you've been adding and subtracting these various numbers, then you've already realized that the rural population declines are too small to offset the urban increases. The world's total population of people will continue to grow substantially in the next half-century. We are now on a trajectory to add nearly 40% more population by the year 2050, raising our number to around 9.2 billion.[54] Who will we be in 2050? In that year, for every one hundred of our future children and grandchildren born, fifty-seven will open their eyes in Asia and twenty-two in Africa, and mostly in cities.

What Kind of Cities Will They Be?

So the people of Earth are rushing into town. "The twenty-first century," declared the United Nations, "is the Century of the City."[55] But what kind of cities will they be? Will they be prosperous or Dickensian? The best of times, or the worst?

There is certainly reason for optimism. The economic downturn of 2008–09 notwithstanding, the long-term trends all point to continued economic globalization, rising urban wealth, and a host of new technologies to help make cities cleaner, safer, and more efficient. It seems plausible to imagine the ascendance of shining, modern, prosperous cities all over the world. Take, for example, the success story of Singapore.

A port city situated on a large island at the southern tip of the Malay Peninsula, Singapore began as a British trading colony in 1819 and remained under colonial rule for one hundred and forty-one years before gaining independence in 1960. Since then, despite its small size (less than 270 square miles), few natural resources, and no domestic fossil fuel supply, Singapore's growth and economic success have been phenomenal.

Between 1960 and 2005 Singapore's population grew rapidly, averaging 2.2% annually or doubling every thirty-six years. Once a calm British trading outpost, Singapore today has nearly five million people and has become a throbbing services, technology, and financial hub for Southeast Asia. It is a global supplier of electronic components and runs the busiest port in the world, with over six hundred shipping lines. Despite having no oil to speak of, it is a major oil refining and distribution center. Singapore is also attracting major foreign investments in pharmaceuticals, medicine, and biotechnology. With a 2008 gross domestic product (GDP) of USD $192 billion, Singapore's economy is bigger than those of the far more populous Philippines, Pakistan, and Egypt.

Geopolitically, Singapore has become one of the most globalized, stable, and prosperous countries in the world. Per capita income is over USD $50,000, higher than in the United States. It has a democratically elected government and ranks second in the world's Index of Economic Freedom.[56] It is a member of the IMF, WTO, UNESCO, Interpol, and many other global institutions. Since the 1970s the performance of its sovereign wealth funds has been legendary. Through heavy global investments they've returned 4%–10% annually, growing a few humble millions into over USD $200 billion today.[57]

Singapore has learned to manage long-standing tensions between its main ethnic groups (Chinese, Malay, and Indian) and religions. Mass transit is abundant, clean, and energy-efficient.[58] There are wonderful parks, theaters, and museums. Singapore's health care is excellent and its life expectancies are the fourth-longest in the world (seventy-nine years for men and eighty-five years for women). Aggressive law enforcement—while also leading to complaints of excessive strictness and a sort of police-state authoritarianism—has made corruption, violent crime, and the trafficking of sex and drugs virtually nonexistent.

Singapore is a good example of how rapid population and economic growth, when properly managed, can grow a city that not only has a large economy but is also technologically advanced, culturally vibrant, and an enjoyable place to live. To borrow a name coined by my UCLA colleague Allen Scott,[59] it is a shining *technopolis*. Writes author Henri Ghesquiere about Singapore's success:[60]

> Rapid growth was matched by enhanced well-being. The quality of life improved for large numbers of people. Singapore succeeded from the perspective of not only growth, but also social development. . . . China's momentous decision in 1978 to reverse five centuries of economic isolation was influenced in part by Deng Xiaoping's visit to Singapore that year. His dream to "plant a thousand Singapores in China" sparked numerous delegations on study tours to the island. South Korea was impressed with Singapore's success in overcoming corruption. The city-state's mastery in keeping urban traffic flowing has fascinated officials from many countries, and its housing program is studied by planners from around the world. Dubai eyes Singapore continually. . . .

Unfortunately, there is no rule saying a city must be a nice place to live in order to attract fast population and economic growth. Parks, good governance, and smoothly flowing traffic are optional, not required. Sometimes cities grow at an astonishing rate, despite being hell on Earth.

Take Lagos, Nigeria. Like Singapore, Lagos is a coastal port city, is built on an island, and was once a British colony. It guards the mouth of a huge swampy lagoon and for centuries has been one of the most important trading ports of West Africa. Over the years it has variously exported slaves, ivory, peppers, and, most recently, oil. Like Singapore, Lagos also won its independence from Great Britain in 1960. Both cities are located a few degrees north of the equator in moist, tropical climates. Both are governed by civilian democracies, although Nigeria's is still young and shaky after years of military rule.

Since independence, Lagos' population has grown even faster than Singapore's—averaging about 5% annually since 1960. Between 2000 and 2010 its population grew almost 50%, from 7.2 to 10.6 million people.

Nigerians are pouring in from the surrounding countryside and villages because there is money to be made in Lagos. The city has now run out of room, filling up its island and spilling across congested bridges to penetrate more than fifteen miles inland. By the year 2025, Lagos is projected to grow another 50% to sixteen million people, making it the twelfth-largest city in the world. With a 2007 gross domestic product of about USD $220 billion—bigger even than Singapore's—Lagos is the economic epicenter of Nigeria and indeed of the entire western African continent.

Similarities between the two cities end there. Unlike Singapore, Lagos has not handled its growing pains well. It is an overcrowded dystopia of traffic jams, squalor, corruption, murder, and disease. Per capita income averages around USD $2,200 per year. Millions live in boats without electricity or sanitation. Four out of ten women cannot read. Police are outnumbered, ineffective, and unpredictably dangerous. The physical infrastructure is simply overwhelmed. Writes the urban geographer Matthew Gandy:[61]

> The sprawling city now extends far beyond its original lagoon setting to encompass a vast expanse of mostly low-rise developments including as many as 200 different slums. . . . Over the past 20 years, the city has lost much of its street lighting, its dilapidated road system has become extremely congested, there are no longer regular refuse collections, violent crime has become a determining feature of everyday life and many symbols of civic culture such as libraries and cinemas have largely disappeared. The city's sewerage network is practically non-existent and at least two-thirds of childhood disease is attributable to inadequate access to safe drinking water. In heavy rains, over half of the city's dwellings suffer from routine flooding and a third of households must contend with knee-deep water within their homes.

The flooding observed by Dr. Gandy is a serious problem. Lagos' extreme growth has pushed development, much of it slum, into the city's last remaining real estate: swampy, low-lying swales that are barely above sea level. Human excrement flows in open ditches. Drainage is so bad that when it rains, the waste floats into people's homes. Fewer than fifteen people per hundred have piped water, most instead rely on shared outdoor taps or wells. Nearly all water sources are regularly contaminated with *E. coli*, streptococcus, and salmonella. Unsurprisingly, disease is rampant, including typhoid fever, yellow fever, Lassa fever, malaria, leptospirosis, shistosomiasis, hepatitis, meningococcal meningitis, HIV/AIDS, and the H5N1 avian flu. Human life expectancies are just forty-six years for men and forty-seven for women.

It gets worse. Two-fifths of the people living in Lagos are victimized by corruption, especially demands for bribes from their public officials.[62] Robberies, assaults, and murders are a constant fact of life. Failed by their police and judiciary, citizens form vigilante militias with names like the "Bakassi Boys," who fight back at criminals with machetes and shotguns.[63] When government officials perceive disorder, they issue orders to shoot on sight. In general, police and soldiers are best avoided in Nigeria, as it is not uncommon for police officers to simply shoot potential suspects rather than arrest them. Nigeria's National Human Rights Commission, a domestic agency charged with monitoring the country's human rights violations, recently compiled a heartbreakingly long list of abuses, including the following three incidents:[64]

[March 2, 2005] "A commercial bus with registration number, XA 344 plying the NorthBank to Wadata route was stopped by Police Constable (PC) Vincent Achuku, and without searching the bus, he demanded (money) bribe from the driver, Godwin Anuka. The driver pleaded with PC Achuku to let him continue with his journey and that he would pay on his return trip. This infuriated PC Achuku and he immediately cocked his service rifle, shot and killed the driver."

[July 28, 2006] "In the early hours of July 28, 2006, officers of the Special AntiRobbery squad . . . went to the compound of Adulkadir Azeez, 70 years old, where he lives with his extended family. . . . The sound of a gunshot woke him up and he went out to find out the cause. Immediately he stepped out of his house, he was shot dead. The policemen then went to his son's house (Ibrahim Abdulkadir), forced open the door and shot him dead too.

The second son of the old man, Shehu Abdulkadir, heard the gunshots, opened his door and saw his father lying dead on the ground. He tried to find out what happened and was also shot dead by the policemen."

[September 15, 2006] "On Friday, September 15, 2006, at about 3.00 P.M., a team of over 200 policemen in eight trucks from Delta State Police Command drove to Afiesere Community. . . . On arrival, they started shooting sporadically. According to media reports villagers including women and children took to their heels to avoid the bullets of the rampaging policemen. Those that could not escape were shot and killed or wounded. The policemen then looted several shops and homes before setting them ablaze. When the policemen eventually retreated around 5.30 P.M. 22 persons lay dead, 60 houses and 15 vehicles were burnt. The police also set ablaze two corpses while some were taken away and dumped in the bush. Five other victims, mostly elderly persons also reportedly died from shock over the incident."

While all is not lost in Nigeria—it completed its first transfer of power between civilian governments peacefully in 2007, and Lagos crime rates fell sharply in 2009—it remains a dangerous place to live. Despite growing Nigeria's gross domestic product to the second largest on the African continent, Lagos is a *slum city* and—like other slum cities in Africa, Asia, and Latin America—is a stark illustration of an urban world we don't want. Clearly, there is far more to building shining technopolises than urbanization and economic growth.

Sub-Saharan Africa is a collection of countries bursting with natural and agricultural resources. Many have functioning or semifunctioning democracies. Yet, colonialism, nonsensical borders, tribal loyalties, HIV, and other problems have bogged it in muddy poverty. Nearly two-thirds of its towns and cities are slums. Recall that at current growth trajectories, these urban populations will triple in the next forty years. The still-unfolding results of the Lagos experiment do not bode well for this new African urbanism. Under the conservative ground rules of our thought experiment, it's hard to envision how so many problems can be eliminated overnight. By 2050 I imagine much of sub-Saharan Africa—the cradle of our species—to be a dilapidated, crowded, and dangerous place.

Shifting Economic Power

Not only is the geography of the world's urban population shifting, so also is its wealth. The economic impact of nearly two billion new urban consumers in Asia has not gone unexamined by economists. Unlike the situation in Africa, there is every indication that the rising Asian cities will be modern, globalized, and prosperous. In a thoughtful, forward-looking assessment the U.S. National Intelligence Council writes:[65]

> The international system—as constructed following the Second World War—will be almost unrecognizable by 2025. . . . The transformation is being fueled by a globalizing economy, marked by an historic shift of relative wealth and economic power from West to East, and by the increasing weight of new players—especially China and India.

China and India—along with Brazil and Russia—are considered such economic giants-in-waiting that they've won their own acronym, the "BRICs" (from the first letters of *B*razil, *R*ussia, *I*ndia, and *C*hina), first coined in 2003 by the global financial services company Goldman Sachs.[66] According to econometric model projections by Goldman Sachs, PricewaterhouseCoopers, the Japan Center for Economic Research, the International Monetary Fund, and others, the BRICs are on pace to displace current economic leaders faster than you might think, thus redrawing the map of global economic power over the next forty years.[67]

The world's three biggest economies today are the United States, Japan, and Germany. But by 2050 most model projections have China and India leaving all other countries save the United States in the dust. The Goldman Sachs model, for example, projects U.S. gross domestic product to rise from USD $10.1 to $35.1 trillion, Japan's from $4.4 to $6.7 trillion, and Germany's from $1.9 to $3.6 trillion by then. In contrast, India's GDP is projected to rise from $0.5 to $27.8 trillion and China's from $1.4 to a whopping $44.4 trillion. Brazil and Russia are projected to rise from $0.5 to $6.1 trillion and $0.4 to $5.9 trillion GDP, respectively.[68] China would thus overtake the United States as the world's largest economy and India would become the third largest.

The 2008–09 global financial crisis only reaffirmed this picture. While the economies of America, Japan, and Germany were shrinking, the economies of Brazil, India, and China grew 2%, 6%, and 9% annually, respectively.[69] By late 2009 Brazil, one of the last countries in and first out of the downturn, was again growing 5% per year and was on pace to become the world's fifth-largest economy even sooner than Goldman Sachs envisioned, overtaking Britain and France sometime after 2014.[70] New post-crisis modeling by the Carnegie Endowment for International Peace reaffirmed that China's GDP would indeed surpass that of the United States, probably by 2032. By 2050 the world's three largest economies would still be China ($45.6 trillion), the United States ($38.6 trillion), and India ($17.8 trillion).[71]

While such growth rates sound dramatic, they are actually less spectacular than those enjoyed by Japan for three decades between 1955 and 1985. If these economic model projections hold—and in accordance with our "The Models Are Good Enough" ground rule let's assume here that they will—the world will move from having not one huge economy but three. Of the original top three, only the United States will remain, in distant second place behind China. The relative clout of deposed Japan, Germany, and others of the original G6 (France, Italy, and the United Kingdom) on the world stage will be diminished.

So if China and India are poised to dethrone the original G6 economies, what does that mean for you? Will Chinese and Indians soon enjoy more lavish lifestyles than Germans and Italians? Will Parisians emigrate to São Paulo, seeking better pay and a happier future for their children?

Almost certainly not. Recall that one of the big drivers of all this economic growth is rising urban population and modernization. The economies of China and India *must* grow to support that. If they didn't, per capita incomes would decrease. The cost of living would have to go down, not up. That's not how things work. Can you imagine all those rural migrants pouring into New York City in the early 1900s driving the price of food and housing *down*?

No. Asia's rising cities demand that the economies of China and India grow many times over, and this will also multiply per capita incomes in these countries. However, that progress in personal wealth will still be relative to the extremely low per capita incomes of today (averaging less than $3,000 per year for both countries in 2010). With the sole exception of Russia,[72] personal incomes in BRIC countries are not expected to surpass those of France, Germany, Italy, Japan, the United Kingdom, or the United States by 2050. An average Indian today makes less than one-thirtieth the income of an average Brit. In 2050 she or he will make less than one-third.[73] That's a tenfold improvement, to be sure, but still a yawning divide. The Goldman Sachs model, for example, projects that the average Chinese worker will earn around USD $31,000 per year in 2050. That is much better than in 2010 ($2,200 per year) but still substantially below the projected 2050 per capita incomes for Italy ($41,000), Germany ($49,000), France ($52,000), the United Kingdom ($59,000), Japan ($67,000), and the United States ($83,000).

At the national geopolitical level, however, new superpowers mean complicated, shifting alliances. Having more superpowers portends intense strategic rivalries for trade, foreign investment, and natural resources. It means having more powerful political leaders in the world, and history tells us that their ideas matter. The choices made by Vladimir Lenin, Joseph Stalin, Adolf Hitler, Mao Zedong, Winston Churchill, Franklin D. Roosevelt, Harry Truman, and George W. Bush will reverberate for years. Running through everything are the fault zones of historical, cultural, and religious divisions. "Bad outcomes are not inevitable," the National Intelligence Council assessment concludes, but "today's trends appear to be heading toward a potentially more fragmented and conflicted world."[74]

The choices of future political leaders cannot possibly be divined here. But what we can foresee is an assortment of growing demographic, economic, and resource pressures that will shape the context and options available to them. We are barreling toward a world with nearly 40% more people and a doubled food requirement by 2050. We are transforming from a poor rural to wealthier urban species. We are in the midst of a historic transfer of money and power from West to East. The bad news, as we saw in Lagos, is that some parts of our world are poorly equipped to deal with these changes. The good news is that in our rush to urbanize, we may have found the golden pliers for defusing Ehrlich's population bomb.

I See Old People

These megatrends have personal consequences. Honestly, for the long haul, I've begun socking away shares of pharmaceutical company stock. Because beginning right about now, the world is starting to fill up with old people.

Demography just might be the most fascinating academic subject you've never studied. Underneath its dull name and dry statistics lie gripping stories of sex and death, of the rise and fall of communities, of why migrants choose to pick up and move, of the futures for our retirements and for our children. It uncovers big surprises like the myth of the American melting pot.[75] Although combing through census and national registry databanks on numbers of births, deaths, and marriages may not sound very fun, a new world is revealed. These data comprise a road map of our future still wired into today.

Consider the "baby boomer" phenomenon—that is, the post–World War II baby boom. Like a snake swallowing a big meal, this age bulge has worked its way through the decades, triggering all manner of economic and cultural transformations along the way. Many of them—like demand for doctors, vacation getaways, and Viagra as the boomers now enter their sixties—have been anticipated for years. Like all baby booms, it had a softer "baby echo" that cropped up a generation later—again, predictable.

"Population momentum" is another example of how demographic futures can be foreseen. Even if a society's average fertility rate[76] suddenly falls, its population will continue to grow twenty years later owing to the abundance of new parents carried forward from when fertility rates were high.[77] This works in the other direction, too, meaning that elderly countries will keep shrinking even if fertility rises, owing to a small cohort of parents born when fertility rates were still low.

The unprecedented explosion of people on Earth happened because births began outnumbering deaths, but there's more to it than that. The "Demographic Transition" concept described in Chapter 1 emerged from what transpired in Europe and the United States. And it appears to now be unfolding in the rest of the world as well. Recall that the Demographic Transition has four stages:

1. High and similar rates of birth and death (e.g., the preindustrial era, with a small and relatively stable total human population); followed by
2. Falling deaths but not births (initiating a population explosion); followed by
3. Falling births (still exploding, but decelerating); and finally
4. Low and similar rates of birth and death (population stabilization at a new, higher total number).

Most OECD countries have now passed through these stages and—except for those allowing high levels of immigration like the United States—have stabilizing or even falling populations. Most developing nations, however, are still in Stage 2 or early Stage 3. Thus, our run-up in global population is still under way.

Once a population enters Stage 3 its net rate of growth starts to slow, and this has generally been happening, beginning at different times and to varying degrees, for most of the world. On average, growth rates in developing countries have decreased from around +2.3% per year in 1950 to +1.8% in 2007. Expressed as "doubling times" (the number of years needed for a population to double), that means we have slowed from doubling our developing world population about every thirty years in 1950 to every forty years in 2007.

As we saw in Chapter 1, urbanization, modernization, and the empowerment of women push fertility rates downward, thus ushering in the final stage of the Demographic Transition. Put another way, the urbanization of society—if also associated with modernization and women's rights—helps slow the rate of growth. There are, of course, exceptions to this tendency, but as these phenomena continue to expand throughout the developing world, the global population explosion so feared by Thomas Malthus and Paul Ehrlich is expected to decelerate. Already, in late-stage, low-immigration developed countries like Japan and Italy, and in regions like Eastern Europe, populations have not only stabilized but are falling. Assuming that fertility rates continue to drop as they are now, we are heading toward a total world population of around 9.2 billion in 2050, at which point we will still be growing but about half as fast as we are today.[78]

One of the most profound long-term effects of women having fewer babies is to skew societal age structures toward the elderly (the pulse of babies from population momentum is only temporary). That is precisely what has now begun, in varying stages, all around the world (improving health care, of course, also extends our life spans, thus increasing the proportion of elderly even more). Demographers agree that we are racing forward to not only a more urban world, but a grayer one. This, too, is unprecedented in the history of humankind. For 99.9% of the time we humans have existed on Earth, our average life expectancy was 30 years or less. Archaeologists have never dug up the prehistoric remains of anyone over 50.[79]

This aging will hit some places faster and harder than others. With a median age of 44.6 years[80] Japan is the world's most elderly country today. In contrast, the median age in Pakistan is just 22.1 years, almost half that of Japan. Pakistan is youthful; Japan is full of geezers. But both places will become grayer during the next forty years. By 2050 the median Pakistani age will rise twelve years to 34. Japan's will rise another decade to 55.

When I was young I remember seeing magazine advertisements targeting people who planned to retire at 55. In forty years fully half of all Japanese will be at least that old. On the following page, a table shows how the graying trend will transpire for a few other countries over the next forty years.[81]

As seen plainly on this table from just sixteen examples, there are large age contrasts around the world today and even larger ones emerging in the future. Korea, Russia, and China will join Japan as the world's great geriatric nations. Mexicans will be older than Americans. Median ages will be higher everywhere, but Korea, Vietnam, Mexico, and Iran will age radically, by fifteen years or more. Only our poorest, least-developed countries—like Afghanistan, Somalia, and the Democratic Republic of Congo—will still have youthful populations in 2050; and even they will be somewhat older than today.

This patchwork of wildly contrasting age patterns around the globe is especially driven by the timing of fertility transitions—when the baby booms and echoes happened, and most importantly, when fertility rates first began to fall.[82] In many OECD countries this began in the late 1950s, so the graying process is nearing its end. Median ages have already become quite high today and by 2050 will average only six years older. The same process

Some World Aging Patterns by 2050

(median age, in years)

Country	2010	2050	Change
Republic of Korea	38.0	54.9	(+17 years)
Russian Federation	37.9	45.3	(+7 years)
China	34.9	45.0	(+10 years)
Germany	44.2	44.9	(+1 year)
United Kingdom	40.0	43.4	(+3 years)
Mexico	27.5	43.1	(+16 years)
Chile	32.1	43.1	(+11 years)
Vietnam	26.9	41.6	(+15 years)
USA	36.5	41.1	(+5 years)
Iran	25.9	40.6	(+15 years)
Brazil	28.5	40.4	(+12 years)
Argentina	30.2	40.3	(+10 years)
India	25.0	38.6	(+14 years)
Saudi Arabia	24.4	36.0	(+12 years)
Iraq	19.6	31.1	(+12 years)
Afghanistan	16.7	23.0	(+6 years)

(*Source:* United Nations Population Division)

is now under way in developing countries, where fertility drops beginning in the 1960s, 1970s, and 1980s will unleash successive waves of aging over the planet for the next forty years.

In his 1982 film *Blade Runner,* director Ridley Scott imagined that my home city of Los Angeles would be filled up with Japanese people by the year 2019. In light of Japan's economic might at the time, it's not hard to see where the idea came from. But Mr. Scott should have consulted a demographer, because I just don't see where all those Japanese settlers will come from. Over the next forty years, Japan is going to lose about 20% of her population.

Is an elderly population a good thing or bad? Clearly, there are some benefits: perhaps a wiser, less violent society, for example. But it also strains health care systems,[83] and from an economic perspective it absolutely raises the burden on younger workers. Economists stare hard at something called the "elderly dependency ratio," usually calculated as the percentage of people aged sixty-five or older relative to those of "working age," between fifteen and sixty-four.[84] By the year 2050, elderly dependency ratios will be higher all around the world. Some places, like Korea, Spain, and Italy, will have elderly dependency ratios exceeding 60%. That's barely sixteen people of working age for every ten elders. Japan, with a dependency ratio of 74%, will have only thirteen.

Elsewhere, the overall dependency ratio will be lower but the transition shock greater. Relative to 2010, dependency ratios will more than quadruple in Iran, Singapore, and Korea. They will more than triple in China, Mexico, Brazil, Cuba, Turkey, Algeria, Thailand, Vietnam, Indonesia, and Saudi Arabia. Many of these places today have large youthful workforces that attract global business in its tireless search for labor. But by 2050, the United States may find itself in the unfamiliar position of being unable to find enough migrant farm laborers from Mexico's aging workforce.

Clearly, the whole concept of "retirement" is about to undergo a major overhaul. People will have to work later in life, at least part-time, and perhaps as long as they are able. This is not necessarily a bad thing, as there is some evidence that most people are actually happier with a phased retirement[85] just so long as they perceive a sense of choice in the matter.[86] On the other hand, a "gray crime wave" has now begun in Japan: Arrests of struggling pensioners over age sixty-five has doubled—mostly for shoplifting and pickpocketing—and the number incarcerated has tripled to over 10% of Japan's prison population.[87] It is also apparent that some big cultural shifts will be needed in the way we treat and value our elderly. "Our society must learn that ageing and youth should be valued equally," writes Leonard Hayflick of the UCSF School of Medicine, "if for no other reason than the youth in developed countries have an excellent chance of experiencing the phenomenon that they may now hold in such low esteem."[88]

Another thing about to undergo a major overhaul is how the countries treat and value foreign immigrants. As the world grays, skilled young people will become an increasingly coveted resource, both for direct immigration and for globalized labor pools abroad. This creates the opportunity for new economic tigers to emerge when today's "youth bulges" mature into "worker bulges" in Turkey, Lebanon, Iran, Morocco, Algeria, Tunisia, Colombia, Costa Rica, Chile, Vietnam, Indonesia, and Malaysia—countries offering a reasonably educated workforce and business-friendly environment.[89] A graying world also bodes well for women's employment in places that currently discourage it, because allowing women into the labor pool is the quickest and easiest way to double it. Countries where women don't work for religious and/or cultural reasons will experience an increasingly powerful economic incentive to abandon that tradition by 2050.

The following point will become particularly important later in this book. In an aging world, those countries best able to attract skilled foreign workers will fare best. The early signs of a migrant planet are already here. In 2008, some two hundred million people—3% of the world population—were living outside their native countries. In most OECD countries the proportion of foreign-born was over 10%, even in countries like Greece and Ireland, where emigrants used to flow out, not in.[90] Foreign workers benefit their homelands as well as host economies: The World Bank estimates overseas remittances to poor countries was USD $283 billion in 2008, constituting a huge share of GDP in countries like Tajikistan (46%), Moldova (38%), and Lebanon (24%).

What about in 2050, when the nursing homes in Mexico, China, and Iran are packed full? Who will be running the computers and caring for the residents? Unless the entire world has entered a full-blown robotic age by then, we will still need young people around to do things. Where on Earth will they come from?

This is harder to project demographically, because those young people haven't been born yet. But based on current population structures, the most youthful countries in 2050 will be the same ones where fertility rates are highest today—in the world's least modernized places. Somalia, Afghanistan, Yemen, the West Bank and Gaza, Ethiopia, and much of sub-Saharan Africa will offer our world's youth in 2050.

It's a critical but open question whether our poorest countries can convert their forthcoming demographic advantages into the new skilled workforces needed to help care for an elderly world. Just having a bunch of young people running around is not enough. Huge improvements in education, governance, and security are also required. Women will have to start attending school and working in places where this is uncommon today. Terrorism must be sufficiently quelled such that the countries that need young work-ers will accept immigrants from the countries that have them. I hope that these things can be achieved, and a global skilled-worker program all worked out, by 2050. I'll be eighty-two years old—and I just can't imagine anything lonelier than being turned over in my bed by a robot.

CHAPTER 3

Iron, Oil, and Wind

All I wanna do is to thank you
Even though I don't know who you are
You let me change lanes
While I was driving in my car
—Lyrics from "Whoever You Are" by Geggy Tah (1996)

I nose my compact SUV out of traffic and into the Mobil gas station at Cahuenga Pass, just off the 101 Freeway in Los Angeles. Perched high above me atop the Santa Monica Mountains are the enormous white letters of the Hollywood sign. The nine letters gleam out proudly over a booming young megacity that barely existed a century ago.

I find an open pump and hop out of the car. I swipe a credit card and tap in my ZIP code. I choose a fuel grade, lift the pump handle from its cradle, and jam it into the tank's orifice. I squeeze the pump's handgrip and feel its metal grow cold as fuel churns from another tank in the ground beneath me to the one in my car. It is a simple, mindless act I have repeated countless times since I was seventeen years old. I give no more thought to the process than I do to washing my hands or drinking a glass of orange juice. But I really should be more appreciative. In L.A. the elixir of life isn't Botox: It's gasoline.

The average man must labor for ten hours a day, for two solid months, to perform as much physical work as one gallon of crude oil. No wonder we've abandoned horses and carriages in favor of oil-powered vehicles. This raw material, from which all gasolines, diesels, and jet fuels are refined, is miraculous stuff. It fuels 99% of all motorized vehicles today. And oil is so much more than just a transport fuel—it is an essential ingredient of nearly everything we make. Our plastics, lubricants, cosmetics, pharmaceuticals, and millions of other products all derive somehow from oil. Our food is grown with oil. So besides what I was pumping into my gas tank, I was sitting in oil while driving and was drinking oil as I sipped coffee from my cup.

Since the Industrial Revolution oil, coal, natural gas, and metals have improved nearly every aspect of human life. Before then, a meager existence was the norm no matter what country one lived in. It is naïve to romanticize the eighteenth century as simpler, happier times—the lives of those farmers and townspeople were a constant struggle. Without fossil fuels and metals our lives would be very different. Indeed, today's urbanization megatrend and gigantic cities would not even exist.

The modern city survives upon constant resupply from the outer natural world, from faraway fields, forests, mines, streams, and wells. We scour the planet for hydrocarbons and deliver them to power plants to zap electricity over miles of metal wire. We take water from flowing rivers with distant headwaters of snow and ice. Plants and animals are grown someplace else, killed, and delivered for us to eat. Wind, rivers, and tides flush out our filth. Without this constant flow of nature pouring into our cities, we would all have to disperse, or die.

This reliance of cities upon the outside natural world is a profound relationship to which their occupants give little if any thought. Unlike a hardscrabble Uzbek farmer, modern urbanites worry little about securing water and food, and instead focus on securing jobs and wealth. But a lack of awareness doesn't make this dependency any less profound. Swedish cities, for example, import at least twenty-two tons of fossil fuel, water, and minerals per person annually.[91] In a single year Portugal's growing city of Lisbon gobbles some 11,200,000 tons of material (things like food, gas, and cement) but excretes just 2,297,000 million tons (things like sewage, air pollution, and trash).[92] That's twenty tons coming in and only four going out for every one of Lisbon's 560,000 residents. The difference—nearly nine million tons—stays in Lisbon, mostly in the form of added buildings and landfills. So not only do cities *feed* on their outside natural resource base, they *retain* and *grow* from it.[93]

Clearly then, our global rush to urbanize does not mean giving the natural world a break. As we saw in the previous chapter, when people move to modern cities, consumption goes up, not down. And cities import all sorts of materials besides food, water, and consumer goods. Roads, buildings, and power plants require serious tonnage of steel, chemicals, wood, water, and hydrocarbons. Even in rural areas, the departing farmers are being replaced by tractors and petrochemicals.[94]

As described in the last two chapters, the developing world will experience extraordinary urban and economic growth over the next forty years. What does this portend for our third global force, demand for natural resources? Do we face oil wars and crazy steel prices? Stump forests and dried-up water wells? Are we about to run out of the raw materials our cities and mechanized farmlands so desperately need?

Are We Running Out of Resources?

The debate over natural resources, and whether we are running out of them, is a contentious and surprisingly ancient debate. Even Aristotle wrote about it. In 1798 Thomas Malthus' first edition of *An Essay on the Principle of Population* argued that the exponential growth of human population, set against the arithmetic growth in the area of arable land, must ultimately lead us to outstrip our food supply, thus inevitably dragging us toward a brutal world of famines and violence.[95] Among Malthus' more odious ideas was that social programs are pointless because they enable poor people to have more babies, thus making the problem worse.

Not surprisingly, Malthus' ideas angered many people in his day and since. John Stuart Mill, Karl Marx, Friedrich Engels, and Vladimir Ilyich Lenin were among his vocal critics, mostly retorting that social inequity, not resource scarcity, is the root cause of human suffering. More than two centuries after the publication of this slim book the battle rages on, pitching modern-day "neo-Malthusians" like Stanford's Paul Ehrlich against opponents like the late Julian Simon at the University of Illinois.[96] The debate has now expanded well beyond food production to include all manner of natural resources.[97]

To enter this debate it is simplest to start off with finite, nonrenewable raw commodities that are essential to modern human enterprise, like metals and fossil hydrocarbons (we will take up water, food, and renewable hydrocarbons later). Are we running out?

Let us tabulate estimates of known geological deposits that we have already discovered and know to be of sufficiently high grade that they could be profitably developed tomorrow if necessary. These quantities are called *proved reserves,* or simply *reserves.* It is then a simple calculation to divide the world's total reserves by their current rate of depletion (i.e., their annual production rate) to see how many years are left until the remaining reserves run out. This simple measure is called the "R/P" (reserve-to-production) ratio or the "life-index" of a resource. On the following page are some examples of global proved reserves (both in total and per capita) and R/P ratios for twenty-two of the Earth's especially useful nonrenewable resources.

Two observations leap from these data. The first is that the absolute abundance of a reserve is not always a good predictor of when it might be depleted. The current world reserve of oil—despite being the second-largest at nearly two hundred billion metric tons (about twenty-four metric tons for every man, woman, and child alive on Earth)—is scheduled to run out in just 42 years at current production rates, whereas the supply of magnesium would appear to last for 4,481 more years, despite having only 1/75th the abundance of oil. Platinum would appear to have 150 years left despite being more than two million times scarcer (just 100 grams for every man, woman, and child).

The second observation is that there is an enormous range in R/P ratios, with some reserves projected to be exhausted as soon as eight years from now and others not for hundreds or even thousands of years. The known proved reserves of magnesium, for example, appear sufficient to carry us to the year 6491 at today's rate of consumption. Interestingly, commodity prices do not necessarily reflect this. For example, one can buy silver and lead much more cheaply than platinum, despite their having shorter index lifetimes.

Proved World Reserves of Some Important Natural Resources

Resource	Proved Reserves (metric tons)	Tons per person	R/P life-index (years)	Important Uses
Coal	847,488,000,000	124	133	energy
Oil	168,600,000,000	25	42	energy
Natural Gas (LNG equiv.)	128,611,000,000	19	60	energy
Iron	70,000,000,000	10	72	steel
Bauxite (Aluminum)	23,000,000,000	3	148	many
Magnesium	2,200,000,000	0.3	4,481	alloys, refractories, agriculture
Copper	470,000,000	0.1	35	construction, electricity, electronics
Titanium	470,000,000	0.1	90	industrial, catalysts, food, aerospace
Zinc	220,000,000	0.03	24	coatings, alloys, supplements
Lead	67,000,000	0.01	22	batteries, pigments
Nickel	62,000,000	0.01	21	stainless steel, superalloys, batteries
Cobalt	7,000,000	0.001	171	superalloys, magnets, chemicals
Tin	6,100,000	0.001	24	plating, electronics, construction
Lithium	4,100,000	0.0006	291	ceramics, batteries, pharmaceuticals
Uranium	3,537,000	0.0005	97	energy
Tungsten	2,900,000	0.0004	63	cutting tools, superalloys, electrical
Cadmium	600,000	0.0001	35	NiCd batteries, pigments, coatings
Silver	270,000	0.00004	14	electronics, jewelry, investment
Platinum (group)	71,000	0.00001	157	catalysts, jewelry, electronics
Gold	43,000	0.000006	17	jewelry, industrial, investment
Tellurium	21,000	0.000003	123	alloys, catalysts, semiconductors
Indium	2,500	0.0000004	8	LCD screens, semiconductors

(*Sources: BP 2008;* British Geological Survey 2005)[98]

Why is this? Can the markets be wrong? Before you rush off to hoard lead ingots, note that there are serious flaws with the use of this simple "fixed-stock" approach to project future resource scarcity. An obvious one is that not all "nonrenewable" resources are irreparably destroyed when used, meaning they can be recycled. This is particularly true for metals. Lead and aluminum are highly recycled today, for example. A second flaw is that the size of proved reserves is not truly fixed but tends to rise over time as new deposits are found, extraction technologies improve, and commodity prices go up. The latter can make a low-grade deposit become economically viable, thus adding it to the list of proved reserves despite no new geological discoveries whatsoever. And to an economist, a big problem with the R/P ratio is its implicit assumption that the cost of production for all those tons is equal around the world, when we know that is not the case.

In principle there is sufficient aluminum, iron, zinc, and copper within the Earth's crust to last humanity for millions of years, if we had the energy and technology and desire to extract such dilute materials and didn't object to mining away vast portions of the planet from beneath our feet. Mineral "depletion," at least in the strictly physical sense, is thus meaningless.[99, 100] The better question, therefore, is not "will we run out of aluminum?" but "to what lengths will we go to get it?"

The above flaws—ignoring recycling, and the tendency for proved reserves to increase over time with advancing prices, technology, and new discoveries—make R/P life-index calculations, like the ones tabled on the previous page, overly pessimistic. However, two other factors tend to make them overly rosy. The first is that governments or companies holding a resource sometimes find it in their best interest to be optimistic when assessing the size of their proved reserves. This is particularly true for oil and is a serious concern with Saudi Arabia, currently the world's largest oil producer.[101] The second problem with life-index calculations is that they imply today's rate of consumption will remain fixed into the future. As we saw in the previous chapter, enormous growth in the global economy and population is projected for developing countries. Resource consumption is expected to rise right along with them, thus making life-index projections too short. In light of these weaknesses, R/P life-index values are best used for illustrating the present-day situation, rather than for making projections into the future.

A more sophisticated approach is to link resource consumption to GDP or some other economic indicator, thus allowing it to rise with projected economic growth. Model studies that add this extra step all indicate serious depletion of in-ground reserves of certain key metals, notably silver, gold, indium, tin, lead, zinc, and possibly copper, by the year 2050.[102] Pressure is also rising on some other exotic metals (besides indium) needed by the electronics and energy industries, notably gallium and germanium for electronics; tellurium for solar power; thorium for next-generation nuclear reactors; molybdenum and cobalt for catalysts; and niobium, tantalum, and tungsten for making hardened synthetic materials. Clearly, we are transitioning toward a world where some industrial metals will become either geologically rare and increasingly recycled, or abandoned altogether in favor of cheaper, man-made substitutes.[103] So while physical mineral depletion won't happen soon—and we will see it coming if it does—perhaps you might stash away a little silver and zinc after all. They could well bring you a tidy payback in forty years' time.

What About Oil?

Much less ambiguous is the long-term outlook for conventional oil. *Conventional* means oil in the traditional sense: a low-viscosity liquid that is relatively easy to pump from the ground.[104] Unlike metals, oil cannot be recycled because we burn about 70% of every barrel as transportation fuel. And unlike metal ores, which are diffused in varying grades throughout the Earth's crust, conventional oil is a pure liquid and found only in a narrow range of geological settings. Therefore, after a new oil field is first developed, over the course of several decades its production will inevitably rise, peak at some maximum, and then decline. This sequence is normal and predictable and observed in all oil fields ever drilled on Earth.[105]

For over one hundred years the United States was the world's dominant oil producer. Then, in October 1970. its domestic production peaked at just over ten million barrels per day—about the same as Saudi Arabia's production today—before beginning to fall.

American oil companies launched an epic search to find new domestic reserves. Within ten years the United States was drilling four times as many wells as during the peak, but its production still dropped anyway—to 8.5 million barrels per day and falling. By December 2009 it was down to just 5.3 million barrels per day.[106] So much for "drill, baby, drill" as the solution to energy supply problems.

This story is not unique to America. Azerbaijan's Baku oil fields—once Russia's biggest supplier and the target of Adolf Hitler's eastern front invasion in World War II—are now mostly empty except for littered hulks of rusting junk. Venezuela's enormous Lake Maracaibo Basin is in decline. Iran's oil production peaked in 1978 and now produces barely half the six million barrels per day that it did then.

Most of the world's oil still comes from giant and supergiant oil fields discovered more than fifty years ago. Many of them have now begun their decline, including Alaska's North Slope region, Kuwait's Burgan oil field, the North Sea, and Canterell in Mexico. Saudi Arabia is so far maintaining production from its massive Ghawar field—currently providing over 6% of the world's oil—but eventually it, too, must decline.[107]

A common debate, which to me is not a very interesting one, is whether world production of conventional oil has "peaked" already or whether that day still lies ahead—say in thirty or forty years. Beyond that time window, the chances of finding huge new discoveries of conventional oil—of sizes needed to maintain even our current rate of oil consumption, let alone meet projected growth in demand—grow dim. New oil is still being found, and exploration and extraction technologies continue to improve, but it is now quite clear that conventional oil production cannot grow fast enough to keep up with projected increases in demand over the next forty years.

The reasons for this go even beyond geological scarcity to include "aboveground" challenges in geopolitics, infrastructure, environmental protection, and an aging industry workforce. Many of the fields awaiting development are in parts of the Caucasus and Africa that are dangerously unstable.[108] It takes decades and enormous investments of capital to develop an oil field, and will cost increasingly more in blood and treasure than energy investors are accustomed to. Further supply tightening derives from the fact that oil producers have a long-term financial incentive in limiting production of what is, after all, a finite resource. A large fraction of the world's oil is now controlled by national rather than transnational oil companies. These companies, notes former U.S. secretary of energy Samuel Bodman, are beginning to wonder why they should produce now, when the same oil could make them even more money in the future.[109]

The world currently consumes some 85 million barrels of oil every day and is forecast to demand 106 million barrels per day by 2030, despite the 2008–09 economic contraction and the creation of new government policies encouraging alternative energy sources.[110] To meet this demand, as another former U.S. secretary of energy, James Schlesinger, recently noted, means that we must find and develop the equivalent of nine Saudi Arabias. The probability of this happening is vanishingly small.

Even if total world oil production can be increased, if production cannot keep up with demand, that is still a supply decline. Disturbing twenty-first-century scenarios of intense competition for oil—even to the point of economic collapse and violent warfare—are described in the books *Out of Gas* by David Goodstein, *Resource Wars* and *Rising Powers, Shrinking Planet: The New Geopolitics of Energy* by Michael Klare, and *Twilight in the Desert: The Coming Saudi Oil Shock and the World Economy* by Matt Simmons.[111] These authors are neither hacks nor alarmists. Simmons is a lifelong Republican and oil industry insider, and is widely respected as one of the smartest data analysts in the business. Goodstein is a Caltech physicist, and Klare has long experience in military policy. "Of all the resources discussed in this book," writes Klare in *Resource Wars,* "none is more likely to provoke conflict between states in the twenty-first century than oil." There is ample empirical evidence to support this, including the 2003 U.S. invasion of Iraq and a 2008 war between Russia and Georgia over South Ossetia, a breakaway republic proximate to a highly strategic transport corridor for Caspian oil and gas. A struggle for control of Sudan's south-central oil fields has contributed to ongoing unrest in a country that has seen perhaps three hundred thousand people killed and two million more displaced since 2003.

It's true that we're always just one borehole away from a huge new oil discovery. But realistically speaking, despite great leaps forward in geophysical exploration technology, we stopped finding those about fifty years ago. All of the world's supergiant fields still producing significantly today were discovered in the late 1960s. World production is still rising, but to achieve it we are expending many times the effort to find fewer and smaller pockets of oil. To make matters worse, not only do these smaller fields hold less to begin with, they also decline more precipitously than big fields after they've peaked.[112] According to Simmons' research in *Twilight in the Desert,* a far more likely scenario than a big find is a big crash in the Middle East—home to two-thirds of the world's conventional oil supply—brought on by years of overstatement about the size of Saudi reserves.

Also more likely than a giant new find are supply problems with the reserves we have already. There are plenty of geopolitical problems with oil besides the nationalization trend described earlier. All oil-importing countries worry incessantly about supply disruptions and vulnerabilities. Oil infrastructure is under constant threat from oil spills and terrorism, for example, at Saudi Arabia's Abqaia facility, where Saudi forces thwarted an Al Qaeda attack in 2007.[113] More than two-thirds of all the oil shipped in the world passes through the heavily militarized bottlenecks of the Strait of Hormuz or the Strait of Malacca. When prices hit one hundred dollars a barrel, the United States sends roughly a half-trillion dollars per year to oil-producing countries—including political foes like Venezuela—just to secure its transportation fuel. Few would dispute that securing stable access to oil supplies is a driving force behind U.S.-led military actions in the Middle East.

In light of all this, world leaders, financial markets, and even oil companies have already decided that it's time to add other options to the energy basket. They know the world is entering a time of unprecedented energy demand just as our great oil fields are aging and new ones are harder to find and more expensive to tap. Future production will increasingly come from new discoveries that are smaller, deeper, and riskier; the remnants of depleted giants; and unconventional sources like tar sands. It seems probable that the world will eventually begin regulating carbon emissions one way or another, at least by a token amount. For all of these reasons the cost of *using* oil—regardless of geological supply—is expected to rise.

Obviously, energy conservation measures are the cheapest and most immediate way to soften this blow, and will comprise a key part of its solution. But however we end up feeding our vehicles in 2050, it won't be the same as how we did it back in 2010. We are moving from a narrow fossil-fuel economy to something much more diverse—and likely safer and more resilient—than what we have today. We will explore this exciting range of possible energy futures next.

————

"You got five minutes?"

It was two o'clock in the afternoon and my weight-lifting neighbor, whose hobby is driving racing cars, was standing at my front door. He was grinning fiendishly.

Moments later, my happy excitement had curdled to pure adrenaline and fear and the feeling that I was about to die. My neighbor tapped the accelerator and there again was that terrifying sensation of heart and lungs being pressed against the back of my chest cavity. My body sank into the open-air cockpit, inches above the mountain curves of Mulholland Drive, as the Tesla Roadster screamed silently around them at ninety miles an hour. Flower-fragrant Southern California air pushed up my nose. Smells like a funeral, I thought weakly, and gripped the windshield frame harder. Some-one was howling, probably me. I was trapped in the fastest roller coaster of my life and there were no rails pinning it to the ground.

It felt like an hour, but true to his word, my maniac neighbor had me back home safely in five minutes. He was on his way to Universal Studios to give the CEO a ride. The day before, it had been Anthony Kiedis, lead singer of the Red Hot Chili Peppers. "Faster than a Ferrari from thirty to sixty, and just two cents a mile!" he said, beaming and waving as he drove off. I wandered inside, collapsed on my couch, and wondered if I might be having a heart attack. That's when I realized that electric cars weren't just for eco-pansies anymore.

It is rapidly becoming obvious that plug-in electric cars will be the great bridging technology between the cars of today and the cars of a hydrogen fuel-cell economy later this century (should there be one[114]). Plug-ins differ from conventional cars and hybrids (like the Toyota Prius, first sold in Japan in 1997) because they are powered mainly or exclusively from the electric grid, not by gasoline. And because plug-ins emit very little tailpipe exhaust (zero for fully electric cars with no hybrid conventional motor), that means urban air quality is about to become cleaner.

One of the biggest reasons to be happy about the phase-in of plug-in electric cars has less to with solving climate change or reducing dependency on foreign oil, and more to do with quality of life for all those new city people. Take, for example, my home. It's only a thousand square feet in size, with one bedroom and one bath, but my wife and I love it. It clings to the Hollywood Hills, high above everything, with sweeping views of the downtown Los Angeles skyline and beyond. Every morning one of the first things I do is step out on the deck to check out the view. It's usually crummy, the skyscrapers and distant mountains obscured by the orange-stained smog of ten million belching tailpipes. But on good days, when winds clear out the fumes, we win a breathtaking vista spanning over fifty miles, from blue ocean in the west to snow-covered peaks in the east. It's stunning, and I'm looking forward to those rare views becoming downright ordinary over the next forty years. The public health benefits of this are obvious. Today, as a resident of Los Angeles, I suffer a 25%–30% higher chance of dying from a respiratory disease than my parents, who live on the Great Plains.[115]

This is not to suggest that electric cars are environmentally benign, because they aren't. All of that new electricity must come from somewhere, and for the foreseeable future it will mostly come from power plants burning coal and natural gas. And while the vehicles themselves emit virtually no pollution, these power plants do.[116] Producing millions of electric batteries also requires mining huge volumes of nickel, lithium, and cobalt. There are many technology hurdles remaining with battery lifetime, disposal, and price. Mileage rates are improving (the Chevrolet Volt goes 40 miles, the Tesla 244 miles as of 2010) but still well below the range of a conventional car. Recharging takes several hours unless a system of battery-exchange service stations can be set up. For these reasons and others most first-generation plug-in electrics will likely be hybrids, with a small gasoline or diesel motor that kicks in when the battery range is exceeded. To the extent that they are driven beyond this range, cars will continue to emit pollution and greenhouse gas from their tailpipes.

There is also the "liquid-fuels" problem: Not all transport can be electrified. There is no foreseeable battery on the horizon that will power airplanes, helicopters, freight ships, long-haul trucks, and emergency generators. These all require the power, extended range, or portability offered by liquid fuels. For these forms of transport, gasoline, diesel, ethanol, biodiesel, liquefied natural gas, or coal-derived syngas will be necessary for decades. However, electrification of the passenger vehicle fleet will help ensure adequate supplies of these liquid fuels. And perhaps one day, our descendants will be grateful that we left them enough oil to still make plastic affordable.

So peering forward to 2050, we find a world more heavily electrified than today, and an assortment of strange new liquid fuels. Where will these new energy sources come from? Will clean renewable electricity replace hydrocarbon-burning power plants? And what about hydrogen power, the fuel of space ships, sci-fi movies, and Arnold Schwarzenegger's specially designed Humvee?

Let's start with the last. First, it is important to remember hydrogen is not truly an energy *source* but, like electricity, an energy *carrier*. Pure hydrogen makes a wonderful fuel but isn't just lying around for the taking.[117] Instead, just like making electricity, it must be generated using energy from some other source.[118] A feedstock material is also needed from which to strip hydrogen atoms. The most common feedstocks in use today are natural gas or water, but others, like coal or biomass, are also feasible sources of hydrogen. Energy is used to crack the hydrogen from the feedstock—for example through electrolysis of water[119]—yielding a portable fuel in gas or liquid form. One kilogram is packed with about the same energy as a gallon of gasoline.

But unlike gasoline, the hydrogen is not then burned in a combustion engine. It is instead converted to electricity on-site, by feeding it into a fuel cell. Fuel cells essentially reverse the hydrolysis reaction, combining hydrogen with oxygen to create electricity and water. The newly made electricity is then used to power the car, appliance, furnace, or whatever, with the water by-product either released as vapor or recycled. Like plug-in electrics, fuel-cell cars release no tailpipe pollution or greenhouse gases (besides water vapor[120]). However, they *are* released at the hydrogen plant, unless fossil fuels or biomass can be avoided as sources of energy or feedstocks. In principle, solar, wind, or hydroelectric power could be used to split hydrogen from a water feedstock, making the entire process quite pollution-free from beginning to end.

Sounds wonderful, and many energy experts and futurists believe that one day we will have a full-blown hydrogen economy. The ultimate dream is to use solar energy to split hydrogen from seawater, thus providing the world with an infinite supply of clean hydrogen fuel—and even some fresh-water as a bonus—with no air pollution or greenhouse gases. But nothing like that will be in place by 2050.

Years of research are needed to resolve a rat's nest of challenges concealed within the previous two paragraphs, with major technology advances and cost reductions necessary in all areas.[121] Basic research in hydrogen manu-facture, transport, and fuel cells is still lacking. The cost of making a fuel-cell vehicle is extremely high. A completely new physical infrastructure is required, including manufacturing plants, pipelines, distribution and bottling centers, and filling stations. Hydrogen is explosive, so there are many safety issues to be resolved, like how to safely pack enough of it into a vehicle to drive three hundred miles, comparable to vehicles today. One way is to use highly pressurized hydrogen, but the collision safety of ten-thousand-psi tanks remains unproven. Early hydrogen supplies are all but certain to be made from fossil fuels, and thus will help little with reducing carbon emissions.

In light of these challenges, most experts agree that a hydrogen econ-omy lies at least thirty to forty years in the future, at which point hydro-gen fuel-cell cars might possibly be the new "next-generation" technology that plug-in hybrids are today. Under the conservative ground rules of our thought experiment, we will assume the world will not convert to a hydro-gen economy by the year 2050.

Running on Moonshine and Wood

Unlike hydrogen, biofuels offer a quicker solution to the liquid-fuels prob-lem. Like gasoline, they are refined hydrocarbons that are burned in an internal combustion engine. They use the same filling stations and, with only slight modifications, the same car and truck engines of today.[122] The only real difference between biofuels and current fuels is that they are made from contemporary organic matter rather than ancient organic matter, and are somewhat cleaner. They emit similar levels of carbon dioxide from the tailpipe as gasoline or diesel, but fewer sulfur oxides and particulates. In principle, when biofuel crops grow back they draw down a comparable amount of new carbon from the atmosphere, thus offsetting their emission of greenhouse gas, but this does not take into account the added emissions of growing, harvesting, and transporting the crop. The biggest appeal of biofuels, therefore, is that they offer a domestic or alternative liquid-fuel source to oil, and potentially less greenhouse gas emission, depending on how efficiently the biofuel can be produced.

The most common biofuel today is ethanol made from corn (in the United States), sugarcane (Brazil), and sugar beets (European Union). Making ethanol is essentially the ancient art of fermenting sugars to make alcoholic drinks, meaning that corn-based car fuel is very similar to moonshine. It is commonly mixed with gasoline, and in Brazil, cars run on flex-fuel mixtures containing up to 100% ethanol. Ethanol has higher octane than gasoline and for this reason was used in early racing cars. In fact, when cars were first being developed about a century ago, their makers strongly considered fueling them with ethanol.[123]

The world's two largest ethanol producers are the United States and Brazil, together producing more than ten billion gallons per year. That may sound like a lot, but it's less than 1% of the liquid-fuels market. The good news is that Brazil is becoming quite expert at making sugarcane ethanol. Production is rising rapidly and is expected to double by 2015.[124] Sugarcane plantations are expanding and, contrary to popular belief, represent little deforestation threat to Amazon rain forests because they are found mostly in the south and east of Brazil.[125] Improved agricultural practices have more than doubled the ethanol yield per unit area, and new genetic methods called marker-assisted breeding suggest further increases of up to 30% in the future. The price Brazilians pay for ethanol has steadily fallen for the past twenty-five years even as the price paid for gasoline has gone up.[126] In 2008, for the first time in history, Brazilians bought more ethanol than gasoline.[127]

The United States is also ramping up ethanol production. The 2007 Energy Independence and Security Act calls for a tripling of U.S. corn-based ethanol production by 2022, a goal reaffirmed by the Obama administration in 2010. Ethanol also comprises a large part of the U.S. Department of Energy's official goal to replace 30% of gasoline consumption with biofuels by 2030. The European Union hopes to derive a quarter of its transport fuels from biofuels by the same year.[128]

Unfortunately, there are tremendous differences in production efficiency among the different plant crops used to make ethanol. Sugarcane is a high-value feedstock, yielding up to eight to ten times the amount of fossil-fuel energy needed to grow, harvest, and refine sugarcane into ethanol. Corn-based ethanol, in contrast, is terribly inefficient, usually requiring as much or more fossil fuel in its manufacture as is delivered by the final product. Therefore the greenhouse gas benefit of corn ethanol over oil is negligible.[129] While often pitched otherwise, American subsidies for it are for objectives other than greenhouse gas reduction. For that goal, a far smarter biofuel investment would be production of sugarcane ethanol in the Caribbean, a potential "Middle East" for ethanol export to the United States.[130]

Another problem is that current technology requires ethanol to be made from simple sugars and starches, putting biofuel crops in direct competition with food crops. The U.S. corn ethanol program was widely blamed in 2007 for a worldwide rise in food prices, because it subsidized farmers to plant fields with corn for fuel rather than with wheat and soybeans for food.[131] This notion that biofuels threaten global food supply reared up again in 2008 in response to a series of food riots in Haiti.[132] While this fear is probably overblown—the share of arable land currently used for biofuel production is only a few percent, and geographic models indicate adequate land does exist for the coexistence of energy and food crops[133]—it is nonetheless disturbing to imagine, in a 2050 world with half again more people than today, converting large swaths of prime farmland to feed cars instead of people.

An attractive alternative would be making ethanol from cellulose, extracted from low-value waste and woody material. Indeed, to make sense any large-scale conversion to biofuels must include cellulosic technology.[134] Cellulose is found in waste products like sawdust and cornstalks, or in grasses and woody shrubs that grow on marginal land not suitable for food crops. It is also the only way to achieve large greenhouse gas reduction through biofuels: Because cellulose requires little or no mechanical cultivation, fertilizers, or pesticides, the amount of fossil fuel needed to produce it is greatly diminished.

At the moment, we do not yet have the technology to produce cellulosic ethanol at sufficiently low price and large scale to penetrate the liquid-fuels market. Woody material contains lignin, a tough polymer that surrounds the cellulose to strengthen and protect the plant. Lignin prevents enzymes from reaching the cellulose to break it down to sugars that can then be converted to ethanol. Current methods for doing this require strong acids or high temperatures, making them uneconomic. But cows and termites, through a symbiotic relationship with gut bacteria, have no problem breaking down cellulose, and promising research is under way to discover how we can too.[135] Another potential source of liquid biofuels is algae (e.g., algenol), which can be grown in non-agricultural, non-forest places like deserts, potentially even from wastewater and seawater.

Whether from increased competition with food crops, or the harvesting of brush and wood for cellulose, a downside of all biofuels is a pressure to expand cultivation, putting even more pressure on natural habitats. Because they consume so much land area, biofuels have the largest "ecological footprint" of any energy source including fossil fuels.[136] Another challenge is purely logistical. Most plant biomass is dispersed over the landscape. How will we secure enough of it, and deliver it to plants at a reasonable cost, without also burning large amounts of fuel in the process? In an echo of hydrogen, this lack of broad-scale processing infrastructure thus remains an open challenge to major production of liquid biofuels.

Of the nonfossil fuel sources of energy, biomass is the world's most important source today, accounting for around 9%–10% of total primary energy consumption. Most of this comes from burning wood and dung for heating and cooking in developing countries. While less than 1% of the world's electricity production comes from biomass, its role is expected to grow across all energy sectors in the next forty years, with total biomass consumption rising 50%–300% by the year 2050.[137] Sugarcane ethanol is already a success, and most experts feel that an economically viable cellulosic technology will be found. If the described challenges to agriculture, land management, and infrastructure can be met, biofuels could possibly supply up to a quarter of all liquid transport fuels by 2050.[138] But this is no small task: With world population growing another 50% over the same period, it means tripling our current agricultural productivity. Total bioenergy use in 2050 would have to approach the level of world oil consumption today.

Was Jack Lemmon's Oscar a Setback for the United States?

On March 16, 1979, the movie thriller *The China Syndrome* opened, starring Jack Lemmon, Michael Douglas, and Jane Fonda. It was about a nuclear accident, compounded by a series of human blunders and criminal acts, at a fictional nuclear power plant in California. By sheer coincidence, just twelve days later a nuclear reactor core was seriously damaged at the Three Mile Island power plant near Harrisburg, Pennsylvania. The level of radioactivity leaked into the environment was too low to harm anyone, but the accident's timing was uncanny. The real accident, although quickly contained, brought immediate attention to the film and it became a box-office smash.

Jack Lemmon won an Academy Award for his performance as the distraught plant manager who barricades himself inside the control room to prevent a criminal cover-up by the plant's owners. I won't spoil the ending, but the story remains gripping to this day. *The China Syndrome* horrified an audience of millions and, together with the accident at Three Mile Island, helped to turn the court of U.S. public opinion against nuclear energy. The last year that a construction permit for a new nuclear power plant was issued in the United States was 1979.[139]

Then, a second, far more deadly catastrophe occurred. On April 26, 1986, nuclear reactor unit No. 4 exploded at the Chernobyl power plant in Ukraine, then part of the Soviet Union. The blast and consequent fire that burned for days released a radioactive cloud detected across much of Europe, with the fallout concentrated in Belarus, Ukraine, and Russia. Two people were killed in the plant explosion, and twenty-eight emergency workers died from acute radiation poisoning. About five million people were exposed to some level of radiation.

Soviet officials initially downplayed the accident. It took eighteen days for then–general secretary Mikhail Gorbachev to acknowledge the disaster on Soviet television, but he had already mobilized a massive response. Soviet helicopters dropped more than five thousand tons of sand, clay, lead, and other materials on the reactor's burning core to smother the flames. Approximately 50,000 residents were evacuated from the nearby town of Pripyat, still abandoned today with many personal belongings lying where they were left. Some 116,000 people were relocated in 1986, followed by a further

220,000 in subsequent years. Approximately 350,000 emergency work-ers came to Chernobyl in 1986–87, and ultimately 600,000 were involved with the containment effort. Today, a thirty-kilometer "Exclusion Zone" surrounds the Chernobyl disaster site, and Ukraine's government expends about 5% of its budget annually on costs related to its aftermath.[140] Although claims of tens or even hundreds of thousands of deaths are exaggerated—by conservative estimates perhaps 8,000 people suffered cancer as a result of Chernobyl[141]—and the failures leading to the explosion are unlikely to be repeated, it was an epic catastrophe from which the Soviet Union and nuclear industry never fully recovered. In the United States and many other countries, what lingering support for nuclear power had remained after Three Mile Island was largely buried alongside the victims of Chernobyl.

Today, that situation appears about to change. In late 2008, the U.S. company Northrop Grumman and the French company Areva, the world's largest builder of nuclear reactors, announced a $360 million plan to build major components for seven proposed U.S. reactors. Twenty-one compa-nies were seeking permission to build thirty-four new nuclear power plants across the United States, from New York to Texas. By 2009 the French firm EDF Group was planning to build eleven new reactors in Britain, the United States, China, and France, and contemplating several more in Italy and the United Arab Emirates. In 2010 U.S. president Barack Obama pledged more than $8.3 billion in conditional loans to build the first nuclear reactor on U.S. soil in over three decades, and for his 2011 budget sought to triple loan guarantees (to $54.5 billion) supporting six to nine more. In a *Wall Street Journal* Op-Ed, U.S. secretary of energy Steven Chu called for building "small modular reactors," less than one-third the size of previ-ous nuclear plants, made in factories and transported to sites by truck or rail. And for the first time nearly two-thirds of Americans were in favor of nuclear power, the highest level of support since Gallup began polling on the issue in 1994.[142]

One reason for all the renewed interest is that nuclear fission is one of only two forms of carbon-free energy already contributing a significant fraction of the world's power supply.[143] Notwithstanding the threatening appearance of billowing white plumes streaming from concrete nuclear towers, they emit no greenhouse gases directly,[144] thus winning the support of a surprising number of climate-change activists. To date, nuclear reactors have been tapped mainly to produce electricity, but they also have potential uses for seawater desalinization, district heating, and making hydrogen fuel.[145] Nuclear power plants are very costly and take years to build, but once established they can provide electricity at prices comparable to burning fossil fuel. In some countries like Japan, nuclear power is actually cheaper than fossil-fuel power.[146] Nuclear advocates point to France, which gets about 80% of its electricity from nuclear plants with no accidents so far. Belgium, Sweden, and Japan also obtain large amounts of electricity from nuclear reactors, so far without major mishap.

Public health remains the single greatest concern with nuclear energy. Although great strides have been made to increase reactor safety,[147] accidents and terrorism remain legitimate threats. Of grave concern is the disposal of radioactive waste, which must be safely interred for tens of thousands of years. The most feasible way to do this is probably subterranean burial in a geologically secure formation. But certifying anything as "geologically secure" for a hundred thousand years is exceedingly difficult. After more than two decades of research and $8 billion spent, the U.S. government recently killed plans to tunnel a long-term nuclear waste repository into Yucca Mountain, a volcanic formation in Nevada. Even in the middle of desert, there was simply too much evidence of fluctuating water tables, earthquakes, and potential volcanic activity to declare the site "safe" for a hundred thousand years.

Finally, there is the issue of fuel supply. Estimated R/P life-index estimates for conventional uranium are under a hundred years, with most closer to fifty years. Therefore, over the long run a shift to nuclear power will require the reprocessing of spent uranium fuel rods from conventional "once-through" nuclear reactors so as to recycle usable fissile material. But spent-fuel reprocessing yields high-grade plutonium, even small amounts of which are the principal barrier to acquiring a nuclear bomb. Therefore, any expansion in nuclear power that involves spent-fuel reprocessing or breeder reactors elevates the threat of proliferating nuclear weapons and creates attractive targets for terrorism.

Nuclear power generates about 15% of the world's electricity today. In a recent analysis of the industry's future, the Massachusetts Institute of

Technology concluded that if aggressive steps are taken to deal with the issues of waste disposal and security, it is feasible to more than triple the world's current capacity to 1,000–1,500 conventional "once-through" nuclear reactors, up from the equivalent of 366 such reactors today.[148] Enough natural uranium is available to support this to at least midcentury or so. Depending on the choices we make,[149] our global nuclear power capacity is projected to either stagnate or grow fivefold, producing as little as 8% to as much as 38% of the world's electricity by the year 2050.

Renewable Carbon-Free Electricity: The Holy Trinity

Besides nuclear fission, there are only three other carbon-free sources of energy positioned to significantly dent the world's power needs by 2050.[150] Unlike nuclear energy (which consumes uranium), they are truly renewable. One of them, hydropower, is already important, generating about 16% of the world's electricity today. The other two sources—wind and solar—provide barely 1% combined. But that breakdown is poised to change.

Hydropower is a mature technology that has already been developed to or near its maximum potential in much of the world. There are only so many large rivers, and even fewer appropriate places to build a dam. Except in Africa, South America, and parts of Asia, most of the good spots have already been taken. Big dams also create many local problems. They pool huge reservoirs, displacing farmland, wildlife, and people. They dramatically change hydrological conditions downstream—a big source of strife between countries sharing transboundary rivers—and fill up with silt, requiring dredging. While "small hydropower" schemes that don't require dams, like waterwheels, have great potential for growth, big dam projects do not. For this reason, regardless of the choices we make,[151] hydropower is expected to lose market share despite doubling in absolute terms. By 2050, it is projected to supply just 9%–14% of the world's electricity.

Wind and solar, in contrast, are the fastest-growing energy sectors today. Although wind power provides barely 1% of the world's electricity, that number hides enormous differences around the globe. Nearly 4% of electricity in the European Union, and nearly 20% in Denmark and the Canadian province of Prince Edward Island, comes from wind.[152] This has partly to do with geography—the mid to high latitudes are windier than the tropics, for example—but much of it is driven by investment.

The wind power trend kicked off in the 1980s in California and in the 1990s in Denmark. Today, Germany, the United States, and Spain are aggressive wind developers and presently lead the world in total installed power capacity, each with fifteen thousand megawatts or more (a typical coal-fired power plant is five hundred to a thousand megawatts; a thousand megawatts might power one million homes). India and China are close behind with six to eight thousand megawatts. Canada, Denmark, Italy, Japan, the Netherlands, Portugal, and the United Kingdom all have installed capacities of one thousand megawatts or more. Altogether, at least forty countries worldwide are now developing wind farms,[153] and all of these numbers are growing quickly.

The reasons for this rapid growth are many. To start, wind is free. Wind turbines are relatively cheap, consume no fuel or water, emit no greenhouse gases, and, aside from the permitting process, can be installed quickly. Because wind farms are comprised of many turbines, it is possible to start small, then grow capacity over time. At present, wind power is one of the cheapest renewable energies, averaging around $0.05 per kilowatt-hour,[154] putting it closest to conventional fossil-fuel electricity prices ($0.02–$0.03/ kWh). The main concerns with wind power are bird and bat deaths, conflicts over land use, and aesthetics. Most wind farms today are on land, but offshore installations are also gathering investors' interest. While it's harder to install turbines and grid connections in the ocean, offshore winds are stronger, so they produce more electricity, and there is less competition for the space. In 2010 the Obama administration approved the United States' first offshore wind farm near Cape Cod, Massachusetts.

The wind power industry has a thirty-year legacy and is now reaping double-digit growth. Depending on the choices we make,[155] our global wind power capacity is expected to grow anywhere from tenfold to over fiftyfold by the year 2050, cornering 2%–17% of the world's electricity market.

That leaves solar energy. The Sun, in principle, offers us more inexhaustible clean power than we could ever possibly use. One hour of sunlight striking our planet contains more energy than all of humanity uses in a year. It absolutely dwarfs all other possible energy sources, even if we add up all of the world's coal, oil, natural gas, uranium, hydropower, wind, and photosynthesis combined. It is nonpolluting, carbonless, and free. Panels of solar photovoltaic cells have been powering satellites for over half a century, and we see their familiar shape all around us—encrusted on streetlights, garden lamps, and pocket calculators. Why, then, is our total world production of solar photovoltaic electricity equivalent to that of just one very large coal-fired power plant?

For all its largesse, sunlight has a fundamental problem. Although vast in total, its energy density is low. Unlike a power-packed coal nugget, sunlight is diffuse, low-grade stuff. Getting significant power out of it requires covering a large area, either with mirrors to focus the Sun's rays, or with panels of photovoltaic (PV) cells that directly convert solar photons into electricity. Both are expensive (especially photovoltaics) and efficiencies are low.

Theoretically,[156] PV cells can convert sunlight to electricity with efficiencies as high as 31%, but most are considerably lower, around 10%–20%. If that sounds pathetic to you, then consider that the efficiency of plant photosynthesis, after three billion years of evolution, is just 1%. Nonetheless, a typical silicon-based solar photovoltaic panel, with 10% efficiency and a manufacturing cost of around three hundred dollars per square meter, produces electricity that costs around thirty-five cents per kilowatt-hour. That's seven to seventeen times greater than coal-fired electricity. So sunlight, despite being far and away the world's biggest energy source, is also the most expensive.

Finding a cheaper way to hijack sunlight is thus the single greatest barrier to the widespread use of solar power. Most photovoltaic panels are made of sliced wafers of extremely pure silicon that are highly polished, fitted with electrical contacts, sealed into a module, and encased in transparent glass. They are heavy, cumbersome, and expensive to make, and become even more costly when the price of silicon goes up. As ardent renewable-energy enthusiast Chris Goodall points out, installing large solar panels on the roof of his Oxford home costs about £12,000, yet the total market value of the electricity they produce after four years is just £300. While it makes sense for governments to subsidize such investments initially, eventually the technology must become competitive with fossil fuels in order to take hold.

That means the cost of PVs must fall to about one-fifth of what they are today, a huge challenge. It's a materials-science problem and there is much exciting research under way, particularly in the area of "thin-film" photo-voltaics that abandon heavy silicon panels in favor of exotic coatings of semiconductors like cadmium telluride, or even carbon nanotubes.[157] The conversion efficiencies of these materials would probably be lower than that of traditional silicon PV cells (8%–12%), but if they could be manufactured cheaply—even printed as shrink-wrap for buildings, for example—the cost of PV electricity would tumble and we could start enshrouding the planet in electricity-making paints and films.

At the moment, photovoltaic paint lies in the sweat-soaked dreams of nanotech graduate students. A safer bet for 2050 lies in the expansion of so-called concentrated solar thermal power, or CSP, technology. Like wind power it has been around for years, and is already providing economically viable electricity from a handful of pilot installations. Unlike photovoltaics, CSP does not attempt to convert sunlight into electrons directly. Instead, in much the way that kids fry ants with a magnifying glass, CSP relies on mirrors or lenses to focus the Sun's rays, heating a fluid like water, mineral oil, or molten salt inside a metal tube or tank. The fluid boils or expands, forcing a mechanical turbine or Stirling engine to move, making electricity. Sound familiar? It's just plain old-fashioned electricity generation[158] driven by a new source. And because CSP plants work best on hot, sunny days—a time when millions of air conditioners drive up the price of electricity—their product commands top dollar. Unlike photovoltaics, CSP requires no silicon wafers, cadmium telluride, or other fancy semiconductors, just a great many polished mirrors, the motorized steel racks to mount them on, and a traditional power plant.

To make the most sense, CSP plants should be located in deserts. Current operations include several in Spain and the U.S. states of California, Nevada, and Arizona. Seventy miles southwest of Phoenix a billion-dollar project is under way to spread mirrors across three square miles of desert, enough to power seventy thousand homes.[159] Other projects are operating or planned in Algeria, Egypt, Morocco, Jordan, and Libya.[160] In terms of sheer untapped potential, these North African countries are the next Saudi Arabia–in-waiting for solar energy (as is Saudi Arabia). The same goes for Australia, much of the Middle East, the southwestern United States, and the Altiplano Plateau and eastern side of Brazil in South America.

So why, then, haven't we plastered CSP plants all over our deserts? One reason is that because there are still so few built, the necessary mirrors and other equipment are still specialty products and thus quite expensive. These costs are expected to fall as the industry grows, but at the moment, with electricity prices of at least twelve cents per kilowatt-hour, CSP is still less economical than conventional power plants. Another challenge is the lack of high-voltage transmission lines connecting hot, empty deserts to the places where people actually live. All the electricity production in the world is worthless if it can't be delivered to customers. This entails running hundreds of miles of high-voltage direct-current (HVDC) power cable, which suffers lower transmission losses than traditional alternating current (AC) transmission lines. HVDC is already used to transmit electricity over great distances in Africa, China, the United States, Canada, and Brazil but, like all major infrastructure, is quite expensive. An undersea HVDC cable between Norway and the Netherlands cost about a million euros per kilometer in 2008.[161] So while doable, channeling solar power from the world's deserts to cities will require major capital investments in infrastructure.

One disadvantage that afflicts not just CSP but all forms of solar and wind energy is energy storage. Few of us marvel that a light beam appears with the simple click of a flashlight button. Yet, imagine if the flashlight were powered not by battery but by hand-crank, with no battery storage whatsoever. Use of this flashlight would require constant hand-cranking (I would simply give up and sit in the dark). Furthermore, for maximum efficiency the turning hand would have to exactly match the electricity requirement at all times: Without battery storage, any excess power generated (i.e., beyond the wattage of the bulb) is lost; any deficit causes the bulb to dim.

Scaling this problem up, we see that meeting society's volatile electricity needs in a nonwasteful manner poses an enormous challenge. Demand fluctuates by the week, hour, and minute in response to all sorts of things, from business cycles to the commercial breaks of popular television programs. Power utilities must constantly adjust their production of electricity accordingly. Too much capacity wastes money as power plants make unused electricity; too little capacity triggers brownouts or rolling power outages.

It's hard enough to predict fluctuations on the demand side. Solar and wind sources—because they wither or die on calm days, cloudy days, and at night—add new volatility on the supply side. In a world powered substantially from wind and solar sources, avoiding brownouts will require vast "smart grids," meaning highly interconnected and communicative transmission networks, plenty of backup capacity from conventional power plants,[162] and new ways to store excess electricity for times of deficit.

Storing excess electricity is challenging. One way is "pumped storage" using water. If excess electricity becomes available, it is used to pump water uphill, from a reservoir or tank, to another one at higher elevation. When electricity is wanted, the water is released from the upper to lower container again, flowing by force of gravity over turbines to make electricity. Pumped storage is relatively efficient, inexpensive, and has been around for a long time, but requires lots of water and reservoirs.[163]

An exciting storage idea is to tap into the batteries of millions of parked electric cars whenever they plug into the power grid. By communicating with the grid, car owners can elect to charge up when electricity demand is low, and discharge back into the grid when demand is high. Google Inc. is actively developing such a "V2G" (vehicle-to-grid) technology through their RechargeIt initiative.[164] In effect, a city's entire motor pool becomes a giant collective battery bank, helping to buffer fluctuations in electricity supply and help protect against brownouts. In return, cars earn a profit by buying electricity when it is cheap and selling when it is expensive. Thus, the notion of a "cash-back hybrid." Jeff Wellinghoff, commissioner of the U.S. Federal Energy Regulatory Committee, estimates that if millions of cars were made available to the grid, cash-back hybrids could earn their owners up to two thousand to four thousand dollars per vehicle.

Solar power is an exciting, fast-evolving field, and is positioned for technological breakthroughs on multiple fronts.[165] With transmission line investments CSP technology has good potential to bloom in well-placed deserts, for example tapping the northern Sahara to supply electricity to Europe. Globally, the solar power industry is over USD $10 billion per year and growing 30%–40% annually, even faster than wind power.[166, 167] Depending on the choices we make,[168] world electricity production from solar sources is expected to grow anywhere from fiftyfold to nearly *two thousandfold* by 2050, cornering some 0%–13% of the world's electricity market.

That zero was not a typo. This is all very exciting and will surely inspire many investor fortunes in the stock market. But if you've been adding up the numbers as we went along, you've already figured something out: Fast-growing as they are, the blunt truth is that the clean, renewable energy sources we'd all love to have—wind, solar, hydro, geothermal, tidal, and (sustainably grown) biomass—are in no position to replace nonrenewable sources by 2050.[169]

Despite blistering growth, by 2050 solar energy will just be starting to substantially dent our energy needs. It takes time to grow from a base of near-zero. Our present capacity is so minuscule that a fiftyfold increase of solar power in the next four decades will still supply about 0% of the world's electricity. Even the most aggressively modeled expansion of solar sources suggests they can meet just 13% of the world's electricity demand by 2050. So buy the stocks if you wish, but in forty years where will the bulk of the world's energy be coming from? Very likely from the same sources they come from today. There is simply no realistic way to eliminate oil, coal, and natural gas from the world's energy portfolio in just forty years' time.

Natural Gas versus the Dirty Temptation

As oil supply tightens we will harden our gaze more than ever upon coal and natural gas, until that distant day when renewable sources can catch up. Both have their handicaps and benefits relative to oil and to each other. Neither approaches the value of oil for making liquid fuels and chemical products. However, these two fossil fuels already dominate the world's electricity generation, with about 40% coming from coal and 20% from natural gas (in contrast, only 7% of all electricity is generated using oil). A transition to electric cars, therefore, would seem a natural one even without renewable and nuclear sources of electricity.

Should current trends continue unabated, coal demand will nearly triple by 2050, at which point it would capture 52% of the electricity market. Natural gas demand will more than double, at which point it would capture about 21%. However, nothing is fixed about these "business as usual" projections. Through aggressive conservation measures, and development of natural gas, nuclear, and renewable sources, for example, global electricity production from coal could be as little as a few percent by then.[170] There are compelling reasons for the world to work toward this goal, as we shall see shortly.

Demand for natural gas is projected to more than double between now and 2050, and it is difficult to imagine any scenario in which we will *not* be aggressively pursuing it (and oil) between now and then. Natural gas is widely used for heating, cooking, and industrial purposes. It comprises about one-fourth of all energy consumption in the United States. It has a growing niche as a gaseous transportation fuel, and various gas-to-liquid technologies have good potential for providing liquid fuels. It is the prime feedstock for making agricultural nitrogen fertilizers. Of the big three fossil hydrocarbons, natural gas is by far the cleanest, with roughly one-tenth to one-thousandth the amount of sulfur dioxides, nitrous oxides, particulates, and mercury of coal or oil. When burned, it releases about two-thirds as much carbon dioxide as oil and half as much as coal. There is also considerable room to improve the efficiency of natural-gas-fired plants, mainly by replacing gas-fired steam cycles with more efficient combined-cycle plants.

The biggest drawback of natural gas, of course, is that it's a gas. Unlike coal and oil, which can be simply dumped into tankers or a train car, it isn't very portable. Getting natural gas from wells to distant markets requires either an intricate pipeline system or construction of a special refinery to chill it into liquefied natural gas (LNG). Because LNG takes up only about one six-hundredth the volume of natural gas, it can then be transported using tankers. At present, LNG comprises only a tiny fraction of world gas markets, but its use is growing fast. It is especially appealing for remote gas fields that would otherwise be uneconomic to develop. However, this does not come cheaply. A joint LNG venture begun in 2010 by Chevron, Exxon Mobil, and Shell off the coast of Australia, for example, was expected to cost roughly USD $50 billion. The project will tap offshore gas fields for Asian markets and, together with other LNG projects, could make Australia the world's second-largest LNG exporter after Qatar, with revenues in excess of USD $24 billion per year by 2018.[171]

A second drawback of natural gas, similar to a big drawback of oil, is that most of it is concentrated in a handful of countries. The world's largest reserves, by far, are controlled by the Russian Federation (about 1,529 trillion cubic feet or 23.4% of world total), followed by Iran (16.0%), Qatar (13.8%), Saudi Arabia (4.1%), the United States (3.6%), United Arab Emirates (3.5%), Nigeria (2.8%), Venezuela (2.6%), Algeria (2.4%), and Iraq (1.7%).[172] China and India, projected to be the first- and third-largest economies by 2050, have only 1.3% and 0.6% of world reserves of natural gas, respectively. These countries will require aggressive imports of foreign gas to meet their needs.

Like oil, gas fields are finite, so our transition to natural gas is something of a bridging solution to our long-term energy problems. But, as the cleanest-burning fossil fuel, with lowest greenhouse gas emissions and greatest room for efficiency improvements, it is by far the most environmentally appealing of the three. There are substantial world reserves remaining, a long history of exploitation, and additional markets for fertilizers and perhaps hydrogen feedstocks. In the coming decades natural gas will be an elite commodity, highly prized wherever it is found. There seems little doubt that natural gas, like oil, is a raw resource we shall pursue to the last corners of the Earth.

Coal, in contrast, is plentiful and found all over the world. Proved reserves of natural gas have R/P life-index lifetimes of only around sixty years, but for coal they are at least twice as long, often up to two hundred years.[173] The largest reserves are in the United States (238.3 trillion tons, or 28.9% of world reserves), Russia (19.0%), China (13.9%), and India (7.1%), but coal is mined all over the planet. Coal fueled the Industrial Revolution and, despite popular perceptions, is the world's single largest electricity source today. Half of all electricity in the United States comes from more than five hundred coal-fired power plants. In China it's 80%, and the country is building about two new plants per week, equivalent to adding the entire United Kingdom power grid every year.[174] Coal can even be gasified to make synthetic natural gas (SNG) or liquid diesel and methanol transport fuels. South Africa has been doing this since the 1950s and currently makes nearly two hundred thousand barrels of liquid coal fuel every day.[175] Under our current trajectory, world coal consumption is projected to grow 2%–4%

annually for many decades, surpassing oil to become the world's number one energy source. Should current trends continue unabated, coal demand will nearly triple by 2050.

It's enough to make you wish there was more oil. Coal is the dirtiest and most environmentally damaging fuel on Earth. Entire mountaintops are leveled to obtain it. Coal mining pollutes water and devastates the landscape, covering it with toxic slurry pools and leaving behind acidic, eroding deposits upon which nothing will grow. I studied one of these places for my rather traumatizing master's thesis. An hour's fieldwork would leave me covered in black grime, hands and clothing stained orange from an acidic creek full of chemical leachate.[176] Coal mining also releases trapped methane, a powerful greenhouse gas and even more powerful explosive inside subterranean mines. Several thousand coal miners are killed each year in China.

Coal is worse than oil and much worse than natural gas when it comes to emissions of greenhouse gas, because its carbon content is the highest of all fossil fuels. To produce an equivalent amount of useful energy, burned coal unleashes roughly twice as much carbon dioxide as burned natural gas. It also releases a host of irritating or toxic air pollutants, including sulfur dioxide (SO_2), nitrogen oxides (NO and NO_2), particulates, and mercury. It makes acid rain. If converted to a liquid, it releases 150% more carbon dioxide than oil fuels. To people hoping to bring our escalating release of greenhouse gases to the atmosphere under control, coal is Public Enemy Number One.

As my University of California colleague Catherine Gautier writes, "Were it not for its environmental impact, coal would be the obvious choice to replacing oil."[177] From a geological perspective, there will be no scarcity of the stuff anytime before 2100.[178] And therein is the problem: From nearly all model projections, coal *is* slated to replace oil. By the year 2030 its consumption in the United States is projected to rise nearly 40% over 2010 levels. In China, which already burns twice as much coal as the United States, consumption is projected to nearly *double*.

Other than banning the stuff, the only thin hope lying between this future and a giant upward lurch in the atmosphere's greenhouse gas concentrations is something called Carbon Capture and Storage (CCS), often called "clean coal" technology. There's no such thing as clean coal, but CCS does appear technically possible and, at first blush, alluringly simple: Rather than send carbon dioxide up the smokestacks of coal-burning power plants, use chemical scrubbers to capture it, compress it to a high-pressure liquid, then pipe the liquid someplace else to pump deep underground. Oil companies already use a similar process to force more petroleum out of declining oil fields. Successful pilot demonstrations of CCS technology are under way in Norway, Sweden, and Wyoming, the longest running for more than a decade without mishap.

The main problem with CCS is one of scale, and therefore cost. First, the "capture" process consumes energy itself, requiring significantly bigger plants burning even more coal to generate the same quantity of electricity. Second, a vast network of pipelines is needed to transport staggering volumes of liquid CO_2 away from the power plants to suitable burial sites (abandoned oil fields or deep, salty aquifers). The United States alone produces about 1.5 billion tons of CO_2 per year from coal-fired power plants. Capturing and storing just 60% of that means burying twenty million barrels of liquid per day—about the same as the country's entire consumption of oil.[179] Small pilot demonstrations are one thing, but a demonstration of CCS at the scale of even one full-sized power plant has yet to be attempted. FutureGen, the only proposed prototype, was scrapped in 2008 when its estimated cost swelled to $1.8 billion (the project has since been revived). Finally, there are no guarantees that the stuff won't leak back out to the atmosphere. A leakage rate of just 1% per year would lead to 63% of the stored carbon dioxide being released within a century, undoing much of the supposed environmental benefit.[180]

Carbon Capture and Storage has become a commonly accepted bullet point among proponents of coal, as if all of the above problems have somehow been worked out. Politicians and many scientists have dutifully lined up behind it. It figures prominently in all of our biggest blueprints for reducing greenhouse gases, including model scenarios of the Stern Report, the Intergovernmental Panel on Climate Change, and the International Energy Agency projections outlined above. CCS is embraced by Barack Obama, Angela Merkel, Gordon Brown, and other leaders of the G8. It is the single strand of hope upon which a thunderous increase in carbon emissions from our coming coal boom might possibly be restrained.

I'm not holding my breath.

CHAPTER 4

California Browning, Shanghai Drowning

"Behold, he withholdeth the waters, and they dry up: also he sendeth them out, and they overturn the earth."

—Job 12:15

In January 2008, the U.S. state of Iowa was on the front pages of newspapers all around the world. Ninety-four thousand voters of the Iowa Democratic Party had just propelled Barack Obama—a freshman Illinois senator who was virtually unknown just two years earlier—over the long-time national front-runner, Senator Hillary Rodham Clinton of New York. The Iowa caucuses are the first major electoral event in the U.S. presidential race and are widely believed to influence its outcome. Iowa's voters had delivered a stunning upset and the opening salvo of one of the most exciting and protracted primary battles in U.S. electoral history. Little did they know that only five months later, their state would be on the front pages of newspapers around the world once again.

Within weeks after the political campaigns had left for other battles in other states, the snow started to fall. Two big storms dumped more than three feet of it around the little town of Oskaloosa. By March, Iowa had tied its third-highest monthly snowfall total in 121 years of record keeping. Then came the rain. April's statewide average was the second-highest in 136 years. Twelve inches deluged the town of Fayette, obliterating its previous record of eight inches set back in 1909.[181] Snowmelt and water ran everywhere, flooding cornfields and swelling streams and rivers. On May 25, a category F5 tornado—the strongest category of tornado and Iowa's first F5 in forty years—leveled a forty-mile swath through tiny Parkersburg, killing eight people, destroying hundreds of homes, and narrowly missing populous Cedar Falls. President George W. Bush declared four counties federal disaster areas and the Federal Emergency Management Agency (FEMA) dispatched thirty-nine relief workers to the state.[182] Forty-eight other tornados followed in the month of June, killing four Boy Scouts and raising the state's tornado fatalities to its highest since 1968.

Then things got nasty. The wettest fifteen days in Iowa history began on May 29. Global food prices soared as farm fields in America's top state producer of corn and soybeans melted away in the rain. In Cedar Rapids, thirteen hundred city blocks were inundated when the Cedar River leapt its banks and climbed eleven feet higher than had ever happened in the city's 159-year existence. In Iowa City, parts of the University of Iowa campus were underwater. When I arrived in mid-July the university's magnificent arts buildings and museum were trashed. Cedar Rapids was piled high with gutted wood, dead cars, and molding drywall. A train dangled crazily from a crushed bridge into the river. The little farming town of Oakville was simply wiped off the map—its former green fields cratered or buried in sand by the flood. There was nothing left but wrecked homes and fields, with plumes of black smoke rising from piles of burning wreckage.

By August, eighty-five of Iowa's ninety-nine counties had been declared federal disasters. FEMA's response team had grown from thirty-nine to fifteen hundred. Two million acres of the world's finest farmland had lost twenty tons or more of topsoil per acre; six hundred thousand acres of bottomland were simply scoured away.[183] The statewide damage estimates had swelled to $10 billion—roughly $3,500 for every man, woman, and child in Iowa—and would later go even higher. By 2009 damage estimates to the University of Iowa alone were approaching one billion dollars.[184] Forty thousand Iowans—almost half the number of voters who in January helped send Barack Obama to the White House—had been displaced from their homes.

Meanwhile, six states and eighteen hundred miles to the west, a very different water-related disaster was unfolding. On June 4, 2008—right in the middle of those wettest fifteen days of Iowa history—Governor Arnold

Schwarzenegger strode to a podium in Sacramento to declare an official state of drought in California, the largest total producer of agricultural products in the United States.

Conditions in the Golden State had deteriorated rapidly in an already dry decade. The year before, rainfall in Southern California had been 80% below average. Statewide snowpack and rainfall levels were so low that farmers had begun abandoning their crops. By October, the extreme dryness had fueled a series of vicious wildfires, killing ten people and forcing almost a million more to evacuate. Thousands of homes were destroyed.[185] By May 2008, northern California was also suffering. In many areas its rainfall, too, fell 80% below normal. Flows in the Sacramento and San Joaquin rivers were critically low. Reservoir levels were down across the state, and Lake Oroville, a key supplier to California's massive State Water Project, was half gone. More than a hundred thousand acres in California's sprawling Central Valley—the very heart of the state's gigantic agricultural engine—went unplanted.

Schwarzenegger issued an executive order setting into motion water-transfers, conservation programs, and other measures to combat the crisis,[186] but the drought deepened. Water levels fell further and more fires burned. Eight months later, in February 2009, he proclaimed a state of emergency. Citing "conditions of extreme peril to the safety of persons and property" and "widespread harm to people, businesses, property, communities, wildlife, and recreation,"[187] he ordered even more draconian measures to be taken. Experts were predicting that field fallowing would rise from one hundred thousand to eight hundred thousand acres—meaning that nearly 20% of the Central Valley's farmland would go unplanted.[188] Suddenly, on top of a historic economic crisis from collapsed housing and global credit markets, California was bracing to lose another eighty thousand jobs and $3 billion in agricultural revenue from drought.

Iowa and California were not alone in their water-related crises. As Schwarzenegger mobilized California, the *southeastern* United States, which is usually moist, was also in historic drought, triggering a wave of outdoor-watering bans, withered crops, and unheard-of water battles between states like Georgia, Tennessee, and the Carolinas.[189] Mexico had been in severe drought, with only limited relief, for fifteen years.[190] Exceptional droughts were under way in Brazil, Argentina, western Africa, Australia, the Middle East, Turkey, and Ukraine.[191] Drought emergencies were triggering food aid in Lesotho, Swaziland, Zimbabwe, Mauritania, and Moldova.[192] By February 2009, precipitation was 70%–90% below normal in northern and western China, threatening 10% of the country's entire cereal production.[193] That same month, extreme dryness primed "Black Saturday," when six hundred blazes killed two hundred people in the worst Australian wildfires in history. By April, crop failures in Chattisgarh state drove fifteen hundred Indian farmers—unable to repay their debts without water—to commit suicide.[194]

Within days of the Iowa floods, heavy rains also struck eastern India and China, killing sixty-five people and displacing five hundred thousand in India. In China, floods in Guangdong and Guangxi Zhuang, Sansui City, and the Pearl River delta killed 176 and displaced 1.6 *million*. While America's eyes were fixed on Sarah Palin, hydrologist Bob Brakenridge at Dartmouth was watching floods from space, using satellites to track them all over the world.[195] In the ten months between Barack Obama's winning the Iowa caucuses on January 3, and the general election on November 4, Brakenridge documented 145 major floods carving destruction around the planet. As Barack Obama took down first Hillary Clinton and then John McCain, those rivers took down lives and property from Taiwan to Togo. They killed almost five thousand people and washed seventeen million more from their homes.

Our Most Necessary Resource

It's hard to imagine anything humans need more than freshwater. If it were to all somehow vanish, the human race would be extinct in a matter of days. If it stopped flowing to our animals and fields, we would starve. If it became unclean, we would become sick or even die. Our societies need water in proper quantity, quality, and timing to preserve civilization as we know it. Too little, or at the wrong time of year, and our food dies off and industries fail. Too much, and our fields dissolve and people drown. For the past ten thousand years the very existence of permanent human settlements has depended upon having a consistent, dependable supply of usable water.

What does the future hold? Are we running low on water, as we must ultimately run low on oil? In the past fifty years we've doubled our irrigated cropland and tripled our water consumption to meet global food demand. In the next fifty, we must double food production again.[196] Is there really enough water to pull that off?

In his book *When the Rivers Run Dry* environmental journalist Fred Pearce describes in vivid, firsthand detail the stark reality of impending water crises in more than thirty countries around the globe. We now withdraw so much water that many of our mightiest and most historic rivers—like the Nile, the Colorado, the Yellow, the Indus—have barely a trickle left to meet the sea.

The good news is that, unlike oil, which is ultimately finite, water is endlessly returned to us by the hydrologic cycle. Except for fossil groundwater, there is no such thing as "Peak Water" in the same sense as "Peak Oil." It always comes back—somewhere—as rain or snow. It may be too much, or too little, or come at the wrong time, but it does come back. The bad news is that in addition to the aforementioned problems of too much, too little, or bad timing, our water sources can also become polluted. Finally, while it's true that there is plenty of water circulating out there someplace, nearly all of it is useless to us.

The Russian hydrologist Igor Alexander Shiklomanov estimates that almost 97% of the world's water is salty ocean, unfit for drinking or irrigation; 1% is salty groundwater, again useless. Of the 2.5% or so that is fresh, most *would* be salty if not for the glaciers of Antarctica, Greenland, and mountains that hold it up on land in the form of ice, rather than letting it run off into the ocean. Fresh groundwater holds about three-quarters of 1%. The minuscule remainder—about eight one-thousandths of 1%—is held in all the world's lakes, wetlands, and rivers combined. Our atmosphere's clouds, vapor, and rain hold even less, just one ten-thousandth of 1% of all water on Earth.[197]

There are three points to be taken from Shiklomanov's numbers. The first is that the most important sources of water for people and terrestrial ecosystems—rivers, lakes, and rain—are actually fleetingly rare forms of H_2O. If all the water in the world was a thousand-dollar bill, these sources would amount to about eight cents. The second point is that relative to rivers, lakes, and rain, far larger volumes of freshwater are frozen up inside glaciers, or stored underground in aquifers. These, too, are critically important to humanity and will be discussed shortly.

The third point—and frankly one that is all too often neglected by policy makers and scientists alike—is that these numbers alone do not tell the whole story when it comes to human water supply. Recall that water, unlike oil, is a circulating resource. It recycles constantly through the hydrologic cycle, in infinite loops of rain, runoff, evaporation, and various storage compartments, like ice. From a practical standpoint the *throughput* of freshwater (or "flux") is just as important as the absolute size of its various containers. The total volume of water held in rivers at any given instant is tiny, but it is replaced quickly, unlike, say, an ancient glacier or slowly oozing aquifer. A water droplet moves down a natural river in a few days, whereas the same droplet moving through glaciers, groundwater, and deep ocean currents could be stuck there for centuries to hundreds of thousands of years. This explains the seeming paradox that despite the world's rivers' instantaneous storage capacity of just two thousand cubic kilometers of water, we pull almost twice that amount from them every year.[198]

This is why rainfall and surface water, despite their diminutive holdings, are so critically important to land-based ecosystems and people. Their fast throughput is what makes them so valuable. But because their storage capacities are so tiny, we are vulnerable to the smallest of variations in that throughput. Unlike an ocean or glacier, the atmosphere and rivers have no meaningful storage capacity from which to draw water in dry times or hoard it in wet times. Therefore, terrestrial life is highly sensitive to floods and droughts, whereas marine life is generally not. Tuna have plenty of worries, but droughts are not one of them. Battling this vulnerability is a prime reason why we have built millions of dams, reservoirs, lakes, and ponds throughout the world. Yet even after all this massive engineering, we still have only enough of these artificial impoundments to store slightly less than two years' water supply.[199]

The other big problem for humans, of course, is that this small bucket of fast-recycling river water is spread very unfairly around the planet. Canada, Alaska, Scandinavia, and Russia are veined with so many permanent streams, rivers, and lakes that most have never been named, whereas Saudi Arabia has no natural ones at all. Water-rich Norway has 82,000 cubic meters of renewable freshwater per person while Kenya has just 830.[200] And to a very large degree, this unfair distribution of surface water is created by the pattern of the global atmospheric circulation itself.

Rainmaker, Land Baker

Just a hundred steps into the rain forest my head was thudding, my shirt drenched, and I couldn't breathe. It wasn't claustrophobia—although I couldn't see well through the green gloom of filtered canopy light—but the wet, steaming heat. It was like inhaling vapors over a teakettle. Something went soft under my foot—I had unwittingly crushed an exotic caterpillar the length of my hand. I excused myself from the group and walked gasping back toward the boat, but was intercepted by an aboriginal man. He was selling tiny clay couples with enormous genitalia, eternally frozen in joyous copulation. Back on the boat, a hot breeze blew down the Amazon River but my skin dripped even faster. The air was totally saturated. I couldn't wait to get back to my air-conditioned hotel room in Manaus.

I must have caught the Amazon on a bad day. Most living things love tropical rain forests. Their wide green sash—plain on any world map, roughly encircling the equator—is bursting with life and contains the vast majority of species, known and as yet undiscovered, on Earth. Rain forests grow there thanks to the condensate downpours dumped by the moist, rising air masses of the Intertropical Convergence Zone (ITCZ). This band of clouds and rain follows the Sun, circling nearly directly overhead, as it sizzles the equatorial oceans and landmasses to evaporate huge quantities of water vapor. The vapor rises, cools, and condenses, deluging the tropics with rain and triggering the Asian and African monsoons as the ITCZ drifts back and forth across the equator each year, endlessly chasing the seasonal march of the Sun. Billions of living things hang on the strength and reliability of these annual rainfall patterns, including us.

To the north and south, straddling the lush equatorial belt and monsoonal areas like the dried-out bun halves of a veggie sandwich, are two huge drought-stricken bands of drylands and deserts. The Sahara, Arabian, Australian, Kalahari, and Sonoran are all found here, huddled at roughly 30° N and S latitude. While not lifeless, these zones are decidedly stark compared with their green equatorial neighbor. They mark the killing fields of the moist ITCZ air masses. Emptied of their rain holdings, the air masses drift north or south before tumbling earthward again, baking the land with crushing dry heat, pressed downward by the weight of still more air falling from above. Like the perpetual circuit of rising and falling wax in a Lava lamp, this sinking air closes the convection loop, flowing from both hemispheres back toward the equator in the form of trade winds. From there, the Sun's rays will moisten and lift the air once again, repeating the cycle. This overall pattern of atmospheric circulation, called the Hadley Cell, is one of the most powerful shapers of climate and ecosystems on Earth.

Despite the harsh aridity, billions of people live in or around those twin subtropical blast zones of sinking dry air, which contain some of our fastest-growing human populations. Pressing hard into the Sahara's southern flank are nearly eighty million people of Africa's Sahel, a population projected to reach two hundred million by 2050.[201] North of the Sahara are the large populations of northern Africa and Mediterranean Europe. Australian cities cling to the coastline of their dusty continent, leaving the continent's vast desert interior mostly uninhabited. But the parched Middle East, southern Africa, and western Pakistan are heavily populated and have some of the youngest, fastest-growing populations in the world.

Phoenix and Las Vegas—two briskly growing cities in the arid southwestern United States—lie in the middle of a Hadley Cell desert. Nineteen million people can survive in Southern California only because there are a thousand miles of pipelines, tunnels, and canals bringing water to them from someplace else. It comes from the Sacramento–San Joaquin Delta and Owens Valley to the north, and from the Colorado River to the east, far across the Mojave Desert. They enjoy green lawns, burbling fountains, and swimming pools in a place where rainfall averages less than fifteen inches per year. A second canal[202] from the Colorado pumps water up nearly three thousand feet in elevation and 330 miles east to Phoenix and Tucson, prompting Robert Glennon, author of *Water Follies,* to observe that we literally move water "uphill to wealth and power."[203] Without this infrastructure and the energy to run it, Arizonans' water supply would more closely resemble that of Palestinians: fifteen dubious gallons a day haggled from the back of a water trafficker's truck.

Which Is Worse?

Even if there were no climate change, the world would still be facing declining per capita water supply because of our growing economy and population. In general, more people means more water demand. Even if we could freeze population growth, advancing modernization means more meat, finished goods, and energy, all of which raise per capita water consumption.[204] Contrary to common perception, population growth and industrialization thus represent an even bigger challenge to the global water supply than does climate change.

Policy wonks and water managers have long sensed this. But hydrologist Charlie Vörösmarty blew it wide open in 2000 when he and his colleagues Pamela Green, Joe Salisbury, and Richard Lammers at the University of New Hampshire compared climate and hydrologic models with long-term population and water-consumption trends.[205] As part of the study, they published three brightly colored maps of projected water demand for 2025. I make my students stare at these maps at least once in my introductory course lectures at UCLA.

One of the maps is quite scary-looking and captures the combined effects of both climate and population trends on human water-supply stress. Most of the world is colored red (indicating less water availability than today) with a few places colored blue (more water availability, mostly in Russia and Canada) and even fewer in green (meaning little or no change). This fearsome red map suggests that by the year 2025 much of humanity's water supply will be worse off, either from population growth, or climate change, or both.

The other two maps separate out the effects of population and climate change. The population-only map is even scarier than the combined map. Nearly all the world is bathed in red, with blue colors even rarer than before. Compared to it, the climate-only map seems almost benign, with roughly equal proportions of blue and red tones and even more in green. In other words, climate changes are expected to both harm and help water availability in different parts of the world, whereas population and economic growth harm it nearly everywhere.[206] So even if our climate-change problems could somehow disappear tomorrow (and they won't), we would still face enormous challenges to water supply in some of the hottest, most crowded places on Earth.

Drinking Sh**

It's hard to imagine the world behind those red maps. To most people—especially living in cities—clean water is like oil and electricity: one of those things upon which they depend mightily yet give barely a passing thought. In my own city of Los Angeles, everyone will gladly pay a hundred dollars a month for cable television, yet would roar in protest if forced to pay that much for life's elixir piped directly into their homes. When Governor Schwarzenegger declared a state of drought emergency, I studied my water bill closely for the first time in my life. For two months of clean drinking water, snared from faraway sources and delivered to my house by one of the world's most expensive and elaborate engineering schemes, I was charged $20.67. I spend more on postage stamps.

If only everyone could indulge such ignorant bliss. While eight in ten people have access to some sort of improved water source,[207] this globally averaged number masks some wild geographic discrepancies. Some countries, like Canada, Japan, and Estonia, provide clean water to all of their citizens. Others, especially in Africa, do so for under half. The worst water poverty is suffered by Ethiopians, Somalis, Afghanis, Papua New Guineans, Cambodians, Chadians, Equatorial Guineans, and Mozambicans.[208] Even their statistics hide the most glaring divide—between cities and rural areas. Eight in ten urban Ethiopians have some form of improved water whereas just one in ten rural Ethiopians do.

As we saw in Chapter 3, cities empower efficient channeling of natural resources to people. It is far more economical to lay water pipes and sewerage in a densely populated area than to spread them across the countryside. For much of the world, even sewers are a luxury. Unbelievably, four in ten of us don't even have a simple pit latrine. Small wonder that waterborne diseases kill even more people than our raging epidemic of HIV/AIDS. As Jamie Bartram of the United Nations World Health Organization writes:

> Far more people endure the largely preventable effects of poor sanitation and water supply than are affected by war, terrorism, and weapons of mass destruction combined. Yet those other issues capture the public and political imagination—and public resources—in a way that water and sanitation issues do not. Why? Perhaps in part because most people who read articles such as this find it hard to imagine defecating daily in plastic bags, buckets, open pits, agricultural fields, and public areas for want of a private hygienic alternative, as do some 2.6 billion people. Or perhaps they cannot relate to the everyday life of the 1.1 billion people without access to even a protected well or spring within reasonable walking distance of their homes.[209]

Most experts agree that getting clean water to the world's poorest people is largely a matter of money. According to the United Nations, the price tag for everyone to have safe, clean drinking water would be about $30 billion per year. But in the poorest countries, building water treatment plants and a network of pipes to move it is still prohibitively expensive, especially for rural areas. Well-intentioned foreign aid often fails to leave the cities of ruling elites. And while small, inexpensive water treatment technologies like ultraviolet purification hold promise, microprojects have failed to attract much interest from the big lenders. Water expert Peter Gleick, cofounder and president of the Pacific Institute, likes to point out that the World Bank and International Monetary Fund know how to spend a billion dollars in one place (on a big dam project, for example) but not how to spend a thousand dollars in a million places. But all too often, a thousand-dollar solution is what's needed most. Getting clean water to people living in our most impoverished places remains an enormous challenge, with no clear solution on the horizon.

Another trend is further clouding the picture. Multinational corporations are increasingly moving to privatize and consolidate water supplies. Over the past decade, at least three—Suez, Veolia Environmental Services (formerly Vivendi), and Thames Water—have expanded into for-profit water delivery ventures all over the developing world. In early 2009 Germany's industrial giant Siemens paid nearly $1 billion for U.S. Filter, the leading supplier of water treatment products and services in North America. Multinational giants like General Electric and Dow Chemical are also jumping into the water business, alongside other companies you've never heard of, like Nalco, ITT, and Danaher Corporation.

The benefit of this water-privatization frenzy is the expansion of modern water treatment and distribution facilities into impoverished places that desperately need them. However, these are for-profit companies, not public municipalities. In return for the new infrastructure, they must charge fees for the water in order to recoup building costs and generate profits for their shareholders. This is a familiar transaction in the developed world, where people are accustomed to paying for water, but is a radical shift in poor countries where municipal water supply—to the extent that it is available—is often free.

Control of life's most essential natural resource by overseas multinational corporations is an abomination to people like Maude Barlow, author of *Blue Gold* and *Blue Covenant.*[210] These books point out that to the poorest of the poor, even a few cents for water is unaffordable, forcing them to drink from polluted streams and ditches, fall sick, and die. Extrapolating the current globalization trend into the future, Barlow imagines the following in *Blue Covenant:*

> A powerful corporate water cartel has emerged to seize control of every aspect of water for its own profit. Corporations deliver drinking water and take away wastewater; corporations put massive amounts of water in plastic bottles and sell it to us at exorbitant prices; corporations are building sophisticated new technologies to recycle our dirty water and sell it back to us; corporations extract and move water by huge pipelines from watersheds

and aquifers to sell to big cities and industries; corporations buy, store, and trade water on the open market, like running shoes. Most importantly, corporations want governments to deregulate the water sector and allow the market to set water policy. Every day, they get closer to that goal.

Opponents of multinational companies are a passionate group, and especially when it comes to water. They protest that water privatization has become a key objective of the World Bank, and even of regional lenders like the African Development Bank and Asian Development Bank, with full buy-in from the United Nations and World Trade Organization. They accuse the World Water Council—purportedly an ideologically neutral platform to promote "conservation, protection, development, planning, management, and use of water in all its dimensions on an environmentally sustainable basis for the benefit of all life on Earth"[211]—as in fact being a subversive global champion of water privatization and business corporations. They organize resistance movements and sit-ins, losing a fight with Nestlé over a Poland Spring bottling plant in Michigan, winning another against Coca-Cola at Plachimada, India; and even street riots to force Bechtel out of Bolivia.[212]

Surveying the debate coolly from arm's length, one can appreciate the benefits of the private-sector model. If countries cannot or will not deliver clean water to their citizens who desperately need it, and neither will the World Bank, then why not let private capital have a go? On the other hand, something does feel creepy about transferring control of life's most basic requirement—clean drinking water—from local to overseas control, to corporations whose fiduciary responsibility lies first and foremost with their shareholders. Paying for water works fine in the developed world, but where people earn a dollar per day? Is water property, or human right? This battle continues on fronts all over the world, with no clear best path forward.

World population will grow by 50% in the next forty years, nearly all of it in the developing world and mostly in places that are already water-stressed now. This new population will also be wealthier and eat more meat, thus requiring higher per capita food production than today. To meet this projected demand for food and feed, we must double our crop production by 2050. Finding enough freshwater to support this, plus more industry, plus billions of new apartments, all while keeping the water clean as it cycles endlessly between our kidneys and the environment, is very likely the greatest challenge of our century.

The Information Revolution

Breakfasts at high-powered NASA meetings in Washington, D.C., were much less glamorous than I'd hoped. Rather than sampling astronaut food in a gleaming high-tech boardroom, I was hunched in a bland carpeted hallway at the Marriott, poking a half-empty platter of stale bagels. But I didn't mind. I grabbed the last poppyseed and a cup of coffee and ducked into the cramped meeting room. My old grad-school roommate Doug Alsdorf, now a professor at Ohio State, was bellowing at us to take our seats. I found one and sat quickly. One of the smartest men I have ever known, radar engineer Ernesto Rodriguez from NASA's Jet Propulsion Laboratory, was preparing to give us another update on our half-billion-dollar idea.

The water crisis is about more than failing crops and unsanitary conditions. It is also about *information*—or more precisely, the lack of it—for effective water management. Water is constantly on the move, but unbelievably, we have hardly any idea of where, when, or how much we have at any given moment. Our knowledge of Earth's hydrology is extraordinarily data-poor. Other than large rivers, few streams are measured. Outside the United States and Europe, the vast majority of water bodies receive no hydrologic monitoring whatsoever. We have basically zero information for small lakes, cattle ponds, and wetlands. Even the water levels behind dams, while monitored by their operators, are seldom released to the broader public in many countries.

Because of this information gap, millions of people have no idea whether next week will bring lower water levels in their river or lake, or a raging flood. Emergency workers don't know when a flood has peaked or how high it will go. Along many rivers even the weather isn't a reliable predictor because upstream reservoirs release water at the command of dam operators, not rainstorms. In a complete reversal of their preexisting natural state, many of today's rivers *shrink*, not swell, as they move downstream. In fits and starts, a gauntlet of diversions and dams sips them to death.

Since construction of the High Aswan Dam almost all flow in the Nile

River is now either diverted for irrigation or evaporates away behind reservoirs.[213] Dams along Africa's Volta River system can hold back or release more than four years' worth of its total river flow. Water passage through the Euphrates-Tigris in the Middle East, the Mae Khlong in Thailand, the Río Negro in Argentina, and the Colorado in North America is similarly controlled. But hydrologic data are seldom released. Many countries even classify them, so their downstream neighbors can't tell if they are complying with international water-sharing agreements.[214]

These are the reasons why our group of scientists and engineers were in that Washington, D.C., hotel room, and in other meeting rooms like it in Rome, San Francisco, Barcelona, Paris, Orlando, San Diego, Columbus, and Lisbon. There are now over five hundred of us in thirty-two countries, working on a bold new idea to globalize *information* about water resources, by measuring it everywhere and all the time, from space. The technology to do it is a satellite called a wide-swath altimeter. It uses a remarkable radar technology that Ernesto Rodriguez invented, called a "Ka-band radar interferometer" or KaRIn (named adorably after Ernesto's wife). We're going to put KaRIn into space, mounted on a satellite called SWOT.[215]

SWOT will point not one but two radars—tethered to each other by a thirty-foot boom—toward the Earth. Like two giant police radar guns they will stare down at the planet, zapping millions of rivers, lakes, coastlines, and other wet spots on its rotating face while hurtling through orbit at over fifteen thousand miles per hour. Even one SWOT satellite will stream three-dimensional water-level maps of the entire world, day and night. This technology will constantly scan the pulse of the planet's plumbing. It will unveil its throbs and ebbs of circulating water in all their complexity for the first time. Then, we will post the data online for free.

Billions care about the fate and availability of their water. Especially where it is scarce, little information is available, and lives depend on it. Our satellite is currently wending its way through the political labyrinth of being approved, built, and launched. We are hoping it can be up and orbiting by 2018. But regardless of SWOT's particular fate, I am confident that by 2050, its successors will have made globalized water resource information transparently available for everyone and everywhere on Earth, as has now been done very successfully with other kinds of satellite data.[216] No more water secrets or scientific question marks. It will completely transform the way we study and manage our most vital natural resource.

Wars over Water?

It has become fashionable to declare water the "next oil," over which the world is bracing to go to war in the twenty-first century. Googling "water wars" yields over three hundred thousand hits; the phrase is showing up in scholarly articles as well as newspaper headlines.[217] "Fierce competition for freshwater," said U.N. secretary general Kofi Annan in 2001, "may well become a source of conflict and wars in the future." His successor, Ban Ki-Moon, in a 2007 debate of the U.N. Security Council, warned of water scarcity "transforming peaceful competition into violence," and floods and droughts sparking "massive human migrations, polarizing societies and weakening the ability of countries to resolve conflicts peacefully."[218]

International relations professor and journalist Michael Klare gets more specific. He expects four rivers in particular—the Nile, Jordan, Tigris-Euphrates, and Indus—to provoke "high levels of tension along with periodic outbreaks of violent conflict."[219] Those four are good picks. They are already oversubscribed, and shared between sworn enemies. The Jordan River's water is divided among Israel, Jordan, Lebanon, Syria, and the occupied Palestinian territories. Tigris-Euphrates water is used by Iraqis, Iranians, Syrians, Turks, and Kurds. The Indus is shared by Afghanistan, China, India, Pakistan, and Kashmir. The Nile and its tributaries are controlled by eight other countries besides Egypt.

Virtually all of the water flowing down these four river systems is in use today. By 2050, depending on the basin, their dependent human populations will jump anywhere from 70% to 150%. This means that for a vast area, from North Africa to the Near East and South Asia, human demand for water is rapidly overtaking available supply. "Now at the dawn of the twenty-first century," Klare warns, "conflict over critical water supplies is an ever-present danger."[220]

Scary stuff. But will the world really go to war over water? Here is a pleasant surprise: History tells us that while international conflicts over water are very common, nearly all of them—at least so far—are peacefully settled. A close reading of history reveals that while water and violence are often associated, countries rarely resort to armed violence *over* water.[221]

Peter Gleick at the Pacific Institute and Aaron Wolf at Oregon State University maintain historical databases of past conflicts and their causes.[222] These reveal a rich soap opera of tensions, conflicting interests, and contentious relations, but not outright war—at least not between sovereign countries or specifically over water resources. Most commonly, the violence they document identifies water as a tool, a target, or a victim of warfare— but not its cause.[223]

Remarkably, successful water-sharing agreements are common even between hydrologically stressed countries that go to war over other things. Wendy Barnaby, editor of Britain's *People & Science* magazine, points out that India and Pakistan have fought three wars, yet always have managed to work out their water disputes through the 1960 Indus Water Treaty.[224] The reason is purely rational: By cooperating, both countries are able to safeguard their core water supply. Water is *too important* to risk losing in a war. Israel's water independence ran out in the 1950s, Jordan's in the 1960s, and Egypt's since the 1970s. But their wars have never been fought over water. It's amazing, because these countries no longer have enough even to grow their food.

Instead, they all import someone else's water . . . in the form of grain.

The Virtual Water Trade

The most skilled diplomats in the world couldn't stop a water war if people were starving. What enables sworn enemies to coexist, with large and growing populations, along a dwindling dribble like the Jordan River? Ten million people living between it and the Mediterranean Sea, with barely enough water to grow a fifth of their food? The answer is global trade flows of *food*.

The single biggest users of water are not cities but farms. Fully 70% of all human water withdrawal from rivers, lakes, and aquifers is for agriculture.[225] Because agricultural products require water to grow, they essentially have water resources "embedded" within them. The export and import of food and animals, therefore, amounts to the export and import of water.

This "virtual water trade" is the globalized-world solution to the ancient problem of having abundant water in some places and not enough in others.[226] From the global perspective, it is also less wasteful. It takes far more water to grow an orange in the baking dry heat of Saudi Arabia than to grow the same orange in humid Florida. Hidden inside Mexico's imports of wheat, corn, and sorghum from the United States is the import of seven billion cubic meters of virtual water a year. Not only does this help Mexico—now in its fifteenth year of drought—it also requires less water overall. To produce that same amount of grain domestically, Mexico would need nearly sixteen billion cubic meters of freshwater per year, almost nine billion more. That single trade relationship saves enough water to flood the entire United Kingdom under an inch and a half of standing water.

The virtual water trade is a little-discussed secret not publicized by political leaders. Most people don't enjoy hearing that their country is food-dependent, or that it uses its water to support others. North America is the world's biggest exporter of virtual water. Many countries—including much of Europe, the Middle East, North Africa, Japan, and Mexico—are net importers. Unbelievably, about 40% of all human water consumption is moved around in this way, embedded in global trade flows of agricultural and industrial products.[227] Without these flows the world would look very different than it does today. Dry places would support far fewer people. Lacking distant markets, large areas of terrific farmland would either surge in population or become abandoned. Global trade may be bad for local economies, bad for energy consumption, bad for resource exploitation, bad for other things . . . but it's also spreading the wealth—of water—around.

Despite its endless recirculation, there are parts of the hydrologic cycle that smell suspiciously like depletion of a finite natural resource. This is especially true for underground sources, collectively called *groundwater.*

Groundwater is a very attractive water source. Unlike rainfall and rivers, which have tiny holding capacity and variable throughput, aquifers hold large volumes and are relatively stable. Humans have dug wells for thousands of years—the Egyptians, Chinese, and Persians had them as early as 2000 B.C. However, wells more than seventy to eighty feet deep are a modern invention, brought about by centrifugal pumps and the internal combustion engine.[228] In water-scarce areas this new technology quickly triggered a water-drilling boom, much like the oil-drilling boom described in the previous chapter. We became a horde of mosquitoes, piercing and probing the planet with steel prosboscises in search of fluids.

Tapping subterranean water meant that farmers could convert drylands and deserts into lush, productive fields virtually overnight. Here's a dirty little secret about the agricultural "green revolution" of the latter half of the twentieth century. The green revolution was brought about not only by new petrochemicals, hybrid seeds, and mechanized agriculture, but also by a massive ballooning in the pumping of groundwater to irrigate crops. In just fifty years the world's irrigated land area *doubled* from 60 million acres in 1960 to 120 million and growing by 2007.[229] Much of that irrigation water came from underground. Today, many farmers in California, Texas, Nebraska, and elsewhere are utterly dependent upon groundwater for their livelihoods.[230]

A common misconception about groundwater arises from photographs of headlamp-wearing spelunkers wading through mysterious dark pools in underground caverns. Actually an "aquifer" is rarely a subterranean river or pool but instead just a geological layer of saturated sediment or bedrock, the best material being porous sand.[231] Water is removed from the aquifer by drilling a hole into the layer and installing a pump to raise water to the surface. This creates a cone of depression in the water table, causing surrounding groundwater to ooze through the porous matrix toward the borehole, providing a continuous water supply. Water raised from deep aquifers is normally reliable, clear, cold, and delicious. Deep aquifers don't flood or go into drought. In some of our driest, most water-stressed civilizations, it is the discovery and tapping of giant aquifers—ancient relicts that took many thousands of years to form—that has watered cities and exploded lawns across deserts from Texas to Saudi Arabia.

The problem is that no one knew or cared where the groundwater came from. In the early days many drillers thought it was infinite, or replenished somehow by mysterious underground rivers. But because aquifers are ultimately recharged by whatever rainfall manages to percolate down from the surface, they refill slowly. If water is pumped out faster than new water can ooze in, the aquifer goes into overdraft. The water table drops and wells fail. Farmers drill deeper, then the wells fail again. Eventually the aquifer is depleted or lowered too far to raise, and becomes uneconomic.

We are now coming to appreciate just how widespread this problem is globally, by measuring small variations in the Earth's gravity field precisely from space. In 2009 researchers using the NASA Gravity Recovery and Climate Experiment (GRACE) satellites discovered that despite natural recharge, groundwater tables in heavily irrigated parts of the Indian subcontinent are falling between four and ten centimeters per year, an unsustainable decline in an area supporting some six hundred million people.[232]

Most irreversible is groundwater overdrafting in our driest places. Not only do these aquifers have very low rates of rainfall recharge—and thus faster overdraft—but they are very often the main or only water source upon which people depend. Once gone, they take thousands of years to refill, or may never refill at all because they are relicts left over from the end of the last ice age. For all intents and purposes fossil groundwater, like oil, is a finite, nonrenewable resource. Eventually, the wells must run dry.

Death of a Giant

The Ogallala is a monster aquifer underlying no fewer than eight states across the western United States.[233] Its existence had been known to High Plains ranchers and dryland farmers since the 1800s, but it wasn't until the 1940s—with the arrival of modern pumps powered by electricity or natural gas—that the spigot could be opened wide. Since then, we have been pumping seven *trillion* gallons of cold, clear water out of the Ogallala Aquifer to irrigate circular center-pivot fields of wheat, cotton, corn, and sorghum across the Great Plains. This soon transformed over one hundred million acres of highly marginal land—much of it abandoned after the 1937 Dust Bowl—into one of the world's most productive agricultural regions. From your airplane window or a Web-browser view from Google Earth, you can see for yourself the green circles stamped out across the Texas and Oklahoma panhandles through eastern Colorado, New Mexico, and Wyoming; and running north through Kansas and Nebraska all the way to southern South Dakota. Those verdant, neatly aligned disks are the telltale fingerprints of the Ogallala Aquifer.

Zoom in with your Web browser and you'll see many of the disks are brown. By 1980 it was common knowledge that wells were falling fast in the Ogallala's southern half. By 2005 large portions had fallen by 50 feet, 100 feet, even 150 feet, in southwestern Kansas, Oklahoma, and Texas. Wells in the wetter northern half were holding up fine thanks to much higher natural recharge rates, but the dry southern states, where the Ogallala water is mostly of Pleistocene age,[234] was in serious overdraft. Wells began sputtering. Texas farmers, accustomed to feeding one or more center-pivot fields from a single well, began drilling several wells to support a single field.

In 2009 a team led by Kevin Mulligan, a professor of economics and geography at Texas Tech, completed a detailed study of just how fast Texas farmers are emptying out the southern Ogallala. Using a Geographic Information System (GIS), his team mapped thousands of wells throughout a forty-two-county area of northern Texas. They used the wells' water-level and flow-rate data to calculate the remaining saturated thickness of the Ogallala, and how fast the water table is falling. From these data they constructed a series of maps projecting the remaining useful life expectancy of the aquifer, for ten, fifteen, and twenty-five years into the future.

The results were shocking. Texas' Ogallala Aquifer is dropping an average of one foot per year and in some places as much as three feet per year. Many areas are careening toward a saturated thickness of just thirty feet, at which point the last wells will begin to suck air.[235] These maps are incredibly precise—all of the thousands of individual wells and the green crop circles they support are shown—so the impending demise of the aquifer is mapped out in a very detailed way. Texas' Parmo and Castro counties are plastered with center-pivot crops today, but their lush surface belies the situation below. Both counties are facing the abandonment of irrigated agriculture within the next twenty-five years.

Might the southern Ogallala be saved by sound conservation measures, like converting to drip irrigation? "We don't see it," snorted Mulligan to my question. It sounds great in theory, but his well data show that in practice, converting center pivots from sprinklers to dripping hoses doesn't slow the speed of the Ogallala's depletion. Instead, farmers just run their new drip systems longer so as to pull out the same volume of water, resulting in the same net drawdown. The hard fact is that there just isn't any way to save an aquifer whose natural recharge is one-half to one inch per year, when it is being drawn down a foot or more per year. Ironically, the single biggest benefit of drip irrigation to farmers isn't delaying the Ogallala's death but ensuring it, by allowing access to its last remaining dregs.[236] These wells are the final straws into a doomed giant once thought to be invincible.

Oil and Water Truly Don't Mix

Everyone knows that it takes water to get food. Less obvious is how much energy it takes to get water (for pumping, moving, purifying, and so on). And hardly anyone grasps how much water is needed to get energy. But like hopeless lovers, water and energy are inextricably intertwined. Pressure on water resources, therefore, is intimately linked to pressures on coal, oil, and natural gas resources. Except for wind and certain forms of solar power, even renewable energy sources demand a lot of water.

Power plants—regardless of whether they run on coal, natural gas, uranium, biomass, garbage, or whatever—use water in at least two important ways: to make steam to turn a turbine and thus generate electricity; and to get rid of excess heat. The single greatest demand for water in the energy sector today is for the cooling of power plants. Over half of all water withdrawals in the United States alone, slightly more than for irrigating crops, are used for this purpose. That's a half-billion acre-feet of water per year (enough to flood the entire country ankle-deep in water) to cool off our power plants. In some parts of Europe the percentage of water withdrawn for energy production is even higher.[237]

The total amount of water needed depends greatly on the fuel used, on plant design, whether the water is recycled, the type of cooling apparatus, and so on. But in all cases the volume of water needed to operate the power plant is large, even greater than the volume of fuel. This is why plants are sited next to water bodies or perched over large aquifers. It's not uncommon to find a coal-fired power plant on a riverbank hundreds of miles from the nearest coal mine: It is cheaper to carry the coal to the water, rather than the other way around. The Three Mile Island nuclear power plant, site of the 1979 accident described in the previous chapter, really *is* on an island, stuck out in the middle of the Susquehanna River.

Power plants bite into water supply by reducing both its quality and its quantity. Water recycled back into a river is hotter than the water withdrawn, sometimes by as much 25°C.[238] For plants located on large bodies of water like the ocean, this doesn't introduce significant environmental harm. Putting hot water into a river or lake, however, degrades aquatic ecosystems for many reasons. Warm water holds less dissolved oxygen, slows the swimming speed of fish, and interferes with their reproduction. Desirable coolwater species like trout and smallmouth bass are replaced by warm-water species like carp.

The second problem is water consumption, meaning irrevocable water loss. Most power plants use "wet" cooling towers—or even open ponds—to deliberately evaporate water into the atmosphere, providing cooling in the same way that evaporating sweat cools your skin. Evaporation losses from power plants are much smaller than the total withdrawal but are still significant in water-stressed areas. In very dry places, it becomes increasingly difficult to guarantee enough water for cooling purposes at all.

In the first study of its kind, Martin Pasqualetti, a professor in the School of Geographical Sciences and Urban Planning at Arizona State University,[239] scrutinized how much water consumption (i.e., evaporation) Arizona's different energy technologies require in order to produce one megawatt-hour of electricity. What he found may surprise you:

Water Losses Embedded in Arizona Electricity Generation

Energy technology	Water consumption (gallons/MWhr)
Hydropower	30,078
Solar CSP	800–1,000
Nuclear	785
Coal	510
Natural Gas	195–415 (depending on technology)
Geothermal	<5
Solar photovoltaics	<1
Wind	0

(Data courtesy M. Pasqualetti, Arizona State University)

From Pasqualetti's data we learn that the water consumption of energy production is not only large, but varies tremendously depending on the type of energy being used. For example, a nuclear power plant evaporates about 785 gallons of water to generate one megawatt-hour of electricity, whereas natural gas power plants evaporate considerably less (especially modern combined-cycle plants, which evaporate about 195 gallons per megawatt-hour). This means that an average house in Phoenix, using twenty megawatt-hours per year, will unknowingly evaporate nearly 16,000 gallons of water if its electricity comes from a nuclear power plant, but only about 3,900 gallons if it comes from a combined-cycle natural gas plant. More virtual water.

To put that number into perspective, 15,000 gallons is roughly what a typical Phoenix household with irrigated landscaping uses in two weeks. So this "embedded" water is not an enormous amount, but still significant in such a dry place. But the big surprise here is that in terms of electricity generation, hydropower, of all things, is the worst water waster,[240] followed by concentrated solar thermal (CSP) technology, then nuclear. Arizona does not grow biofuel crops, but other studies show biofuels are even worse than hydro in terms of water consumption.[241] Thus biofuels, hydropower, and nuclear energy, while hailed for being carbon-neutral (or nearly so), are worse even than coal when it comes to water consumption. Of the renewables, only wind and solar photovoltaics are truly benign—something, Pasqualetti points out, that would make solar photovoltaics more cost-competitive if the price of the saved water was taken into account.

The water-energy nexus works both ways. Examined in the opposite direction, energy is needed at every step along the way to deliver clean water to a house. Take again, for example, our typical Phoenix home, which consumes about an acre-foot of water per year. It requires two megawatt-hours of electricity—roughly 10% of the home's total energy use—to pump that acre-foot uphill from the Colorado River some two hundred miles away, purify it, and pressurize it locally. But those megawatt-hours never appear on any electric bill; they are embedded within the water bill itself. Remarkably, almost all the cost of providing drinking water to Phoenix households is for the energy embodied within it, not for the water.

"Indeed," says Pasqualetti, "water and energy are married to one another.

Water is needed in electrical generating stations if they are to run efficiently. Energy, on the other hand, is needed to provide our houses with safe drinking water. How much of each commodity is needed to provide the other is something not well appreciated by the public."[242]

It is something not well appreciated by politicians and planners either. Instead of recognizing this marriage between energy and water, their respective planning and regulatory agencies are almost always totally separate entities. "Energy analysts have typically ignored the water requirements of their proposed measures to meet stated energy security goals. Water analysts have typically ignored the energy requirements to meet stated water goals," concluded a recent Oak Ridge National Laboratory report.[243] Historically we have gotten away with this thanks to cheap water, cheap energy, or both. That cushion will continue to narrow as supplies of both tighten out to 2050.

One of the most widely anticipated outcomes of climate change is that the Hadley Cell circulation will weaken slightly and expand. This appears not only in a broad range of climate model projections for the future, but also from historical data extending three decades into the past.[244] The effect of this is the spawning of more clouds and rain in the tropics, but even drier conditions and a poleward expansion of the two desert blast zones straddling both hips of the equator. Precipitation futures are notoriously difficult to project, but this is one of those things about which all the climate models agree. Put simply, many of the world's wet places will become even wetter, and its dry places even drier.

Rainfall will increase around the equator, but decrease across the Mediterranean, Middle East, southwestern North America, and other dry zones. Rivers will run fuller in some places and lower in others. One highly regarded assessment tells us to get ready for 10%–40% runoff increases in eastern equatorial Africa, South America's La Plata Basin, and high-latitude North America and Eurasia, but 10%–30% runoff declines in southern Africa, southern Europe, the Middle East, and western North America by the year 2050.[245] Through the language of statistics, these models are telling us to brace for more floods and droughts like the ones in Iowa and California.

The Great Twenty-first-Century Drought?

Part of the explanation for the many floods and droughts that happened around the world in 2008 was that it was a La Niña year, meaning that sea surface temperatures (SSTs) in the eastern half of the tropical Pacific Ocean cooled off. This triggered, among other things, dry conditions over California, contributing to its ongoing drought (her counterpart, El Niño, is associated with warm SSTs and wetter conditions there). Through connections between the sloshing ocean and the atmosphere, this "Little Girl" had impacts on human water supply that reverberated worldwide.

My UCLA colleague Glen MacDonald, an expert in the study of prehistoric climate change, is deeply concerned that something like the 2008 La Niña could happen again—but persisting for *decades* rather than months. In fact, MacDonald and his students believe the American Southwest, in particular, could be struck by a drought worse than anything ever seen in modern times. From shrunken tree-rings and other prehistoric natural archives, they have assembled a growing body of evidence that the region suffered at least two extended "Perfect Droughts" (coined by MacDonald to describe periods when Southern California, northern California, and the upper Colorado River Basin all experienced drought simultaneously) during medieval times.[246] These Perfect Droughts were as bad as or worse than the Dust Bowl but lasted much longer, persisting as long as five to seven decades (the Dust Bowl lasted barely one). These prehistoric data tell us that this heavily populated region is capable of experiencing droughts far worse than anything experienced since the first European explorers arrived.

One reason for these massive prehistoric droughts was that between seven hundred and nine hundred years ago temperatures rose. The increase was similar to what we are beginning to see now but not so high as what climate models are projecting by 2050. The reason for the medieval temperature rise (fewer volcanic eruptions plus higher solar brightness) was different from what's happening today, but it nonetheless provides us with a glimpse of how our planet might respond to greenhouse warming.[247]

Not only did the medieval climate warming increase the drying of soils directly, it may also have altered an important circulation pattern in the Pacific Ocean, by shifting relatively cool water masses off the western coast of North America for many decades at a time (this would be a prolonged negative phase of the so-called "Pacific Decadal Oscillation," an El Niño–like oscillation in the northern Pacific that currently vacillates over a 20-30-year time scale). This likely created pressure systems driving rain-bearing storm tracks north, rather than south, across the western United States, triggering drought conditions in the American Southwest. Should the projected rise in air temperatures cause the Pacific circulation to behave like this again, the prolonged medieval megadroughts could return. Similar connections between shifting sea-surface temperatures and geographic rainfall patterns over land exist for the Atlantic and Indian oceans as well.

MacDonald points out that by the time Schwarzenegger declared a state of emergency in 2009, most of the southwestern United States was actually in its eighth year of drought, not third. "Arguably, we are now into the great Twenty-first Century Drought in western North America," he mused to me. "Could we be in transition to a new climate state? Absolutely. Should we be worried? Absolutely." His concerns are echoed by Richard Seager at Columbia University's Lamont-Doherty Earth Observatory. In a widely read *Science* article,[248] Seager and his colleagues showed consensus among sixteen climate models that projected greenhouse warming will drive the American Southwest toward a serious and sustained baking. Their result, of course, is dependent on the group of models analyzed, and the simulation is imperfect because today's coarse-scale climate models don't represent mountainous areas very well (e.g., the Rockies, which produce most of the region's snowpack water). But if these model projections prove correct, then the drought conditions associated with the brief American Dust Bowl could conceivably become the region's new climate within years to decades.

Risky Business

"Stationarity Is Dead," announced another *Science* article in 2008, sending a cold shiver through the hearts of actuaries around the world.[249] A hydrology dream team of Chris Milly, Bob Hirsch, Dennis Lettenmaier,

Julio Betancourt, and others had just told them that the most fundamental assumption of their job description—reliable statistics—was starting to come apart.

Stationarity—the notion that natural phenomena fluctuate within a fixed envelope of uncertainty—is a bedrock principle of risk assessment. Stationarity makes the insurance industry work. It informs the engineering of our bridges, skyscrapers, and other critical infrastructure. It guides the planning and building codes in places prone to fires, flooding, hurricanes, and earthquakes.

Take river floods, for example. By continuously measuring water levels in a river for, say, twenty years, we can then use the stationarity assumption to calculate the statistical probability of rarer events, e.g., the "fifty-year flood," "hundred-year flood," "five-hundred-year flood," and so on. This practice, while creating enormous misunderstanding with the public,[250] has also made us safer. Hard statistics, rather than the whims of developers or mayors, are used to design bridges and for zoning. But flood prediction, and most other forms of natural-hazard risk assessment, rest on the core assumption that the statistics of past behavior will also apply in the future. That's stationarity. Without it, all those risk calculations go straight out the window.

A growing body of research is showing that our old statistics are starting to break down. Climate change is not the sole culprit. Urbanization, changing agricultural practices, and quasi-regular climate oscillations like El Niño all influence the statistical probabilities of flooding. However, the dream team's paper and others like it[251] tell us that climate change is fundamentally altering the statistics of extreme floods and droughts, two things of enormous importance to humans. "In view of the magnitude and ubiquity of the hydroclimatic change apparently now under way," they wrote, "we assert that stationarity is dead and should no longer serve as a central, default assumption in water-resource risk assessment and planning. Finding a suitable successor is crucial for human adaptation to changing climate."[252]

Unfortunately, we have no good replacement for stationary statistics yet, certainly nothing that works as well as they once did. Moreover, there has been hardly any basic research done in this area since the 1970s. We can't just invent a completely new branch of mathematics and train a new generation of water experts in it overnight. "Water resources research has been allowed to slide into oblivion over the past thirty years," Lettenmaier growled later in a separate editorial. "Certainly the profession has been slow to acknowledge these changes and acknowledge that fundamentally new approaches will be required to address them."[253] So even as we're beginning to grasp the enormity of this problem, we presently have no clear replacement for our old way of doing things. Until we find one, risks will be harder to predict and to price. We can expect insurance companies to react accordingly. In 2010, after failing to win a nearly 50% rate increase from state regulators, Florida's largest insurance company abruptly canceled 125,000 homeowner policies in the state's hurricane-prone coastal regions, saying the recent series of devastating hurricanes had rendered its business model unworkable.[254] Get ready for higher premiums, uninsurable properties, and failed or overbuilt bridges.

Nonreturnable Containers

Changing drought and flood statistics are not the only way that rising greenhouse gases harm our water supply. All of our reservoirs, holding tanks, ponds, and other storage containers are trifling compared to the capacity of snowpacks and glaciers. These are free-of-charge water storehouses, and humanity depends upon them mightily.

Snow and ice hoard huge amounts of freshwater on land, then release it in perfect time for the growing season. They do this by bulking up in winter, then melting back in spring and summer. They are the world's hugest water-management system and, unlike a dam reservoir, displace no one and cost nothing. Glaciers (and permanent, year-round snowpacks) are especially valuable because they outlast the summer. This means they can hoard extra water in cool, wet summers, but give it back in hot, dry summers, by melting deeply into previous years' accumulations. Put simply, glaciers sock away water in good years when farmers need it least, and release water in bad years when farmers need it most. Glaciologists call these "positive mass-balance" and "negative mass-balance" years, respectively, and they are a gift to humanity. Glaciers keep the rivers full when all else is dry. They are the ultimate sunny-day fund.

If you read the news, then you already know that many of the world's glaciers are beating a hasty retreat, whether through warmer temperatures, less precipitation, or both. Ohio State University's glaciologist power-couple Lonnie Thompson and Ellen Mosley-Thompson have been photographing the deaths of their various study glaciers since the 1970s. Some of these are even wasting away at their summits, which is a death knell for a glacier. There are ski resorts in the Alps trying to save theirs by covering them with reflective blankets. Most glaciologists expect that by 2030, Montana's Glacier National Park will have no glaciers left at all.

Seasonal snowpack, which does not survive the summer, cannot carry forward water storage from year to year like glaciers do, but it is also a critically important storage container. It creates a badly needed time-delay, releasing water when farmers need it the most. By holding back winter precipitation in the form of snow, the retained water flows downstream to farmers later, in the heat of the growing season. Without this huge, free storage container, this water would run off uselessly to the ocean in winter, long before growing season. Rising air temperatures harm this benefit, both by increasing the prevalence of winter rain (which is not retained) and by shifting the melt season to earlier in the spring. Because the growing season is determined not only by temperature but also the length of daylight, farmers are not necessarily able to adapt by planting sooner. By late summer, when the water is needed most, the snowpack is long gone.

This seasonal shift to earlier snowmelt runoff portends big problems for the North American West and other places that rely on winter snowpack to sustain agriculture through long, dry summers. California's Central Valley—the biggest agricultural producer in the United States—depends heavily on Sierra snowmelt, for example. But the long-term projection for health of the western U.S. snowpack is not good. It has already diminished in spring, despite overall *increases* in winter precipitation, in many places.[255] By late 2008, Tim Barnett at the Scripps Institute of Oceanography and eleven other scientists had definitively linked this phenomenon to human-caused climate warming. This is not good news, they wrote in *Science,* warning of "a coming crisis in water supply for the western United States" and "water shortages, lack of storage capability to meet seasonally changing river flow, transfers of water from agriculture to urban uses, and other critical impacts."[256]

High-profile research like this does not go unnoticed by policy makers. One response is to build more reservoirs, canals, and other engineering schemes to store and move water. China is now planning fifty-nine new reservoirs in its western Xinjiang province to retain water from glacier-fed rivers. In 2009, U.S. Interior Secretary Ken Salazar announced $1 billion in new water projects across the American West, with over a quarter-billion going to California alone.[257]

Thus begins our new technological race—to adapt to a shrinking water storage capacity, once provided for free by snow and ice. But it is important to understand that *no amount of engineering can replace that storage.* Think back to I. A. Shiklomanov (p. 86), his huge container of ice, and trifling container of surface water. Even if we quadrupled the world's reservoirs, they wouldn't come remotely close to replacement. And even if they did, we'd still end up with less water: Unlike snow and ice, water evaporates like crazy from open reservoirs.

We can't hold it all back. More of the world's water is leaving the mountains to run to the sea.

Into the Sea

It's abnormal to be thinking about melting glaciers when standing on a nice sunny beach during holiday break. But this was no ordinary beach and no ordinary holiday. It was Christmas 2005, and I and other members of the Smith family were staring dumbly at the bones of what had once been my aunt and uncle's house, a dozen blocks inland from the Mississippi coast. With the ease of a kid blowing foam across a cup of hot chocolate, Hurricane Katrina had thrown a wall of water—a *storm surge*—right through their lovely Biloxi neighborhood.

The place was a deserted war zone. Houses smashed to splinters, cars crushed and tossed into swimming pools. Nearer the beach, there were no house bones at all, just smooth rectangles of white concrete, scrubbed and gleaming to show where million-dollar homes had once stood. It was four months since the hurricane but the place was abandoned. No one was hauling away debris, no sound of hammering nails. All was silent except for the songbirds, cheeping and squabbling amid the wreckage. To them it was just another beautiful day on the American Gulf Coast.

In devastated New Orleans, ninety miles to the west, we saw a similar abandonment of entire neighborhoods. There were blocks and blocks of leaning houses, trashed and dark except for the colorful graffiti of rescue-worker symbols. The hieroglyphs recorded each house's history in spray paint—the date searched, any noted hazards, whether any human bodies had been found. Living in one home was a pack of feral dogs.

So that is why, while standing on a gorgeous sunny beach, I was thinking about glaciers. In smashing my uncle's former home, Hurricane Katrina had made the dry statistics of my field feel real—on a personal, visceral level. Although glacial melt hadn't caused Katrina, I was thinking about the indelible control the world's ice holds over our coastlines. When the glaciers grow, oceans fall. When they shrink, oceans rise. Oceans and ice have danced in this way, embraced in lockstep, for hundreds of millions of years. From my geophysical training I knew this. From my own research and that of colleagues, I knew how quickly the world's glaciers were retreating. And for miles inland behind me, and hundreds of miles along the coast in either direction, the ground on which I stood lay barely above the surf. I had understood all this before in abstraction, but this endless plain of destruction made it real.

Global sea levels are now steadily rising nearly one-third of a centimeter every year, driven by melting glacier ice and the thermal expansion of ocean water as it warms.[258] There is absolutely no doubt about this. There is absolutely no doubt that it will continue rising for at least several centuries, and probably longer. Sea-level rise really is happening. The big unknowns are how fast, whether it will progress smoothly or in jerks, and how high the water will ultimately go.

We shall explore the scary possibilities of fast sea-level rise in Chapter 9; for now, let's stick to conservative models and what has been measured thus far. In the 1940s, global average sea level was about ten centimeters lower than today, but was rising more than 1 millimeter per year (a brisk rate at the time). It is currently rising 2–3 millimeters per year, and that number is projected to grow by around 0.35 millimeters for each additional degree Celsius of climate warming.[259]

Depending on whose model you like, this means we are looking at around 0.2–0.4 meters of sea level rise by 2050, or calf-deep. The state of California has just begun damage assessment and planning for 0.5 meters by that time,[259] around knee-deep. And 2050 is just the beginning. By century's end, global sea level could potentially rise from 0.8 to 2.0 meters.[261] That's a lot of water—up to the head of an average adult. Much of Miami would be either behind tall dikes or abandoned. Coastlines from the Gulf Coast to Massachusetts would migrate inland. Roughly a quarter of the entire country of Bangladesh would be underwater.

When oceans rise, all coastal settlements face challenges. Higher sea levels expand the inland reach and statistical probability of storm surges like the one Hurricane Katrina blew into the Gulf Coast. Decidedly unhelpful is a two-in-three chance that climate warming will make typhoons and hurricanes more intense than today, with higher wind speeds and heavier downpours.[262] And just as we saw for water supply, there are other, nonclimatic actors that make the problem even worse. In fact, all four of our global forces are conspiring to place some of the world's most important cities at risk.

Most of the world's largest and fastest-growing urban agglomerations—like Mumbai, Shanghai, and Los Angeles—are globalized port cities on the coasts. Their populations and economies are rising fast. Demographers and economic models tell us they will grow even more over the next forty years.

Particularly in Asia, many of these great cities are located on "megadeltas," enormous flat protrusions of mud and silt that grow where large rivers drop off their carried sediment upon entering and dissipating into the ocean. These piles of sediment are ferociously attacked by the ocean's waves and storm surges, but the rivers keep dumping more. Like giant conveyer belts of cement, they keep trundling material to the river mouths—often from thousands of miles inland—to overwhelm the ocean's defenses. Over centuries to millennia, the rivers grow the land out.

These deltas have always attracted humans. Farmers love their thick, rich soils that are also flat, well-watered, and have few rocks. Ships can ply both oceans and continental interiors. The river brings in freshwater for towns and cities, then carries their wastes off to the sea. A delta's flat terrain is appealing to build on; the surrounding swamps and forests are teeming with fish and wildlife.

The problem, of course, is that the very existence of deltas is maintained by the constant sedimentation from flooding and back-and-forth migration of their rivers. They are full of low-lying swales that inundate readily. As human settlements grow, there is increasing pressure to expand into these dangerous areas. This happens not only with deltas but urbanizing river floodplains as well, like Cedar Rapids in Iowa. Flood damages therefore rise as development pushes into low-lying swamps considered too dangerous before. The reason Katrina spared New Orleans' historic French Quarter is that it was the first place to be colonized: Even in 1718 people knew to perch their houses on that crescent-shaped sliver of natural levee, piled a few feet higher than the nearby swamps where the Upper Ninth Ward would drown nearly two centuries later.

As delta cities grow and their rivers become oversubscribed or polluted, they start pumping their groundwater resources. Groundwater removal—from what is essentially a pile of wet mud—causes the delta sediments to compact and settle, lowering the delta's elevation closer to that of the sea. Even in the absence of groundwater pumping, some settling is normal. In a natural system, this settling is compensated by fresh blankets of silt laid down by floods. But the dikes and levees built to protect delta cities also prevent these fresh reinforcements from arriving. Farther upstream, dams thrown across the river snare the delta's lifeblood of new sediment. Dam operators groan and search their budgets for dredging money. The conveyor belt is cut. Hundreds of miles downstream, the ocean starts taking back the land.

Important delta cities are found all over the world. They face the triple threat of rising oceans, sinking land, and sediment-starved coastlines. Without replenishment their coasts are washing away, bringing ocean wave energy and storm surges ever closer to the sinking cities. When combined with projected trends of rising sea level, population, and economic power, this puts some of the world's most populous and prosperous places in harm's way.

The risk assessment study on the next page was recently commissioned by the OECD.[263] The study considered all 136 of the world's major port cities holding one million people or more. As of 2005, about forty million people living in these cities were considered to be living in places at direct

Top Twenty World Port Cities Most Vulnerable to Global Sea Level Rise, Hurricanes, and Land Subsidence

Urban agglomeration	Exposed in 2005		Exposed in 2070s		Factor increase	
	Population	Assets ($US B)	Population	Assets ($US B)	Population	Assets
Mumbai (India)	2,787,000	46	11,418,000	1,598	4.1	34.6
Guangzhou—Guangdong* (China)	2,718,000	84	10,333,000	3,358	3.8	39.9
Shanghai* (China)	2,353,000	73	5,451,000	1,771	2.3	24.3
Miami (USA)	2,003,000	416	4,795,000	3,513	2.4	8.4
Ho Chi Minh City* (Vietnam)	1,931,000	27	9,216,000	653	4.8	24.3
Calcutta* (India)	1,929,000	32	14,014,000	1,961	7.3	61.3
New York—Newark (USA)	1,540,000	320	2,931,000	2,147	1.9	6.7
Osaka—Kobe* (Japan)	1,373,000	216	2,023,000	969	1.5	4.5
Alexandria* (Egypt)	1,330,000	28	4,375,000	563	3.3	19.8
New Orleans* (USA)	1,124,000	234	1,383,000	1,013	1.2	4.3
Tokyo* (Japan)	1,110,000	174	2,521,000	1,207	2.3	6.9
Tianjin* (China)	956,000	30	3,790,000	1,231	4.0	41.6
Bangkok* (Thailand)	907,000	39	5,138,000	1,118	5.7	28.9
Dhaka* (Bangladesh)	844,000	8	11,135,000	544	13.2	64.5
Amsterdam* (Netherlands)	839,000	128	1,435,000	844	1.7	6.6
Hai Phòng* (Vietnam)	794,000	11	4,711,000	334	5.9	30.2
Rotterdam* (Netherlands)	752,000	115	1,404,000	826	1.9	7.2
Shenzhen (China)	701,000	22	749,000	243	1.1	11.2
Nagoya* (Japan)	696,000	109	1,302,000	623	1.9	5.7
Abidjan (Côte d'Ivoire)	519,000	4	3,110,000	142	6.0	36.7

*delta city

(Source: R. J. Nicholls, OECD, 2008)

The first grizzly/polar bear hybrid ever seen in the wild was killed in 2006 on Baffin Island, Canada, by Idaho business owner and hunter James Martell. A second-generation hybrid was shot in 2010. (photo courtesy J. Martell)

Bridge and train destroyed by 2008 Iowa floods, Cedar Rapids, Iowa. (photo by author)

In August 2005 some 320 kilometers of the U.S. Gulf Coast were innundated by a storm surge up to ten meters high, pushed onshore by Hurricane Katrina. Surge heights of more than two meters were felt as far as the Florida panhandle. In parts of Mississippi the surge penetrated ten kilometers inland, eradicating homes and businesses like these near Biloxi. At least 1,200 people were killed. (photo by author)

New potato farmers near Narsaq, Greenland. (photo by John Rasmussen)

Russia has a long history of maritime accomplishments in the Arctic. In May 1987 the nuclear icebreaker *Sibir* reached the North Pole in late winter, a feat never accomplished before or since (all other North Pole surface ship voyages have been in summer). This photo shows her crew literally "surrounding the world" by encircling all lines of longitude emanating from the most northerly point on earth. (photo by Dr. Ivan Frolov)

Artur N. Chilingarov was awarded the Hero of Russia medal for leading a 2007 expedition that planted this titanium Russian flag on the ocean floor of the North Pole. The daring feat required two small submarines to dive 4,300 meters beneath the sea ice. (photo courtesy of ITAR-TASS News Agency, Russian Federation)

The author lived and worked on the Canadian *CCGS Amundsen* icebreaker, shown here off the coast of Labrador. (photo by author)

Summer barge traffic is busy on the Mackenzie River. The aboriginal-owned Northern Transportation Company Limited has a large modern shipyard in Hay River, N.W.T.
(photo by author)

Canadian grain being loaded in the Port of Churchill for consumption in Europe.
(photo by author)

How a deeply frozen permafrost landscape looks from the air. For scale, note ATV tracks scarring the tundra vegetation. This is the North Slope of Alaska near the town of Barrow. (photo by author)

Reindeer are a common sight in northern Finland. Unlike other NORC countries, Norway, Sweden, and Finland have excellent highways and other critical infrastructure extending far north into their Arctic territories. (photo by author)

This Siberian apartment building was destroyed when thawing permafrost reduced the structural strength of the underlying soil. Within days after the first cracks appeared, the building collapsed. (photo by Dr. Vladimir Romanovsky)

Shipping is the most economical way to resupply remote settlements. Like most other aboriginal communities in North America, this village of Sanikiluaq, Nunavut, has an extremely youthful demographic age structure. (photo by author)

Sports commentators Parminder Singh and Harnarayan Singh now broadcast *Hockey Night in Canada* in Punjabi, poised to become the country's fourth most-spoken language. Of the eight NORC countries, Canada excels at attracting and welcoming skilled foreign immigrants. (photo courtesy *Toronto Star*)

Meltwater departing the Greenland Ice Sheet for the global ocean. Where's Waldo? See author for scale. (photo by Dr. Richard Forster)

The Alberta Tar Sands. Mining operations of Syncrude Canada, Ltd., with stacked sulfur blocks and upgrader plant in the background. (photo by David Dodge)

"Boots on the ground." The author on the shore of the Arctic Ocean, with barge traffic in the background. (photo by Ben Jones)

risk from flooding. The total economic exposure to flooding—in the form of buildings, utilities, transportation infrastructure, and other long-lived assets—was about USD $3 trillion, or 5% of global GDP.[264] Under current trajectories of population growth, economic growth, groundwater extraction, and climate change, by the 2070s the total exposed population is forecast to grow more than threefold, to 150 million people. The economic exposure is forecast to rise more than tenfold, to USD $35 trillion, or 9% of global GDP. Of the top twenty major at-risk cities, exposed human populations could rise 1.2- to 13-fold, and exposed economic assets 4- to 65-fold, by the 2070s. Three-quarters of these major cities—nearly all of them in Asia—are found on deltas. Clearly, we are about to begin paying great attention to a new kind of defense spending. It's called coastal defense.

Imagining 2050

The trends I've described—rising water demand; oversubscribed and/or polluted water sources; reduced time-delays and free storage from snow and ice; sharper floods and droughts that are also harder to predict and insure against; the competitive marriage of water to energy; and booming port cities on increasingly risky coasts—all stem from our four global forces of demographics, natural resource demand, globalization, and climate change.

Whether for-profit multinational corporations offer the best solution for tackling water quality problems in impoverished countries remains an open question that is heatedly debated. However, global trade flows of "virtual water" embedded in food, energy, and other goods are already smoothing out some stark water inequities around the world. Compared with other irritants, international water disputes have seldom led to war. Continued economic integration could foment even better water management across borders—especially when nudged along by free hydrologic data measured from space and posted openly on the Internet. Finally, the not-so-far-fetched possibility that new international trade flows in water—not just virtual but actual, physical water—could emerge as a partial solution for some water-stressed places that will be explored further in Chapter 9.

Looking ahead to the next forty years, it's not hard to see where the big pressure points lie. Joseph Alcamo directs a research institute at the

University of Kassel dedicated to exploring different possible futures for humanity's water supply. To do this they built WaterGAP,[265] a sophisticated computer model incorporating not only climate change and population projections but also other factors like income, electricity production, water-use efficiency, and others. WaterGAP is thus a powerful tool for simulating a range of possible outcomes depending on the choices we make.

A typical, "middle-of-the road" WaterGAP scenario is shown here for 2050.[266] Regardless of how the WaterGAP model parameters are twiddled,

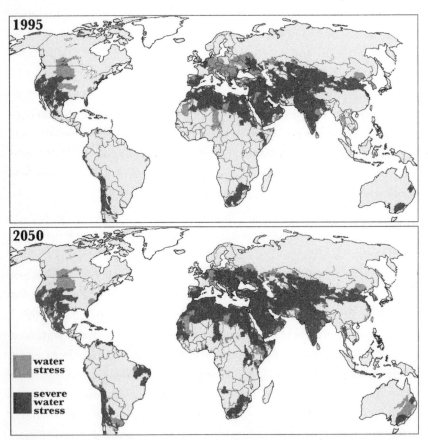

Global water stress to human populations, 1995 versus 2050. "Severe" stress (dark gray) is where humans withdraw more than 40% of available water; "moderate" stress (medium gray) is where humans withdraw 20% to 40%. (Data courtesy of Joseph Alcamo and Martina Flörke, Center for Environmental Systems Research, University of Kassel)

the big picture is clear: The areas where human populations will be most water-stressed are the same areas where they are water-stressed now, but worse. From this model and others, we see that by midcentury the Mediterranean, southwestern North America, north and south Africa, the Middle East, central Asia and India, northern China, Australia, Chile, and eastern Brazil will be facing even tougher water-supply challenges than they do today. One model even projects the eventual disappearance of the Jordan River and the Fertile Crescent[267]—the slow, convulsing death of agriculture in the very cradle of its birth.

Computer models like these aren't built and run in a vacuum. They are built and tuned using whatever real-world data scientists can get their hands on. Take, for example, the western United States. In Kansas, falling water tables from groundwater mining is already drying up the streams that refill four federal reservoirs; another in Oklahoma is now bone-dry. These past observed trends, together with reasonable expectations of climate change, suggest that over half of the region's surface water supply will be gone by 2050.[268] Kevin Mulligan's projection of the remaining life of the southern Ogallala Aquifer requires no climate models at all—it simply subtracts how much water we are currently pumping from what's left in the ground, then counts down the remaining years until the water is gone.

In the United States, the gravest threat of all is to the Colorado River system, the aorta of water and hydropower for twenty-seven million users in seven states and Mexico. It supplies the cities of Los Angeles, Las Vegas, Tucson, and Phoenix. It irrigates over three million acres of highly productive farmland. Global climate models almost unanimously project that human-induced climate change will reduce Colorado River flows by 10%–30%[269] and already, its water is heavily oversubscribed.

More water is legally promised to the Colorado's various shareholders than actually flows in the river.[270] Its left and right ventricles are Lake Mead and Lake Powell, two enormous reservoirs created by the Hoover and Glen Canyon dams, respectively. They haven't been full since 1999. A bitter combination of high demand, high evaporation, and falling river flows has thrown the Colorado River system into a massive net deficit of nearly one million acre-feet per year, enough water for eight million people. By 2005, Lake Powell was two-thirds empty and almost to "dead pool" (the elevation of its lowest outlet, below which no water can be released by the dam and it ceases to function).[271] This desiccation stranded marinas and boat docks on dry land and left a white bathtub ring some ten stories high on Lake Powell's newly exposed canyon walls. "It was as though in four years . . . Lake Powell had simply vanished," wrote James Lawrence Powell of his namesake in *Dead Pool.*

I'm glad humanity has a decent track record with things like settling water disputes with courts rather than missiles, and exporting food from the places that have water to the places that don't: If any of these model forecasts are correct, we're going to need it. Humans currently withdraw about 3.8 trillion cubic meters of water annually, and are projected to require more than six trillion in the next fifty years. To serve India's expected 2050 population of 1.6 billion, even with improved water efficiency, will require a near-tripling of its water supply. Farmers, energy utilities, and municipalities are all in competition for water. Put it all together and the numbers don't add up. Something will have to give.

The survival of California's thirsty dry cities—like Los Angeles and San Diego—seems all but guaranteed. Their populations and economies are growing briskly. Despite annual sales of over USD $30 billion, California agriculture still contributes less than 3% to the state's economy—and cities use far less water than irrigated farms. Even with climate changes and a projected 2050 population of about 20 million, there will still be ample water for Angelenos and San Diegans to drink and shower and cook. Ample water for California farmers, however, is far less assured.

Forced to choose, cities will trump agriculture. Farmers will either lose or sell their historic water rights. Croplands will return to desert. The first signs of an urban takeover have already begun: After years of lawsuits, farmers of California's Imperial Valley were forced to sell two hundred thousand acre-feet of their yearly Colorado River water allocation to San Diego in 2003. That fallowed twenty thousand acres of farmland. By early 2009 the Metropolitan Water District—supplier of twenty-six cities throughout Southern California—was trying to buy *seven hundred thousand* acre-feet more.[272]

Cities versus farmers: the real Water Wars.

PART TWO

THE PULL

CHAPTER 5

Two Weddings and a Computer Model

My adoptive groomsman, whom I'd just met the night before, cracked open the church door and peeked anxiously out at the parking lot. It was a sorry mess of black asphalt, lingering slush, and streaming water. Some early guests were sitting in their cars, peering through their headlights for a dry way into the church. It was early afternoon but very dark. I'd expected dim—we were, after all, just three hundred miles shy of the Arctic Circle in the middle of winter—but not this. The expected reflective blanket of fluffy white snow was gone. My dress socks were wet and cold. We'd strategically timed our wedding day for the prettiest, whitest, most winter-wonderland month of the year. But instead, in the middle of February, some five hours north of Helsinki, a thousand miles northeast of London, and almost twenty degrees of latitude north of Toronto—there was only a steady downpour of rain.

More precisely it was our *first* wedding day, taking place across the Atlantic for my new European family and friends. Our *second* wedding day— for American families and friends—was a month later in the sunny desert resort of Palm Springs, California. Mid-March is peak tourist season in Palm Springs, with infallible blue skies and flawless temperatures hovering in the 70s. We had booked all outdoor venues for the day's events. Our tremulous queries about tents and patio heaters—just in case of a weird- weather repeat—were politely but firmly dismissed. The weather here is *always* perfect in March, we were told. That's why people pay twice as much to come then.

123

You know what happened next. A line of fat squalls sprayed cold rain onto our guests' unprotected heads. By the time the lasagna came out, the temperature had plunged fifteen degrees. We did manage to scrounge up four patio heaters somehow, around which the jacketless masses could huddle. We were shocked and upset—again—by freaky weather. But just like our sub-Arctic celebration, the crowd's good spirits soon prevailed. Both ceremonies went on as planned. Cakes were cut, dances were danced, and good times were had by all.

I shouldn't have been so surprised. While there will always be some weird weather happening somewhere, my wedding experiences were consistent with everything we know about the statistics of climate change. I had described such phenomena many times (though as probabilities, not specific occurrences) to thousands of students in my lectures at UCLA. From my research and travels to the NORCs, plenty of people had told me about bizarre rains in the depths of winter. After a while I'd even become bored with it—one can only listen to so many bizarre-weather stories before it just isn't new information anymore.

In the previous chapter, we explored how the statistical norms of flood and drought frequency are changing, and how they might become more intense in the future. Now it is time to discuss rising air temperatures in the North—even in the dead of winter and at very high latitudes. Indeed, this phenomenon is a central interest throughout the rest of this book.

Four facts about global climate change need to be made very clear.

The first is that any process of climate change—both natural and man-made—unfolds erratically over time. In fact, its behavior is not unlike that of the stock market.

As every investor knows, long-term trends in the stock market are overprinted with short-term fluctuations. We don't normally assume that share prices will move smoothly up or smoothly down. Instead, and usually within days, we expect they will reverse, before reversing again, and so on. Wise investors accept this short-term volatility as being largely unpredictable, yet bank on the existence of an underlying long-term trend to guide their overarching portfolio strategy. They say that while short-term markets react to unpredictable things like profit-taking, news reports, and God-knows-what, a long-term trend is more fundamental. And indeed, they are right. Throughout modernity the long-term trend has been for stock values to rise. Its underlying driver is growth of the real economy, fueled by the steady rise of human population and prosperity.

The long-term trend for the Earth's climate, for at least several centuries, is rising air temperatures in the troposphere (lower atmosphere). Its underlying driver is radiative forcing commanded by the steady rise of carbon dioxide and other greenhouse gases produced as by-products of human activity. Because carbon dioxide, in particular, can linger in the atmosphere for many centuries this buildup is, for all intents and purposes, permanent.[273] Over the long haul, the world's global average temperature must go up. As shown in Chapter 1, the physics of this has been known since Svante Arrhenius' work in the 1890s.

Beyond this broad, average trend, however, the warming process gets more complicated. Our planet is not simply a dry rock with a sunlamp shining on it. The additional heat trapped by greenhouse gases is absorbed, released, and moved around the planet by sloshing ocean currents and turbulent air circulation patterns. Living things breathe air in and out, and store or release carbon—a fundamental ingredient of CO_2 and CH_4 greenhouse gases—in their tissues. When the ground is bare, it absorbs sunlight, causing local heating. When snow-covered it reflects, causing local cooling. Volcanic eruptions punch aerosols into the stratosphere, shading and cooling the planet for a few years until they dissipate.[274] The energy output of the Sun waxes and wanes slightly. All of these little and not-so-little natural mechanisms and feedbacks are to climate change what profit-taking, insider trading, and short sellers are to the stock market. They muck up the underlying greenhouse forcing trend, overprinting it with shorter fluctuations that rise and fall, then rise again.[275] If not for this natural variability, we'd have caught on to the deeper greenhouse signal even sooner than we did.

Any competent financial planner will tell you that the road to secure retirement is paved with market drops. Any competent climate scientist will tell you that our road to a hotter planet will be paved with cold snaps, even record-breakers. But unfortunately, when it comes to communicating this to the general public, we scientists have done a poorer job of it than financial planners. Perhaps it's not surprising, therefore, that so many people will glance outside at the bitter cold and scoff at global warming—even as they log on to E-Trade to buy up the latest stock market dip.

The second important fact about climate change is that its geography is neither always global nor always warming. To be sure, it is *mostly* global and *mostly* warming. But because of the many complex natural mechanisms and feedbacks that inject themselves into the process, the final climatic manifestations of greenhouse forcing vary greatly in spatial pattern. Climate change is not only erratic in time, like the stock market, but also in geography. A globally averaged temperature increase of one degree Celsius does not mean temperatures rise everywhere around the globe by one degree Celsius. That's just the average. Some places will heat up a great deal, others won't or might even cool. Summing them all together gets you to the +1°C global average. But that seemingly small number masks some stunning differences around the world.

Consider the map below. It is a projection of our future temperature changes by the middle of this century. Some places are warming hugely but other places hardly at all.[276] Why is this? Has some climate model gone haywire?

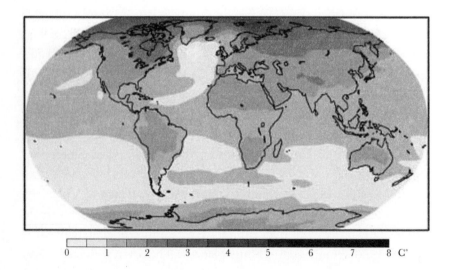

0   1   2   3   4   5   6   7   8 C°

This map is not an oddball, but just one of a family of nine related maps released by the latest IPCC Assessment.[277] They all show irregular geographic patterns and appear together on the following page in a three-by-three grid. From left to right they plot out a three-stage timeline for our century, with average, smoothed-out temperature changes apparent by 2011–2030, by 2046–2065, and by 2080–2099. Like the single map on page 126, each one is actually produced from not one but many climate models—much like a stock index—thus capturing where the models robustly agree rather than the quirks of any particular climate model over another.

Each of the three rows corresponds to a different concentration of greenhouse gas in the atmosphere. That, in turn, rests on all sorts of things, from political leadership to energy technology to gross domestic product. Rather than try to predict which outcome will actually transpire, the IPCC instead calculates outcomes for numerous possible social paths (called SRES scenarios[278]), of which three are shown here. The first outcome (top row) may be described as a highly globalized world, with population stabilizing by midcentury and an aggressive transition to a modern information and service economy. This scenario (known to climate scientists as "B1") is labeled "optimistic" on the figure.[279] The second outcome also assumes a stabilizing population and fast adoption of new energy technologies, but with a balance of fossil and nonfossil fuels. That future (called "A1B" by climate scientists) is labeled "moderate." The third outcome assumes a very divided world with high population growth, slower economic development, and slow adoption of new energy technology. This future (called "A2") is labeled "pessimistic."

The third important fact about global climate change is revealed by comparing these three rows of maps. They show that, regardless of technology path, we are already locked in to some degree of warming; but by century's end, the actions or inactions taken now to curb greenhouse gas emissions really will matter enormously. By 2080–2099 the "pessimistic" world is indeed a cauldron compared to the "optimistic" one, with temperatures rising 3.5°–5.0°C (9°F) across the conterminous United States, Europe, and China, rather than 2.0°–2.5°C (4.5°F). While these numbers may seem small, in fact there is a huge difference between the two outcomes. A 2.5°C rise in average annual temperature is actually huge, equivalent to the difference between a record cool and record warm year in New York City. So even

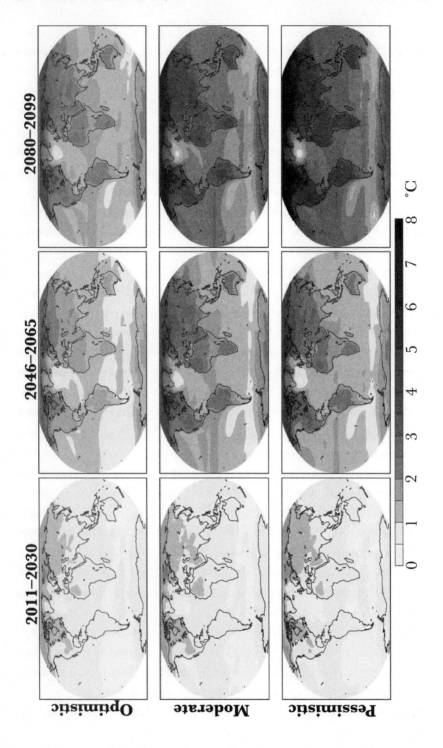

in the "optimistic" world, what is today considered an extreme warm year in New York will become the norm; and the new extremes will be unlike anything New Yorkers have ever seen.

The "pessimistic" numbers are even more alarming. They approach the magnitude of average temperature contrast between the world of today and the world of twenty thousand years ago during the last ice age, when global temperatures averaged about 5°C (9°F) cooler. Many areas of North America and Europe were under ice, sea levels were more than 100 meters (330 feet) lower, and Japan was actually connected to the Asia mainland.[280]

All of these maps are conservative in that they awaken no hidden "climate genies" that give climate scientists nightmares.[281] Instead, they chart out the plain vanilla, predictable intensification of the greenhouse effect, covering a realistic range of options lying well within control of human choices.

The fourth important fact to take from these nine maps is that the irregular geography of climate change presented in the first single map is not at all random. Important spatial patterns remain broadly preserved in all model simulations, for all carbon emissions scenarios, and across all time frames. Temperature increases are higher over land than over the oceans. A bull's-eye over the northern Atlantic Ocean stubbornly refuses to warm up. And without fail, regardless of which emissions path is followed, or what time slice is examined, or what climate models are run, all of the model projections—and measured observations too—consistently tell us something big. Again and again, they tell us that global climate change is hugely amplified in the northern high latitudes.[282]

Even our "optimistic" scenario projects that the northern high latitudes will warm 1.5–2.5°C by midcentury and 3.5–6°C by century's end, more than double the global average. Our "pessimistic" scenario suggests rises of +8°C (14.4°F) or more. Global climate change will not raise temperatures uniformly around the world. Instead, the fastest and most furious increases are under way in the North.

There is another robust trend expected for the northern high latitudes. For much of the world it is very difficult to project future precipitation patterns with confidence. Cloud physics and rainfall are more complicated and tougher to model than greenhouse physics, especially at the coarse spatial resolution of today's climate models. To the frustration of policy makers, model projections of future rainfall often lack statistical confidence, and even disagree as to whether it will increase or decrease. But not in the North. If there is one thing that the climate models all agree on,[283] it's that precipitation (snow and rain) will increase there, especially in winter. It *must* increase, in obedience to physics[284] and rising evaporation from open lakes and seas as they become unfrozen for longer times during the year.

The plainest manifestation of this will be snowier winters and higher river flows. Across southern Europe, western North America, the Middle East, and southern Africa, river flows are projected to fall 10%–30% by 2050. However, they will *increase* by a similar amount across northern Canada, Alaska, Scandinavia, and Russia.[285] This has already happened in Russia. Through statistical analysis of old Soviet hydrologic records, one of my own projects helped to confirm rising river flows there, including sharp increases in south-central Russia beginning around 1985.[286]

Recall the bleak future of stressed human water supply all around the planet's dry latitudes from Chapter 4? That future is not shared by the North. It is water-rich now and, except for Canada's south-central prairies and the Russian steppes, will become even more water-rich in the future.

Uncapping an Ocean

To most people, there is nothing visceral about computer model projections of average climate statistics decades from now. But in September 2007 we got a taste of what the real world inside those maps might look like. For the first time in human memory, nearly 40% of the floating lid of sea ice that papers over the Arctic Ocean disappeared in a matter of months. The famed "Northwest Passage"—an ice-encased explorers' graveyard—opened up. From the northern Pacific, where the United States and Russia brush lips across the Bering Strait, open blue water stretched almost all the way to the North Pole.

There was an error-riddled media frenzy about a melting "ice cap" at the North Pole,[287] then the story faded. But climate scientists were shocked to the bone. The problem wasn't that it had happened, but that it had happened *too soon*. Our climate models had been preparing us for a gradual contraction in Arctic sea ice—and perhaps even ice-free summers by 2050—but none had predicted a downward lurch of this magnitude until at least 2035. The models were too slow to match reality. Apparently, the Arctic Ocean's sea-ice cover could retreat even faster than we thought.

Two months later several thousand of us were milling around the cavernous halls of San Francisco's Moscone Center at our biggest yearly conference,[288] nervously abuzz about the Arctic sea-ice retreat. In a keynote lecture, the University of Colorado's brilliant, ponytailed Mark Serreze drove home the scale of the situation. When NASA first began mapping Arctic sea ice with microwave satellites in the 1970s, he intoned, flashing a political map of the lower forty-eight United States on the screen, its minimum summer sea-ice extent[289] hovered near 8 million square kilometers, equivalent to all of the lower forty-eight U.S. states minus Ohio. POOF! Ohio vanished from the big projection screen. Since then its minimum area had been declining gradually, up until this year when it suddenly contracted abruptly, like a giant poked sea anemone, to just 4.3 million square kilometers. POOF! POOF! POOF! Gone was the entire United States east of the Mississippi River, together with North Dakota, Minnesota, Missouri, Arkansas, Louisiana, and Iowa. A murmur rolled through the hall—even scientists enjoy a good animated graphic over tables of numbers any day.

After Serreze's talk we milled around some more, wrangling over things like "model downscaling," "cloud forcing," and "nonlinear dynamics." Some were revising the old projections for an ice-free Arctic Ocean from 2050 to 2035, or even 2013. Others—including me—argued for natural variability. We thought the 2007 retreat could just be a freak and the sea ice would recover, filling up its old territory by the following year.

We were wrong. The excursion persisted for two more years, with 2008 and 2009 also breaking records for the Arctic summer sea-ice minimum. They were the second- and third-lowest years ever seen, and had followed right on the heels of what happened before.[290]

Ice Reflects, Oceans Absorb

The broader impacts of amplified warming—more rain and snow, and reduced summer sea ice at the top of our planet—extend far beyond the region itself. They will drive important climatic feedbacks that flow out to the rest of the world, influencing atmospheric circulation, precipitation patterns, and jet streams. Unlike land ice, melting sea ice does not directly affect sea level (in accordance with Archimedes' Principle[291]), but its implications for northern shipping and logistical access are so profound they are the subject of the following chapter. Perhaps most importantly of all, an open ocean releases heat, causing milder temperatures to penetrate even the much larger frigid landmasses to the south. Indeed, the loss of sea ice is the single biggest reason why the geographic pattern of climate warming is so magnified in the northern high latitudes.

Look again at the nine maps (p. 128) charting different temperature outcomes for the coming decades. In every one, the epicenter of climate warming is the Arctic Ocean, radiating (relative) warmth southward like a giant mushrooming umbrella. You are looking at the power of the ice-albedo effect, one of the stronger self-reinforcing climate feedbacks on Earth.

Albedo is the light-reflectivity of a surface. Its values range from 0 to 1 (meaning 0% to 100% reflective). Snow and ice have high albedo, bouncing as much as 90% of incoming sunlight back out to space. Ocean water has very low albedo, reflecting less than 10% and absorbing the rest. Just as a white T-shirt feels cool in the Sun but a black T-shirt feels hot, so also does a white Arctic Ocean stay cool while a dark one heats up.

Compared to land glaciers, sea ice is thin and flimsy, an ephemeral floating membrane just 1–2 meters thick. The greenhouse effect, by melting it back somewhat, thus unleashes a self-reinforcing effect even greater than the greenhouse warming itself. It's rather as if when struck by blazing hot sun, one discards a white shirt and puts on a black one. By responding in this way to small global temperature changes, sea ice thus amplifies them even more.[292]

While its global effect is small, the ice-albedo feedback is uniquely powerful in the Arctic because it is the only place on Earth where a major ocean gets coated with ephemeral floating sea ice during the summer. Antarctica, in contrast, is a continent of land, thickly buried beneath permanent, kilometers-thick glaciers. For this and several other reasons, climate warming is more amplified in the Arctic than the Antarctic.[293, 294]

As an ice-free Arctic Ocean warms up, it acts like a giant hot-water bottle, warming the chilly Arctic air as the Sun crawls off the horizon each winter.

The sea ice that does eventually form is thin and crackly, allowing more of the ocean's heat to seep out even during the depths of winter. Winters become milder, the autumn freeze-up happens later, and the spring thaw arrives earlier. The warming effect is highest over the ocean and from there spills southward, warming vast landscapes across some of the coldest terrain on Earth.

Dr. Smith Goes to Washington

I first met National Center for Atmospheric Research (NCAR) climate modeler David Lawrence in Washington, D.C. We had been brought to the Russell Senate Office Building to brief U.S. Senate staffers on the ramifications of thawing Arctic permafrost. It was exciting. The Russell is the Senate's oldest building and the site of many historic events, including the Watergate hearings. Its hallways are white marble and mahogany, with important-looking people clacking around in dark power-suits. Just a few yards from our briefing room were the offices of Senator John Kerry and former senator John F. Kennedy. Moments before we got started, the moderator pulled us aside to whisper that Senator John McCain might show up. He didn't, but it was cool just wondering if he would.

After the briefings and a pleasant lunch reception were over, Dave and I headed out to a local pub for a beer before catching our flights home. Over microbrews, he described his next big idea: figuring out how much northern landscapes might warm up, based purely on the ice-albedo feedback from reduced summer sea-ice. I told him he was on to something. It was critical to separate out the ice-dependent feedback from overall greenhouse gas forcing, I pointed out. That way, if the ice shrank faster than expected, we'd know what the immediate climate response could be— even ahead of the longer-term cumulative effect of greenhouse gas loading. We drained our pints and left. I promptly forgot all about the conversation until eighteen months later when I ran into Dave at a conference. Whipping out his laptop, he showed me a preliminary model simulation of his big idea.[295]

My eyes widened. I was gazing at a world with northern high latitudes plastered everywhere in vivid orange—a pool of spreading warmth as much as five, six, or seven degrees Celsius (8° to 12°F) higher—spreading south-ward from the Arctic Ocean. All of Alaska and Canada and Greenland were bathed in it. It grazed other northern U.S. states from Minnesota to Maine. Russia's vast bulk was lit up from one end to the other. Only Scandinavia and Western Europe, already warmed by the Gulf Stream, were untouched. Then I looked closer and saw what time of year it was.

November . . . December . . . January . . . February. The warming effect was greatest not in summer but during the *coldest months of the year*. I was staring at a map of the relaxing grip of winter's iron clench. It was an easing, a partial lifting, of the Siberian Curse.

The Siberian Curse

The Siberian Curse is the brutal, punishing winter cold that creeps across our northern continental interiors each year. Western Europe and the Nor-dic countries, steeped in tropical heat carried north from the Gulf Stream, are largely spared. But from Russia to Alaska, and tumbling south through Canada into the northern U.S. states, the Curse descends each winter. The name was popularized in a book by Fiona Hill and Clifford Gaddy of the Brookings Institution,[296] but the concept is as ancient as life itself. When it arrives, the birds depart, the ground cracks, frogs freeze solid in their mud beds. At the extreme end, if temperatures plunge to −40°F (or −40°C, the Fahrenheit and Celsius temperature scales converge at this number) steel breaks, engines fail, and manual work becomes virtually impossible. Human enterprise grinds to a halt.

Regardless of country, all NORC northerners seem to hold something in common when it comes to this special temperature: "Minus forties," as such days are known, are universally despised. The shutdown of activity it commands has been described to me by restaurateurs in Whitehorse, Cree trappers in Alberta, truck drivers in Russia, and retirees in Helsinki. And while they otherwise express varying opinions about the problems or ben-efits posed to them by climate change, the one sentiment they all seem to agree on is relief that "minus forties"are becoming increasingly rare.

The most crushing cold rolls each year through eastern Siberia. On a typical January day in the town of Verkhoyansk, temperatures average around −48°C (−54°F). That is far colder than the North Pole, even though Verkhoyansk lies fifteen hundred miles south of it. Such frigidity stirs up images of hardy Russians bundled in furs, trudging home with some firewood or vodka to beat back the elements. A less familiar image is Verkhoyansk in July, when average daytime temperatures soar to nearly +21°C (+70°F). Our same Russian friends now stroll in short-sleeved shirts and halter tops, licking delicious precast ice-cream cones that taste like pure vanilla cream.

"So . . . what are you doing this summer?" I am asked this question twenty or so times per year. Invariably—after responding I'm going to Siberia, or Iceland, or Alaska—I win a puzzled look, followed by a nodding smile and the advice to not forget my parka and snow boots. When I explain I'll actually require sunscreen, DEET, and plenty of white T-shirts, I get another puzzled look.

In summer, even on the high Arctic tundra, there is muggy heat, hordes of buzzing insects, and water running everywhere. Yes, there are stunted trees, tundra mosses, and no raccoons, but these things are the result of cold winters, not summers. In summer the sun circles the sky day and night. Everything is bathed in heat and light. The ground thaws, flowers bloom, and rodents teem. While driving through Fairbanks, Alaska, I noticed people starting softball games at midnight. The place simply explodes with pent-up life in fantastic overdrive.

There is now overwhelming evidence that northern winters are becoming milder and growing seasons are getting longer. From weather station data, we know that air temperatures rose throughout the northern high latitudes during most of the last century, and especially after 1966. There was a short cooling snap lasting from about 1946 to 1965, but even then large areas of southern Canada and southern Eurasia continued to warm. After 1966, temperatures took off sharply, especially in the northern Eurasian and northwestern North American interiors, where annual air temperatures have been rising at least 1° to 2°C per decade on average. That's about *ten times faster* than the global average, and it's being driven almost completely by warmer springs and winters.[297]

The New Arrivals

As you might imagine, the biological response to this has been brisk. By the 1990s, a greening up of northern plant cover was spotted by satellites. Down on the ground, trees grew taller and barren tundra began sprouting up shrubs.[298] All of this is consistent with the temperature increases recorded by weather stations. Not surprisingly, ecosystem models project plant growth to continue rising right alongside the projected increases in air temperature and growing season length. Even under the "optimistic" emissions scenario shown earlier, Arctic net primary productivity (a measure of overall plant biomass growth) is projected to almost double by the 2080s.[299]

Wildlife is also on the move. From my travels and interviews the appearance of "southern" creatures in northern places was a prevailing theme. I heard repeatedly about raccoons, white-tailed deer, beavers, and even a mountain lion spotted in places they'd never been seen before. My uncle, a longtime outdoorsman in northern New York State, noticed gray squirrels and opossums moving in, along with some crazy disruptions to the spring harvest of maple syrup. The Mountain Pine Beetle, normally kept in check by winter-kill, is now devastating Canadian forests. Other biological examples published in the scientific literature include the common buzzard *Buteo buteo* wintering near Moscow, nearly a thousand kilometers north of normal; a northward shift in Japan's Greater White-fronted Goose, *Anser albifrons;* and Sweden's Brown Hare, *Lepus europaeus,* infiltrating the territory of (and possibly hybridizing with) *Lepus timidus,* the Mountain Hare. Red foxes are displacing Arctic foxes. Beavers are pushing north, and model projections suggest they will also become denser inside their current range.[300]

By midcentury *Ixodes scapularis*—the Lyme-disease-carrying tick—is projected to expand northward from its current toehold in southern Ontario to much of Canada. By century's end the smallmouth bass, today found only near the U.S. border, is projected to live all the way to the Arctic Ocean. In the North Sea—one of the world's most productive fisheries—nearly two-thirds of all fish species have either shifted north in latitude or sunk down to cooler water depths. Even lowly plankton is on the move: In the past forty years Atlantic warm-water species have pushed northward a staggering ten degrees of latitude—almost seven hundred miles—supplanting cold-water species that are in turn retreating north.

The Displaced

The 2007 sea-ice contraction triggered a new wave of public consternation about the future of polar bears, including an environmentalist push in the United States to classify them under the Endangered Species Act. This gesture, ultimately rebuffed by both the Bush and Obama administrations, was largely symbolic (far more polar bears live offshore of Canada, Russia, and Greenland than Alaska, and these countries are certainly not beholden to the U.S. Endangered Species Act), but the concern for these magnificent animals is valid. They exist naturally only in the Arctic[301] and are uniquely adapted to live out their lives roaming on top of a frozen ocean. Their home is on the floating sea ice, hunting ringed seals, napping, and occasionally cavorting or mating with one another. Some females go onto land to give birth, but they otherwise spend as much time as possible out on the ice. Unlike other bears they do not hibernate through the winter. The lean time for polar bears is in summer, when the ice disintegrates and retreats. Forced ashore, they mostly fast and wait until it returns.

There is growing evidence that the waiting and fasting periods are getting longer, leading to skinny bears, strange behavior (like wandering into towns), and even cannibalism. In 2004 biologists confirmed three occurrences of polar bears deliberately hunting and eating each other. In one case a large male bear pounded its forepaws through the den roof of a female, savagely bit into her head and neck, then dragged her off in a trail of blood to be devoured. Her cubs were buried and suffocated in the rubble. Such behavior had never been seen before during the scientists' thirty-four years of research in the area.[302]

The problem is that the bears' favorite prey, ringed seals, also require sea ice. They spend their time either resting on top of it (and watching out for polar bears), or swimming beneath it looking for Arctic cod. The Arctic cod lurks under and along the edges of the ice, watching out for ringed seals while chasing amphipods, copepods, and krill. Those little creatures in turn graze on tiny flagellates and diatoms that grow on the underside of the ice, and also bloom profusely in the water alongside its melting edge. This entire food chain—from microscopic phytoplankton to a thousand-pound polar bear—is inextricably linked to the presence of sea ice. Walruses, bearded seals, and other species also use sea ice, though none so specifically as do polar bears, ringed seals, and Arctic cod.

Despite growing evidence of stress (like bear cannibalism), none of these species is in immediate risk of extinction. But there is little question that if the summertime sea-ice fades completely, then these amazing creatures will fade right along with it. Government scientists, in a report to aid the Bush administration's decision on the proposed Endangered Species Act listing, estimate that two-thirds of the world's polar bears will be gone by 2050.[303]

From these and other indications worldwide, climate change is forcing a massive ecological reorganization of the planet, with both extinctions and expansions now under way. Depending on the emission scenario used, one model projects that anywhere from 15% to 37% of the world's species will be committed to climate-change extinction by 2050.[304] If these numbers hold true, they are devastating—roughly comparable to the impacts of defor-estation and other direct forms of habitat loss. When combined with all the other species extinctions since the last ice age, they will mark the sixth great mass extinction on Earth—and the first since the Cretaceous-Tertiary extinction that ended the dinosaurs some sixty-five million years ago.

The mechanisms for climate-change extinction are many. Amphibians and wetland species are especially vulnerable to droughts. As temperatures rise, polar and alpine species have literally nowhere left to go once pushed off the brink of the northernmost coast or highest mountain peak. A less direct mechanism is the decoupling of codependent species within a food web (called "match-mismatch" by ecologists) when their respective pheno-logical cycles fall out of whack. Imagine birds migrating to their accustomed nesting area only to find that the caterpillar hatch they were planning to gorge on has already come and gone, for example. Another is that warmer temperatures tend to enable insect pests, invasive species, disease, and robust "generalized" species (like rats and raccoons) to outcompete specialized ones. Yet another is that the projected rate of climate change is so rapid that some sedentary species (like trees) may not be able to relocate quickly enough, or their escaping climatic comfort zone will shift to a place incompatible for other reasons, like terrain or soil. Some climates, especially in alpine and polar areas, will simply cease to exist. By century's end, under a high carbon emissions scenario, 10%–48% of the world's land surface is projected to "lose" its extant climate completely, and 12%–39% will develop new "novel" climates that don't exist in the world today (mostly in the tropics and subtropics).[305] These changes will have powerful impacts on world ecosystems and could even render some local conservation efforts obsolete. Finally, because ecosystems and food webs have so many complex interconnections, there will be rippling effects we don't yet know about. All of this is piled on top of an ongoing raft of familiar ecological threats, including habitat destruction, invasive species, and pollution.

Compared with other places, habitat loss and pollution are less severe in Alaska, northern Canada, the Nordic countries, and eastern Russia, where vast boreal forests, tundra, and mountains hold some of the wildest and least-disturbed places left on Earth.[306] However, northern ecosystems also have far simpler food chains and fewer species than, say, the Amazon rain forest. Indeed, much of it is a colonizing landscape, still in the early stages of soil formation and biological expansion after being encased and pulverized by glacier ice just eighteen thousand years ago.

When imagining 2050, I anticipate that a globally unfair assortment of some winners and many more loser species will be very apparent by then. Already the world's plants and animals are in the midst of their biggest extinction challenge in sixty-five million years. Out of perhaps seven million eukaryote species found on Earth, nearly half of all vascular plants and one-third of vertebrates are confined to just twenty-five imperiled "hot spots," mostly in the tropics and comprising just 1.4% of the world's land surface.[307]

Even in the far North, a specialized ecosystem adapted to frigid cold will be under attack by advancing southern competitors, pests, and disease. It is possible that the vast boreal forest—girdling the northern high latitudes from Canada to Siberia—might convert to a more open, savannah-like state.[308] But total primary productivity—meaning plant biomass, the bottom of the food chain—will be ramping up. Certain mobile southern invaders will enjoy growing viability in a vast new territory that is larger, less fragmented, and less polluted than where they came from. Longer, deeper penetration of sunlight into the sea (owing to less shading by sea ice) will trigger more algal photosynthesis, again increasing primary productivity and reverberating throughout the Arctic marine food web. The end result of this can only be greater overall ocean biomass, more complex food webs, and the invasion of southern marine species at the expense of northern ones.

The ecology of the North is imperiled and changing. But it will be anything but lifeless.

Hunters on Thin Ice

People rely on sea ice too. For millennia the Inuit and Yupik (Eskimo) peoples have lived along the shores of the Arctic Ocean and even out on the ice itself, hunting seals, polar bears, whales, walruses, and fish. It is the platform upon which they travel, whether by snowmobile, dogsled, or on foot. It is the foundation on which they build hunting camps to live in for weeks or months at a time.

These hunters have watched in astonishment as their sea-ice travel platform—dangerous even in good times—has thinned, become less predictable, and even disappeared. People's snowmobiles and ATVs are crashing through into the freezing ocean. Farther south, they are crashing through the ice covering rivers and lakes. In Sanikiluaq, Canada, I learned that weaker ice and a two- to three-month shorter ice season is impairing people's ability to catch seals and Arctic char. In Pangnirtung a traditional New Year's Day bash celebrated out on the ice has become unsafe. In Barrow, two thousand miles west on the northernmost tip of Alaska, I learned hunters are now taking boats many miles offshore, hoping to find bits of ice with a walrus or bearded seal.[309]

This is a serious matter. In the high Arctic, eating wild animals is an essential part of human survival and culture. In Barrow I was welcomed into the home of an Inuit elder, who explained that three-quarters of his community relies on wild-caught food.[310] I was struck by this because Barrow is one of the most prosperous and modernized northern towns I have seen. There is a huge supermarket with most everything found in the supermarkets of Los Angeles. But groceries are two or three times more expensive because there is no road or rail to Barrow, so everything must be flown or barged in. Most people at least supplement their diet with wild food; many crucially depend on it. Alongside Pepe's Mexican Restaurant (which has surprisingly good food and is apparently visited by members of the Chicago Bulls basketball team) I saw plenty of bushmeat in Barrow. My host's kitchen and backyard were festooned with racks of drying meat and fish; in his driveway was a dead caribou. Another driveway had two seals, yet another a massive walrus. In the Arctic, obtaining "country food" is not for sport—it is as important to people's diet as thin-crust pizza is to New Yorkers.

Of all northern peoples, the marine mammal-hunters living along the Arctic Ocean coast are suffering the most from climate change. Less sea ice means more accidents and fewer ice-loving animals to eat. It means faster shoreline erosion from pounding by the waves and storms of the open ocean. The Alaskan village of Shishmaref has lost this battle and will need to be relocated farther inland. But even in coastal towns, nearly everyone I meet bristles at the notion of being cast as a hapless climate-change refugee.

Even as they express frustration at having their lives damaged by people living thousands of miles away—and think it only fair that those damages be repatriated—they also point to their long history of adaptation and resilience in one of the world's most extreme environments. They are not sitting around idly in despair, or gazing forlornly out at the unfamiliar sea. They are buying boats, and organizing workshops, and setting about catching the fat salmon that are increasingly moving into their seas.

There is more to this story than climate change. Later, we will discuss some profound demographic, political, and economic trends now under way that promise to be just as important to northerners' lives in the coming decades.

Greenland's Fine Potatoes

One of the more vivid media images of 2007 was one of happy Greenlanders tending lush green potato fields against a backdrop of icebergs melting away into the ocean. The diminished sea ice was wreaking havoc on seal hunting—Greenland's finance and foreign affairs minister observed that subsistence hunting crashed by 75%—but people were beginning to plant potatoes, radishes, and broccoli. "Farming, an occupation all but unheard of a century ago, has never looked better," trumpeted *The Christian Science Monitor*. By 2009 some fields were doing so well that Danish scientists started studying them, to learn why Greenland's potatoes were growing even better than southern ones.[311]

What could be a more iconic symbol of the world in 2050 than seal hunters turned farmers in one of the coldest places on Earth? But in terms of sheer caloric output, any climate-triggered boons to agriculture will not be realized on the narrow, rocky shores of Greenland, or indeed any other place in the Arctic. Similarly to what we saw for certain wild organisms, the pressure is a gradient from south to north, not a leap to the top of the planet. Summers there will always be brief, and its soils thin or nonexistent. A short-lived vegetable garden is one thing, but when it comes to producing major crops for global markets, any significant increases will be realized at the northern margins of present-day agriculture. There will be no amber fields of grain waving along the shores of the Arctic Ocean.

In 2007 I watched some of the world's top agronomists and plant geneticists debate how best to save our temperate crops from the rising heat, droughts, and pathogens forecast for the coming decades.[312] Their solution was part biotech—genetic modifications, for example—and part ancient practice: Move over, water-guzzling corn, here come the best drought-tolerant sorghums and millets . . . from Ethiopia! Without adaptation, the group concluded, the prospect of food insecurity in the low latitudes was a serious threat.

I was particularly impressed with presentations by Stanford University's Dave Lobell and Marshall Burke, who used twenty different climate models to statistically map where the food insecurities were most likely to emerge. Apparently, by the year 2030 South Asia, Southeast Asia, and southern Africa are especially vulnerable.[313] By 2050, agricultural projections for sub-Saharan Africa get even worse, with average crop production losses of –22, –17, –17, –18, and –8% for corn, sorghum, millet, groundnut, and cassava, respectively.[314] By century's end, things become still rougher, with one study concluding it is more than 90% likely that future growing season temperatures in the tropics and subtropics will exceed anything we've ever seen before, with bad implications for food crops. "With growing season temperatures in excess of the hottest years on record . . . the stress on crops and livestock will become global in character," wrote the paper's authors. "Ignoring climate projections at this stage will only result in the worst form of triage."[315]

In contrast to these studies, a broad pattern of rising crop yields in Canada, some northern U.S. states, southern Scandinavia, the United Kingdom, and parts of Russia have been repeatedly demonstrated by climate-change model simulations for years. Already these countries are major producers of wheat, barley, rye, rapeseed, and potatoes. As early as 1990 it was apparent that regardless of what climate model was used, the northern U.S. states of Michigan, Minnesota, and Wisconsin would likely benefit from rising average temperatures, even if corn, wheat, and soybean production in the rest of the country declined.[316] Similar north-south asymmetries in crop yield (with gains in the north and declines in the south) were later demonstrated for Europe and Russia.[317] The general idea is that in the marginal northern fringes of present-day agriculture, rising temperatures and longer growing seasons will boost current crops and perhaps allow introduction of new ones; in marginal southern fringes, rising temperatures and drought frequency should harm them.[318]

Other questions revolve around the relative importance of temperature versus moisture stress on plants, soil quality, strength of CO_2 fertilization, and whether extreme events (heat waves, flooding) might be even more important determinants of future food supply than the long-term temperature and precipitation statistical averages produced by climate models.[319] It is also an oversimplification to assert that Russian and Canadian agriculture, for example, will universally benefit from warmer air temperatures. Russia's current agricultural heartland lies in its dry southern steppes, where crop declines may not be fully offset by gains in the north.[320] The same holds true for Canada's western prairies. But relative to the rest of the world, the NORCs—especially the northernmost U.S. states, parts of Canada and Russia, and northern Europe—count among the few places on Earth where we can reasonably expect to see rising crop production from climate change.

Please pass the potatoes.

CHAPTER 6

One if by Land, Two if by Sea

In August 2007 the Russian nuclear-powered icebreaker *Rossiya* broke a path to the North Pole, the research vessel *Akademik Fyodorov* trailing closely behind. An opening was cut through the sea ice and two tiny submarines lowered by crane into the freezing water. Their crews then dove 4,300 meters—more than two and a half miles beneath the ice—to the floor of the Arctic Ocean. A robotic arm collected samples and planted a titanium tricolor Russian flag directly into the yellow mud of the northernmost spot on the planet. "The Arctic is ours," declared Artur Chilingarov, the polar explorer, oceanographer, and Duma politician who led the expedition and also went down in one of the subs.[321] Vaguely remembered for rescuing a stuck polar ship in the 1980s, he became an instant celebrity; President Putin later awarded him a gold Hero of Russia medal.

For the next several months, the world proceeded to go crazy about Russians staking out the North Pole. Western politicians spluttered in outrage. "This isn't the fifteenth century," Canada's foreign minister Peter MacKay told a crowd of television reporters. "You can't go around the world and just plant flags and say: 'We're claiming this territory.'"[322] Media reports framed the story as a thinly veiled grab for natural resources, citing a recent comment by U.S. Geological Survey (USGS) scientist Don Gautier, who had ballparked that the Arctic could hold up to one-fourth of the last undiscovered hydrocarbons remaining on Earth. The presumption was that Russia had fired the opening salvo in a new sovereignty race for vast riches of untapped oil and gas—resources desperately needed to support the world economy in the coming century—thought to lie beneath the frigid seafloor of the Arctic Ocean.

Despite being closer to the *Rossiya* than just about anyone else on Earth, I had no idea what was going on. I was cut off from the outside world, steaming north through an empty ocean a thousand miles north of Toronto. At the moment the titanium Russian flag was inserted, I was probably either sleeping or hosing off stinky plankton nets. It was several days before I even heard about it.

I was living aboard the CCGS *Amundsen,* a smaller icebreaker of the Canadian Coast Guard, which was headed for Hudson Bay and ultimately the Northwest Passage. My daily routine revolved in a painted metallic world less than a hundred meters long and twenty wide, with erratic rotating shifts of sleep, work, and cafeteria. We had launched with great fanfare from Quebec City just six days before the Russian flag-planting incident.

I hadn't fully grasped what a big deal these scientific icebreaker cruises are. A crowd milled alongside the ship and news crews swarmed the ship's officers and chief scientists. I spotted Louis Fortier, the director of Arctic-Net[323] who had invited me along, surrounded by television cameras. He pumped my hand and told me to enjoy myself before being spun around for another interview. A crane lifted the gangplank and the expedition's first rotation—forty scientists, thirty-five crew members of the Canadian Coast Guard, and a handful of journalists—waved at the mass of people standing onshore. Horns blared, a gleaming red helicopter circled overhead, and the two crowds yelled good-byes over the widening slice of water. As we pulled away down the St. Lawrence Seaway, I was surprised to see a few camera crews (and Louis) still milling around on deck. Were they joining the expedition, too, I wondered? Twenty minutes later my question was answered. The ship's helicopter, which had been buzzing around the ship, landed on the aft helipad and ferried them back to Quebec City.

That first night at sea, there was quite a party. Off-duty crew ditched their crisp military blues to mingle with the scientists in shorts, T-shirts, and halter tops. The room steamed, a stereo thumped, and everyone got at least mildly inebriated. American icebreaker cruises are dry, but the Canadians open a beer bar two nights a week. This early in the expedition, the selection was astonishing. I bought two bottles of Kilkenny and set out to learn more about the rare caste of scientist called oceanographers. I found one and we shouted back and forth about marine stratification, ocean sampling, the sexual habits of right whales (quite promiscuous), and the sexual habits of cruise scientists (apparently, also so). It was a great time. But by the third beer, when she had touched my arm twice, I figured it was time to leave.

Three weeks later, after a grueling round-the-clock schedule of moving, anchoring, crane operating, water sampling, and laboratory work, we disembarked in Churchill, Manitoba. A new rotation of scientists and crew were waiting excitedly to board the ship. It felt strange to give up my tiny cabin, familiar narrow hallways, and new friends to a bunch of strangers. But our rotation was just the first of many. The *Amundsen* was in her first leg of a historic 448-day journey, the longest scientific cruise ever undertaken in the Arctic. Over the next fifteen months she would cycle through some two hundred people and shock the world by gliding easily through the Northwest Passage. At a cost of $40 million, the expedition was Canada's biggest contribution to the 2007–2009 International Polar Year.[324] While less splashy than the titanium Russian flag, Canada, too, was asserting its presence in the new Arctic Ocean.

Who Owns the North Pole?

Unlike the *Amundsen* expedition, Chilingarov's dive to the North Pole was privately funded and really just a daring stunt. But that didn't stop the flag-planting from triggering an international commotion. Russia's response was that the flag was merely symbolic: The United States once planted a flag on the moon—did anyone seriously consider that a declaration of legal sovereignty? Her *real* claim to the North Pole was not from a flag, but from the geological samples collected by this and many other Russian expeditions in the Arctic. These data would prove that the Lomonosov Ridge—an underwater mountain chain, rising some three thousand meters above the seafloor, that bisects the Arctic Ocean—was geologically attached to Russia's continental shelf. This would win her sovereignty of a huge chunk of ocean floor—possibly including the North Pole—in accordance with the United Nations Convention on the Law of the Sea (UNCLOS).

UNCLOS and geology are critically important to this story, as we shall see shortly. But in late 2007 the world's eyes were transfixed by that flag, not sediment samples. The great global economic contraction was still a year away. Energy demand was soaring and resurgent Russia, fueled by hundred-dollars-a-barrel oil and Putin's steely gaze, was growing increasingly assertive on the world stage.

Two months later, when the news hit about the record-shattering low in the amount of summertime Arctic sea ice,[325] the image of uncorked shipping lanes, vast new energy reserves, and Russians planting flags in a brand-new ocean proved too much to resist. Arctic fever went viral. Headlines and pundits declared that a new colonial race for the frontier—a "mad scramble" for control of the Arctic Ocean and its vast presumed resources—had begun.[326]

The perception that vast quantities of valuable natural resources lie awaiting in the North is not without merit. Most of its land surface has yet to be prospected for minerals; the Arctic Ocean seafloor is among the least mapped on Earth. Some of the world's biggest mines are dug into Alaska and Siberia; one of the purest iron ores ever found was recently discovered on Canada's Baffin Island.[327] The discovery of diamonds in the Northwest Territories in 1991 sparked the biggest North American staking rush since the Klondike and propelled Canada from having no diamonds at all to becoming the world's third-largest producer almost overnight. No one really knows what the new Arctic Ocean biology will be, but a longer open-water season can only mean more photosynthesis, more complex food webs, and the prospect of valuable new fisheries there. There are staggering volumes of gas hydrate—a sort of solid methane dry-ice that accumulates in the pore spaces of ocean sediments and permafrost—which no one has yet figured out how to recover but is plausibly a coveted fossil fuel of the future.

The plainest prize of all is natural gas and oil. The Arctic's broad continental shelves are draped in thick sequences of shale-rich sedimentary rock, an ideal geological setting for finding oil and gas. Prospects for natural gas are particularly high. In 2008 and 2009, the U.S. Geological Survey released new assessments concluding that about 30% of the world's undiscovered natural gas and 13% of its undiscovered oil lies in the Arctic, mostly offshore in less than five hundred meters of water.[328] These numbers are huge considering the region as a whole covers just 4% of the globe. The USGS assessments conclude it is more than 95% probable that the Arctic holds at least 770 trillion cubic feet of gas, with a fifty-fifty chance it contains more than double that. To put these numbers into perspective, the total proved gas reserves of the United States, Canada, and Mexico combined is about 313 trillion cubic feet of gas. The global economy consumes some 110 trillion cubic feet per year.

Between the 2007 and 2008 sea-ice retreats, the Russian flag-planting, and the new USGS hydrocarbon assessments, it didn't take long to hear rumbles about an arms race—or even outright war—over the Arctic Ocean. "There is simply no comparable historical example of a saltwater space with such ambiguous ownership, such a dramatically mutating seascape, and such extraordinary economic promise. Without U.S. leadership . . . the region could erupt in an armed mad dash for its resources," offered Council on Foreign Relations (a prominent American think tank) analyst Scott Borgerson, writing in *Foreign Affairs*. "The rapid melt is also rekindling numerous interstate rivalries and attracting energy-hungry newcomers, such as China, to the region. The Arctic powers are fast approaching diplomatic gridlock, and that could eventually lead to the sort of armed brinkmanship that plagues other territories."[329] Nikolai Patrushev, secretary of the Russian Security Council, asserted, "The Arctic must become Russia's main strategic resource base," and "it cannot be ruled out that the battle for raw materials will be waged with military means."[330] The prestigious *Jane's Intelligence Review* concluded, "Military competition is likely to increase, with Russia and Canada increasing their deployments and exercises, while there appears little opportunity for diplomatic resolution of the disputes."[331]

Could competition for hydrocarbons really spark a military buildup in the Arctic? Militarization has happened there before, after all. During the Cold War, it was a place where American and Russian forces played cat-and-mouse war games with spy planes and nuclear-armed subs, and built remote outposts to detect long-range bombers. It was a theater of military intrigue and brinkmanship, the stuff of spy novels and movie thrillers like *Ice Station Zebra* with Rock Hudson and *K-19: The Widowmaker* with Harrison Ford.

The end of the Cold War marked the end of the thriller plots, and Arctic countries quickly downsized their militaries and lost interest in the region. Canada canceled its plan to buy as many as a dozen nuclear-powered submarines. The United States canceled a new class of Seawolf attack subs designed to fight beneath the sea ice. Most dramatically, the former Soviet Union simply parked its northern fleet in Murmansk and walked away.[332] But by 2009, nearly two decades later, a military revival was stirring. All eight NORC countries—Russia, the United States, Canada, Denmark, Iceland, Norway, Finland, and Sweden—were either rebuilding their militaries and coast guards or at least pondering new security arrangements in the region.

Prime Minister Stephen Harper was speaking often about reasserting Canada's sovereignty over her northern territories and the Northwest Passage,[333] and backing it up with new ice-strengthened patrol ships, a military training base in Resolute Bay, and a $720 million icebreaker. Norway was acquiring five new frigates armed with Aegis integrated weapons systems, and nearly fifty American-made F-35 fighter jets. Russia had refurbished its northern fleet and announced plans to expand it with new attack submarines, nuclear-powered ballistic missile submarines, and enough ships to man five or six aircraft carrier battle groups by the 2020s. Russia had also resumed long-range bomber patrols along the airspaces of Canada, Alaska, and the Nordic countries for the first time since the Cold War. On the eve of U.S. president Barack Obama's first visit to Canada, two Canadian Air Force jets were scrambled—perhaps overzealously—to meet an approaching Russian bomber.[334] Even Iceland, nearly bankrupted by the global financial crisis, was pondering how to bolster its security. Finland, Denmark, and Sweden were considering new alliances with each other, or even possible membership in NATO.[335]

The United States—dubbed the "reluctant Arctic power" by political scientist Rob Huebert at the University of Calgary[336]—was not growing its northern military power as noticeably. Its *Polar Star* icebreaker was out of service; a replacement was scrapped from the Obama administration's omnibus stimulus bill.[337] However, America had never downsized its northern forces as much as the other Arctic countries after the Cold War. It still maintained some twenty-five thousand army, air force, and coast guard personnel in

Alaska and had even begun conducting naval exercises offshore.[338] One of the United States' two controversial missile defense complexes (intended to shoot down incoming ICBM missiles) was installed at Fort Greely in Alaska. Perhaps most telling of all was a presidential directive quietly issued in January 2009, during the final days of the Bush administration. This little-noticed document sharply redefined U.S. policy in the Arctic for the first time since the end of the Cold War.

This "National Security Presidential Directive/NSPD 66, Homeland Security Presidential Directive/HSPD 25," or, more compactly, "Arctic Region Policy,"[339] was crafted exclusively for the Arctic, a significant change because all previous directives had lumped it and Antarctica together. Equally significant was its elevation of "National security and homeland security needs" to priority position #1 (out of six)—a return to Cold War prioritization. To political scientists, these changes are significant and signal a growing American strategic interest in the region.

War in the Arctic?

We've seen that current trends in rhetoric, defense spending, and written policy all point to a renewed militarization of the North. That is the trend. But what about *war*? Huebert believes that the world is beginning to per-ceive the Arctic as the "next Middle East" in terms of fossil hydrocarbon energy.[340] Is it also the next Middle East in terms of fault lines for conflict? After all, jostling militaries imply heightened risk of incident; and conflicts needn't even be about the Arctic to erupt there—the region could also become an expanded theater for global tensions and antagonism, as hap-pened during the Cold War.

This last scenario is certainly not the case today. Whether it develops in the future depends on the choices of future political leaders and thus lies outside the bounds of our thought experiment. But what of intrinsic pres-sures within the Arctic *itself*? Is the "mad scramble" so fevered, the oil and gas assessments so compelling, the retreating ice and new shipping lanes so transformative, that extreme tension or violent conflicts in the region become inevitable?

There are good reasons to think not. One is a persistent trend of northern cooperation over the past two decades. A second is a legal document of the United Nations that is fast becoming the globally accepted rulebook on how countries carve up dominion over the world's oceans.

The story of the first begins October 1, 1987, with a famous speech delivered in Murmansk by then–Soviet leader Mikhail Gorbachev. Standing at the gateway of his country's strategic nuclear arsenal in the Arctic Ocean, Gorbachev called for transforming the region from a tense military theater to a nuke-free "zone of peace and fruitful cooperation." He proposed international collaborations in disarmament, energy development, science, indigenous rights, and environmental protections between all Arctic countries.[341] The choice of Murmansk, the Arctic's largest and most important port city and the heart of the Soviet Union's military and industrial north, was highly symbolic. Just as the sea ice would experience a record-breaking melt exactly twenty years (to the day) later, the Cold War thawed first in the Arctic.

Four years after the Murmansk speech the Soviet Union dissolved. The Russian Arctic, which had been totally closed off from the world, plunged into a horrible decade of decimated population and economy, but new opportunities to interact with outsiders opened. After a half century of iron-walled separation, aboriginal Alaskan and Russian relatives become reacquainted across the Bering Strait. Siberians, if they had the money, could travel abroad, while western scientists—including myself—could enter and work in formerly closed parts of the Russian North. New international collaborations and foreign cash[342] were a rare bright spot for many suffering Siberians. All around the Arctic, new collaborations and groups were born. Aboriginal groups, most notably the Inuit, began to organize politically across international borders.

In 1991 all eight NORC countries—the United States, Canada, Denmark, Iceland, Norway, Sweden, Finland, and Russia—signed a landmark agreement to cooperate on the region's pollution problems, and scheduled regular meetings to actually accomplish something.[343] Five years later the

Arctic Council was formed,[344] an intergovernmental forum whose membership includes not only the NORCs but other observer countries and interest groups as well. While voiceless in matters of security, it is now the premier "Arctic" polity in the world. In sum, the 1990s were a time of unprecedented cooperation between northern countries operating at many different levels.[345]

Despite the hype about mad scrambles and looming Arctic wars, that cooperative spirit has persisted. The early twenty-first century saw the Arctic Council release the influential "Arctic Climate Impact Assessment" (a consensus science document, modeled after the IPCC assessments) requiring collaboration and sign-off from all of its members.[346] A multitude of international collaborations was completed, without drama, during the International Polar Year. A major study of the region's current and future shipping potential—again requiring international cooperation and sign-off by the eight NORC countries—was completed in 2009.[347] The list of other examples of successful cooperation and integration between supposed adversaries, on things like search-and-rescue, environmental protection, aboriginal rights, science, and public health, is long.

To be sure, the thorniest matters—national security, sovereignty, and borders—have been (and continue to be)[348] scrupulously avoided. But unlike the past, the Arctic today is no longer a place of suspicious neighbors, armed to the teeth, who don't talk to each other. Instead, there is in place a remarkably civil international network, one that is working cooperatively and effectively at many levels of governance.

The Rule of Law

The second reason to doubt the eruption of an Arctic War lies in UNCLOS, the United Nations Convention on the Law of the Sea. Contrary to popular perception the Arctic is not a ruptured piñata. On land, its international political borders are uncontested. For the Arctic Ocean, there are now clear procedural rules for laying claim to its seabed, and indeed any other seabed. Most importantly, just about every country in the world seems to be following them.

UNCLOS was negotiated over a nine-year period from 1973 to 1982 and has emerged as one of the most sweeping, stabilizing international treaties in the world. As of 2009 it was ratified by 158 countries, with many more in various stages of doing so. Of the eight NORC countries, seven have ratified UNCLOS. The one glaring holdout—the United States of America—is obeying all UNCLOS rules and sending signals that it will eventually ratify the treaty. It therefore constitutes one of the most agreed-upon rulebooks in international law and is a highly effective agent of order.

The cornerstone of UNCLOS is the creation of an Exclusive Economic Zone (EEZ) extending from a country's coastline for 200 nautical miles (about 230 statute miles) outward into the ocean. A country has sole sovereignty over all resources, living and nonliving, within its EEZ. It has the right to make rules and management plans and collect rents for the management and exploitation of these resources. The invention of these zones has greatly reduced "tragedy-of-the-commons" overfishing and other resource pressures and disputes in the world's coastal oceans.

That's not to say UNCLOS is perfect. Now, disputes break out over island specks because they anchor a claim to a 200 nm radius circle on the surrounding seafloor. Tiny Rockall—literally a barren rock peeking out of the North Atlantic—has been claimed by the United Kingdom, Ireland, Iceland, and Denmark. Denmark is also tussling with Canada over Hans Island, another speck sitting between the two countries in the Nares Strait off Greenland. The convoluted coastlines of Russia and Alaska open a doughnut hole of high seas in the midst of their Exclusive Economic Zones, into which Japan, South Korea, Taiwan, and Poland pour fishing trawlers.[349] Finally, border disputes arise over how the two-hundred-nautical-mile extension is to be drawn with respect to other boundaries. Canada, for example, extends the ocean border as a straight-line extension of its land border with Alaska, whereas the United States draws the line at right angles to its coastline. This creates a smallish disputed triangle (about 6,250 square miles) of overlapping claims to the Beaufort Sea. In the Barents Sea, Norway and Russia had quite serious overlapping claims but announced resolution of the conflict in 2010.[350] These are not insignificant disputes but, relative to the mess of conflicts existing prior to UNCLOS, manageable ones.

Beyond the two-hundred-nautical-mile limit are the high seas, their resources controlled by no one. However, UNCLOS Article 76 allows a special exception. If a country can prove, scientifically, that the seafloor is a geological extension of its continental shelf—meaning that it is still attached to the country's landmass, just underwater—then the country may file a claim with a special U.N. commission to request sovereignty over that seabed even beyond the two-hundred-nautical-mile limit.[351] Article 76 lays out a clear and orderly procedure for doing this. Because the Arctic Ocean is small, has unusually broad continental shelves, and is mostly encircled by land, it is unique among the world's oceans in that a great deal of it could potentially be carved up into these extended zones. Russia, Denmark, Canada, Norway, and the United States[352] are the sole countries with frontage on the Arctic Ocean. These five countries are thus well positioned to win control over large tracts of its seafloor and any hydrocarbons or minerals it may contain.

The key words here are *scientific* and *orderly*. The case for an Article 76 claim must be documented exhaustively with petabytes of scientific data. Foremost is a detailed mapping of seafloor bathymetry—its topographic relief—from multibeam hydrographic sonar. Seismic surveys, using explosives or blasts of compressed air to send shockwaves into the seabed, trace out the deeper subterranean geology. Sediment samples, like the ones grabbed by Chilingarov's tiny submarines at the North Pole, are used to establish geological provenance. And so on.

All of this takes years of costly research, but there is a process to it and eventually it gets done. Norway submitted its EEZ extension claim in 2006 and was approved in 2009.[353] The United States, Canada, Denmark, and Russia are still busily mapping, with Russia closest to being done. Canada will file by 2013 and Denmark by 2014. Because the United States hasn't ratified UNCLOS yet, it will likely be last to file but has already gathered much of the required hydrographic sonar and other data from the *Healy* icebreaker, work led by Larry Mayer at the University of New Hampshire.

After all this work and expense, small wonder that these five countries—Russia, Denmark, Canada, Norway, and the United States—recently banded together to issue the "Ilulissat Declaration," an assertion that existing international laws are perfectly sufficient for working out their territorial disputes in the region. Everything's cool, no new Arctic treaties are needed—or wanted. Now, would everyone else—like the European Union—kindly butt out?[354]

These five powers are signaling that UNCLOS is the law of the Arctic Ocean, just like any other ocean. They are heeding its procedures for sovereignty claims to its seafloor, in the same manner as the many other Article 76 claims inching forward around the world. None are interested in relinquishing their existing right to make these claims. There will be no new "Arctic Treaty" with shared international governance, as exists in Antarctica. And centuries of legal precedent tell us that once the boundary lines get set, they will stay set. In Southern California today, property lines set by early *ranchos* have persisted through centuries of rule first by Spain, then Mexico, then the United States.

So how will it all pan out? The bathymetric and geological data are still being collected, but Russia has the longest coastline and the broadest continental shelf in the Arctic Ocean. This optimal geography will win sovereignty over very large tracts of seabed and most of the natural gas promised by the U.S. Geological Survey assessment.[355] Canada is positioned to more than double its offshore holdings northwest of Queen Elizabeth Island. The United States will expand dominion in a triangular wedge extending due north of Alaska's North Slope, winning control of some of the Arctic Ocean's most promising oil-bearing rocks. Norway has secured chunks of the Norwegian and Barents seas and will share claims with Russia in the gas-rich Barents Sea.

But Norway has no shot at the North Pole and neither does the United States. That prize—if it is a prize—hinges on the Lomonosov Ridge mentioned earlier in the chapter. This thousand-mile undersea mountain chain, roughly bisecting the abyss of the central Arctic Ocean, is the only hope for a continental shelf extension claim extending as far as the geographic North Pole. Russia, Denmark, and Canada are busily mapping it.[356] But the importance of the North Pole seabed, a distant, heavily ice-covered area and unpromising oil and gas province at that, is primarily symbolic. Personally speaking, if it's really necessary for any country to control the North Pole, then it seems only fair that it be Russia. Russia's first hydrographic surveys date to 1933. No surface ship ever reached the North Pole until the Soviet nuclear icebreaker *Arktika* accomplished it in 1977. By the end of 2009 the feat had been accomplished just eighty times: Once each by Canada and Norway, twice by Germany, thrice by the United States, six times by

Sweden, and *sixty-seven* times by Russia. Her *Sibir* icebreaker completed the first (and only) voyage to reach the North Pole in winter back in 1989. As far as I'm concerned, Russia has earned it.

Whether the science will reveal the Lomonosov Ridge to be geologically attached to Russia, or to Greenland (Denmark), or to Canada, or none of them, is unknown. What *is* known is that no one is bristling missiles over this. And there's little reason to think that anyone will.

The Five-Century Dream

The dream is a northern shipping route between the Atlantic and Asia's Far East, a quest spanning over five centuries since the English, Dutch, and Russians first began looking for it. The only alternative, until the Suez and Panama canals were built, was to sail all the way around the southern horns of either Africa or South America. Many intrepid souls died looking for a shorter route over the North American or Eurasian continents. Probing northwest (the Northwest Passage), their ships got stuck like bugs to flypaper in Canada's perennially frozen northern archipelago, en route to the Bering Strait. Others died trying northeast (the Northern Sea Route), attempting to trace Russia's long northern coastline to reach the Bering Strait from the other direction. Both routes have now been traversed many times but neither is a viable commercial shipping lane. However, a small amount of international traffic is stirring between Canada's port of Churchill (in Hudson Bay) and Europe, and occasionally Murmansk.

Since the 2007 and 2008 sea-ice convulsions, the prospect of global trade flows streaming through the Northwest Passage, the Northern Sea Route, or even straight over the North Pole has become one of the most breathlessly touted benefits of global climate change. After all, those fifteenth-century navigators were geographically correct: Even after the Panama and Suez canals were made, the shortest shipping distances between Asia and the West would still lie through the Arctic Ocean.[357]

Lest we get carried away with visions of colorful sailboat regattas in the Arctic Ocean, keep in mind just how formidable sea ice is to the maritime industry. Only the largest heavy class of icebreaker like the *Rossiya* can break through it confidently.[358] Canada has just two heavy icebreakers, the United States three. Russia—by far the world leader in this domain—is expanding its fleet to around fourteen. Seven are nuclear-powered, the largest and most powerful in the world.[359] But icebreakers are costly and few. They require a strengthened hull, an ice-clearing shape, and serious pushing power, features not possessed by normal ships.[360] There are barely a hundred of them operating in the entire world. The world's other vessels, of course, number in the hundreds of thousands, but cannot navigate safely through sea ice.

However, there is a very real possibility that by 2050, if not sooner, the Arctic Ocean will become briefly free of sea ice in September, by the close of the northern hemisphere summer. The ice will always return in winter (much like the Great Lakes today), but this is nevertheless a radical transformation, one that will dramatically increase the seasonal penetration of shipping and other maritime activities into the region. For part of the year, it would change from being the domain of a handful of heavy icebreakers to that of thousands of ordinary ships.

One doesn't need a fancy climate model projection to appreciate this. It's already obvious today. On the following two pages, consider the seasonal cycle of shipping activity that already happens each year in the Arctic. When sea ice expands in winter, ships retreat. When it shrinks in summer, they advance.

Note the profound restriction that sea ice imposes upon shipping activity. Few, if any, vessels dare to enter the ice pack, but there are thousands of them poking and probing around its southern periphery (there were at least six thousand ships operating in the Arctic in 2004, the year that these two maps capture).[361] In January, sea ice confines them to the Aleutian Islands, northern Fennoscandia, Iceland, and southern Greenland. Even the icebreakers retreat then. Only Russia did any serious icebreaking—to and from Dudinka, a port for the Noril'sk mining complex on the Yenisei River. But in July, when the ice melts, the ships pour in.

The Arctic Ocean will never be ice-free in winter, but summer shipping will last longer and penetrate more deeply. If it really does become ice-free by late summer, it should be briefly possible to sail a ship right over the top of the world.

Not all shipping companies are thrilled about the prospect of this. Take,

The importance of sea ice to maritime activity. Shipping activity in the Arctic Ocean in winter . . .

for example, Northern Transportation Company Limited, northern Canada's oldest Arctic marine operator. Since 1934 NTCL has been providing cargo transport down the Mackenzie River and all across North America's western Arctic coast, from Prudhoe Bay, Alaska, to Taloyoak in Nunavut. The bulk of their business is cargo transport to villages, oil and gas operations, mines, and offshore energy exploration. The company's vice president,

... and in the summer of 2004. (Shipping data source: AMSA[362])

John Marshall, was kind enough to show me around their port in Hay River, on the shore of Great Slave Lake.

I was impressed. There were a hundred barges in operation, acres of other vessels parked, and a Syncrolift to raise huge ships entirely out of the water. Workers were swarming all over the barges to load them up and move them out. The company moves fast to capitalize on their short shipping season—only about four months—before the ice returns in October. But when I bounced the long-term climate model projections for sea ice off my host, I was surprised to learn he hopes to never see their simulations materialize. A longer shipping season on the Mackenzie would be wonderful, but an open Northwest Passage would allow competition in from the east. The sea ice blocking that passage, Marshall told me, was keeping his southern competitors out.[363]

If the Arctic Ocean becomes ice-free in summer, it will also affect maritime activities in at least one other important way. It spells the disappearance of so-called "multiyear ice," the more obstructing of two forms of sea ice currently present there. "First-year" ice, as the name implies, is baby ice, less than twelve months old. It is one or two meters thick and relatively soft, owing to inclusions of salty brine and air pockets. While definitely dangerous, it is easily cleared by icebreakers and will not generally gore a properly handled vessel with an ice-strengthened hull. Importantly, first-year ice is also less damaging to the drilling platforms and other infrastructure needed to produce offshore oil and natural gas.[364] But multiyear ice is hard and can grow up to five meters thick.[365] It is utterly impassible to most ships and can foil even a Russian nuclear-powered icebreaker.

In a world where all sea ice melts away each summer, multiyear ice will go extinct and icebreakers will go where they please. Ships with fortified hulls—and even ordinary vessels—would be somewhat safer.[366] From a regulatory standpoint, this could lead to ships of a lower polar class being permitted to enter and operate in the Arctic.[367] The Northern Sea Route (especially) and the Northwest Passage would become viable lanes. For a brief time window each year, it would become feasible to cross right over the North Pole in ice-strengthened ships. A dream come true.

Dream On

So by 2050 will global trade flows be pouring through the Arctic Ocean, as they do today through the Suez and Panama canals?

Impossible. Those operate 365 days per year with no ice whatsoever. At best the Arctic Ocean will become ice-free for a few days to a few weeks in summer and even then, there is no such thing as a truly "ice-free" Arctic

Ocean. From autumn through spring, there will be expanding first-year ice cover, slowing ships down even with icebreaker escort. In summer, there will always be lingering bits of sea ice floating around, as well as thick icebergs calved from land-based glaciers into the sea (a glacier iceberg sank the *Titanic,* not sea ice). The Arctic Ocean will *always* freeze in winter—or at least we'd better hope so. If it doesn't, that means our planet has become 40°F hotter and a lifeless scorched rock. Superimposed over all of this is ever-present natural variability, making the start and end dates of a part-time shipping season impossible to know with certitude.

The global maritime industry cares about many other things besides geographic shipping distance. It also cares about shipping time, cost, and reliability. To be sure, routes are shorter across the Arctic Ocean, but the travel speeds, owing to the danger of ice, are lower.[368] If the region's emerging regulatory framework demands that only polar-class ships be allowed in, then those vessels will cost considerably more than ordinary single-hulled ships. And how attractive will a short, unpredictable shipping season really be for today's tightly scheduled global supply chains? What about the relative lack of emergency and port services, environmental liability for oil spills, or fees charged by Russia and Canada should they reaffirm their positions that the Northwest Passage and Northern Sea Route are not international straits?[369] Might the Suez and Panama canals lower their prices in response to the new competition? There are many other factors controlling the profitability of transnational shipping lanes besides a shorter geographic route, available for an uncertain few weeks to a few months out of the year.

In imagining 2050, I do see many thousands of boats in the Arctic, but not humming through global trade routes as dreamed of in the fifteenth and early twenty-first centuries. Doubtless some international trade will be diverted through the region as the summer sea-ice retreats northward. It is happening now through the Aleutian Islands, Murmansk, Kirkenes, and Churchill. But few of the vessels I envision are giant container ships carrying goods between East and West.[370] The thousands of ships I see are smaller, with diverse shapes, sizes, and functions. They are not using the Arctic as a shortcut from point A in the East to point B in the West. Instead, they are buzzing all around the Arctic itself.

Look again at the maps of what actually happened in 2004. The action was not *through* the Arctic, but *in* the Arctic. There were tankers, tugs, barges, bulk carriers (for ore), small cargo ships, and fishing boats. There were coast guards, oil and gas explorers, science expeditions, and many pleasure cruises. They were bringing in supplies to villages and mine out-fits. They were fishing, hauling out ore, or looking for hydrocarbons. They were moving goods up and down rivers and through the Bering Strait. They were bringing tourists from all over the world to see one of the last truly wild places on Earth.[371]

With less sea ice, this diverse maritime activity will intensify. It will operate longer and penetrate deeper. It will become more economic to use boats to take food and heavy equipment north, and bring raw natural resources south to waiting markets. Mines located near a coast or an inland river will become increasingly viable. Already, South Korean shipbuilders, like Samsung Heavy Industries, are developing polar LNG carriers spe-cially designed to work there. When those vast new offshore gas deposits are eventually developed, these ships will cruise right up to the wellheads. They will gorge on liquefied natural gas, then turn around and carry it to anywhere in the world.

Ten "Ports of the Future" Poised to Benefit from Increased Traffic in the Arctic

Port	Country	Current Pop.	Likely sector(s)
Archangel'sk	Russia	356,051	timber, trade, metals, energy
Churchill	Canada	923	agricultural exports, trade
Dudinka	Russia	25,132	metals, Northern Sea Route traffic
Hammerfest	Norway	9,261	energy
Kirkenes	Norway	3,300	metals, tourism, NSR traffic
Murmansk	Russia	336,137	metals, energy, trade, military
Nuuk	Greenland	17,834	tourism, energy services
Prudhoe Bay	USA	5[1]	energy
Reykjavík	Iceland	120,165	fisheries, trade
Tromsø	Norway	53,622	energy, fisheries

[1] (+ several thousand temporary workers)

Shipping is the world's cheapest form of transport. As its penetration grows and intensifies, we will see a growing maritime economy in the Arctic. On the opposite page is my qualified guess at ten ports that bear particularly close watching in the coming years. Other possible sleepers include Tuktoyaktuk, Iqaluit, and Bathurst Inlet in Canada; Nome in Alaska; Ilulissat in Greenland; and Varandey, Naryan-Mar, and Tiksi in Russia.

When the *Amundsen* docked in Churchill, I knew exactly what to do. While everyone else was milling around, saying farewells or asking for directions to the town's famous Portuguese bakery, I dashed straight to the train station to ask if the tracks were OK. Just as I'd feared, they weren't. I went immediately to the airport and scooped up one of the last seats on a flight to Winnipeg. I felt guilty because I had beaten out my former friends and comrades, who I knew could be stranded a week or more. But I had just been to Churchill six weeks before, and I knew they would enjoy themselves.

Churchill is famous for being the polar bear capital of the world—thousands of tourists descend on the town each October to watch them from heated buses out on the snowy tundra—but the place is even more incredible in summer. The snow is gone, the weather warm, and some three thousand white beluga whales move into the bay to feast on capelin and have babies. You can see the belugas distantly from the shore, but for eighty dollars a Zodiac tour will take you right out to them. The boil of white bodies leaping all around me, many with little gray calves hugging their backs, is one of the most spectacular sights I've ever seen in my life.

Churchill's other industry is shipping. It is the only northern deepwater seaport in Canada. It is also the closest port to her western provinces, where most of the country's agriculture takes place. Wheat, durum, barley, rapeseed, feed peas, and flax from the prairies are loaded into train cars and sent to Winnipeg, where a spur line runs north for a thousand miles to Churchill on the shore of Hudson Bay. But despite its geographic advantage the port has never done very well. In 1997 the port, grain elevator, and 810 miles of railroad were bought for a pittance from the Canadian government by

Denver-based OmniTRAX Inc., one of the biggest privately held railroad companies in North America. As part of the deal the company poured some USD $50 million in repairs and upgrades to its facilities and rail line.

When I first visited Churchill ten years after OmniTRAX took over, the port still wasn't running at full capacity. Its general manager and Churchill's mayor both offered that the reason was at least partially political.[372] There was also a lingering perception that the Churchill facility could not handle steel hoppers (the industry standard) even after the necessary upgrades had been made. But the biggest problem of all was the rail line linking the port to Winnipeg. Even after millions of dollars in improvements, it was still unreliable. Allowable speeds were slow, and the tracks had to be closed often for repairs. The reason was not bad design, but thawing permafrost.

On Shaky Grounds

Permafrost is permanently frozen ground. It is ubiquitous around the Arctic and high elevations of the world, and extends surprisingly far south in the cold eastern interiors of Canada and Siberia (see maps on pages x–xiii). The topmost part thaws inches deep each summer, but beneath this so-called "active layer," the soil stays hard and frozen year-round. As such, it offers a solid base on which to build roads, buildings, pipelines, and other infrastructure—so long as it *always stays frozen*. The trick is to not warm it up.

An entire subfield of civil engineering is devoted to building things on top of permafrost without somehow warming it. Houses are raised up off the ground on pilings, roads and railroad tracks are perched atop thick pads of insulating gravel, and so on. Oil pipelines require very careful design because flowing fluid generates a surprising amount of heat, and a ruptured pipeline is an environmental disaster. The world's latest permafrost engineering feat, completed in 2006 at a cost of USD $4.2 billion, is China's Qinghai-Tibet Railroad crossing the Tibetan Plateau from Golmud to Lhasa.

But no amount of clever engineering can stop regional permafrost from thawing from milder, snowier winters (snow insulates the ground). When that happens, unless the geological substrate is firm bedrock, the built structures are compromised. The substrate returns to the structural strength of wet mud, or peat, or whatever else it is geologically composed of. The ground slumps, roads buckle, and foundations crack.[373] Pipelines and train tracks become kinked and wavy when they ought to be straight. Even slight undulations force trains to slow down greatly or risk derailment. The sluggardly speeds I'd noticed for parts of the Hudson Bay Express, the otherwise lovely two-night passenger train voyage from Winnipeg and Churchill, was because of this. Deeper kinks require closing down the tracks for repairs. That's what triggered the line's closure six weeks later, when I bailed on the train (and my *Amundsen* shipmates) and caught a flight instead.

Fortunately for OmniTRAX, only the last leg of its long railroad to Churchill lies over permafrost. But other built structures around the Northern Rim are not so lucky. From borehole thermometry and other measurements, we know that permafrost temperatures are generally rising.[374] The endgame of this process is ground slumping, tilted trees, sinkholes, and other disturbances.

Already we see evidence of this from space. Using satellites, my UCLA colleague Yongwei Sheng and I mapped out a strange phenomenon now transforming vast tracts of western Siberia. This region famously holds thousands of wellheads supplying natural gas to international markets in Ukraine and Europe. Less famous are the tens of thousands of lakes that dot its surface like so many spilled marbles. By comparing recent satellite pictures of this region with those from the early 1970s, we discovered a landscape mutating as the underlying permafrost thaws, with many of these lakes disappearing into the ground. [375]

Theoretically, if all permafrost were to go away entirely, about half of the world's northern lakes and wetlands might conceivably vanish.[376] But permafrost thaw is a slow process, so that won't happen anytime soon. Deep permafrost can extend hundreds of meters downward and requires centuries or millennia to defrost. But significant reductions are expected by 2050, with climate models projecting 13%–29% less permafrost area by then, and the depth of seasonal thawing increasing roughly 50%.[377] These numbers are worrisome because from a practical standpoint, the settling and buckling problems commence even when permafrost first starts to thaw. Also troubling is the fact that permafrost ground is commonly stuffed with chunks and lenses of pure ice, which drain out, exacerbating the slumping.

Already in Russia, damages to the Baikal-Amur Mainline (BAM) Railroad have more than tripled. The number of threatened buildings ranges from 10% of all structures in Noril'sk to as high as 80% in Vorkuta.[378] At the center of this book is a photograph of an apartment building destroyed by thawing permafrost. Just days after the first wall cracks appeared, this building collapsed.

The big message here is that climate warming presents a severe challenge to current and future physical infrastructure in northern permafrost areas. The structural strength of many soils will be reduced, threatening existing structures and making new ones more expensive to engineer and maintain. Some permafrost landscapes will slump, collapse, or suffer hydrological changes, rendering them even less appealing for human activities than they are now.

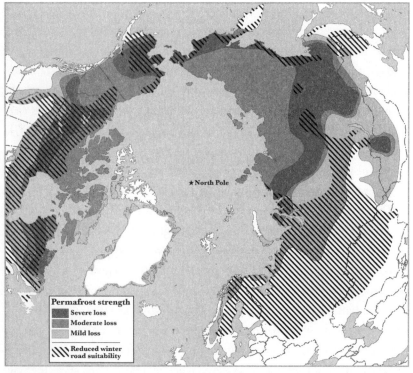

Projected losses by 2050 in (1) the structural integrity of permafrost soils, a threat to buildings and other permanent infrastructure; and (2) suitably freezing temperatures for the construction of temporary winter roads over wet or soft areas.

The map[379] on the previous page illustrates the scale of this problem by midcentury. Part of it derives from a new model of permafrost load-bearing capacity developed by Dmitry Streletskiy, Nikolay Shiklomanov, and Fritz Nelson at the University of Delaware. Dark tones indicate reduced bearing capacities (structural strength) of permafrost soils associated with a middle-of-the road carbon emissions scenario, i.e., the "moderate" (SRES A1B) scenario described in Chapter 5. Widespread losses in Alaska, northern Canada, and most of Siberia suggest that problems of reduced ground strength to support pilings, building foundations, and other heavy installations will be particularly severe there.

The hatched lines on the map are unrelated to permafrost. They illustrate another sort of change that will occur, in places where the ground surface freezes less long and hard during winter than it does now. The repercussions of this are quite different from the threat to infrastructure posed by warming permafrost, as we shall see next.

Ice Road Suckers

The second way in which rising temperatures will make remote northern landscapes less accessible is by reducing our ability to travel on them using winter roads.

Winter roads, also variously called ice roads, snow roads, temporary roads, and other names, are a remarkably well-kept secret. As their name suggests, they are temporary features, requiring a hard, deeply frozen surface to work. Winter roads are used extensively in Alaska, Canada, Russia, and Sweden and are also used in Norway, Finland, Estonia, and several northern U.S. states. In truly remote areas they are the only kind of road at all. Yet, despite their importance, these transient travel lanes rarely show up on maps. Before the popular television series *Ice Road Truckers* was produced, few people even knew they existed. But in many parts of the North—especially wet, boggy areas—they are the only way to economically resupply villages, run construction projects, harvest timber, find oil and gas, or do just about anything. Away from rivers and coastlines the only other option is to use airplanes and helicopters, which are extremely expensive.

In contrast to its biological life, economic activity on northern landscapes springs to action in winter, after the ground freezes and ground vehicles can be brought in. With remote distances and low population densities, the cost of permanent roads is rarely justified. In contrast, even the most expensive of winter roads—built up like an ice-skating rink by repeatedly glazing it with water—costs 99% less to build.[380] So in many remote areas, the road network is not fixed but an ephemeral ghost, expanding briefly each winter, then melting away again in the spring.

One famous winter road, featured in the first season of *Ice Road Truckers*, is the Tibbitt-Contwoyto ice road built each year in Canada's Northwest Territories. It begins near the city of Yellowknife and runs six hundred kilometers northeast into Nunavut, supplying a string of highly lucrative diamond mines. This road traverses bog and lakes and can exist only for about two traffic-jammed months out of the year.[381] During the other ten, the mines can be reached only by air.

Since 2003 one of the richest diamond strikes served by this road has been the Diavik Diamond Mine owned by Rio Tinto, a multinational mining conglomerate. At Diavik's headquarters in Yellowknife, manager Tom Hoefer explained that the Diavik mine yields four to five carats of diamonds per ton of ore, one of the highest grades ever found (the world average is one carat per ton). To get at the diamonds, the company spent $400 million just to dike back an overlying lake that was in the way.[382] Together with one of its neighbors, this mine currently generates about half of the NWT's gross domestic product. But despite its high grade, without the Tibbitt-Contwoyto road, this mine would be uneconomic. "If we didn't have this winter road we wouldn't have these mines," Hoefer told me. "It's as simple as that."[383] Imagine trying to bring in all the heavy equipment, construction materials, and thousands of tons of cement mix by airplane. It just couldn't be done.

For every Tibbitt-Contwoyto there are thousands of lesser winter roads vital to some economic activity or another. In Siberia I saw many long piles of deep sand running across the taiga. They are dormant winter roads and will lie there, useless and undrivable, until the deep freeze of winter returns so they can be graded again. Giant north-flowing rivers like the Ob',

Yenisei, and Lena in Russia, and Mackenzie River in Canada become ice highways in winter. In High Level, Alberta, I visited Tolko Industries—a major softwood producer for the U.S. building industry—and learned that their wood harvest relies on a fourteen- to sixteen-week winter road season. To the consternation of the company, that season has been gradually shortening over time. "We will lose our shirt" if the roads go away, their forester told me.[384]

Most resource extraction operations in the North already face tight profit margins from chronic labor shortages, long distances to market, and an environment that is both too harsh and too delicate. For industries where an entire year's worth of profit must be made in a matter of weeks, even a few days lost is a serious blow. Because northern climate warming is greatest in winter, it uniquely targets this sector. Warm winters mean shorter winter road seasons and/or lighter allowable loads. Deeper snow means more insulation of the ground, further reducing the depth and hardness of its freezing. For all but the most lucrative operations, many industries will become increasingly uneconomic and finally abandoned.

The significance of this goes beyond the major *Ice Road Trucker*–type ice highways that are rebuilt in the same place each year. It means reduced access everywhere. Take, for example, off-road oil and gas exploration on the North Slope of Alaska. To avoid damaging thin tundra soil and vegetation,[385] this can be done only in winter, when its soft, moist surface freezes hard. There's simply no other way to drive on this environmentally sensitive ecosystem without tearing it apart. But since the 1970s the North Slope's permissible off-road travel season has declined from over two hundred days per year to just over one hundred days,[386] effectively cutting the energy exploration season in half.

Put simply, this is not a good century to be out working the land in remote interiors of the North. In permafrost, permanent structures will become even trickier to build and maintain than they are now. Despite ways of prolonging the life of winter roads,[387] there's no getting around the fact that milder winters and deeper winter snow will shorten their seasons, making many of them pointless to build for all but the most lucrative projects—the NWT diamonds,[388] for example, or natural gas pipelines.

Already we see delayed openings and earlier closures harming smaller outfits operating on tight margins.

Extraction industries will favor projects nearer the water. Looking ahead, our northern future is one of diminishing access by land, but rising access by sea. For many remote interior landscapes, the perhaps surprising prospect I see is reduced human presence and their return to a wilder state.

CHAPTER 7

The Third Wave

"Canada: A few acres of snow."
—*Voltaire* (1694–1778)

Number One ($596 billion per year)
—Rank of Canada among U.S. trading partners (2008)

The preceding chapters imagine a 2050 world in which global population has grown by nearly half, forming crowded urban clots around the hot lower latitudes of our planet. Mighty new poles of economic power and resource consumption have risen in China, India, and Brazil. People are urban, grayer, and richer. Many places are water-stressed, uninsurable, or battling the sea. Some have abandoned irrigated farming altogether; their cities rely totally on global trade flows of energy and virtual water to even exist.

We have a diverse basket of new energy sources but still rely heavily upon fossil fuels. Natural gas is especially lucrative and under aggressive development in all corners of the world. Among these is the Arctic Ocean, where investment capital is flowing north as the peaceful settlement of seafloor claims, diminished sea ice, new maritime port facilities, and specialized LNG tankers have made offshore gas extraction increasingly economic. The NORCs' relative water riches are envied by all. Milder winters have encouraged billions of southern organisms to press northward, including us. But in remote continental interiors, many small villages and extraction industries have been abandoned, even as new ones flourish along the coast.

These broad pressures and trends portend great changes to the northern quarter of our planet, making it a place of higher human activity and strategic value than today. But history tells us that the pace and pattern of human expansion will not be uniform. There are many differences among the NORC countries, like steep temperature contrasts and an uneven geography of natural resources. Disparities abound in their historical patterns of settlement and infrastructure. Demographic trajectories, and national views on foreigners and aboriginal rights, vary greatly. The decisions of past political leaders on how to develop their frontiers still carry legacy today, as do current attitudes toward economic globalization and trade.

How much do these different preexisting conditions across northern countries matter? Many of the global and regional forces described thus far will be shaped by them. Their contrasts bring finer detail to the broad outlines of the 2050 thought experiment drawn thus far, and are the subject of this chapter and the next.

Quick! Hazard a guess: Of the following six countries, which has the fastest population growth rate out to 2050—China, Brazil, Canada, Iceland, Mexico, or Norway?

If you picked China, Brazil, or Mexico you guessed wrong. In terms of percent growth (not sheer numbers) you may be surprised to learn that none even makes the top three. Canada, Iceland, and Norway are all growing faster with population increases of 20% or more expected by 2050 (see table on page 173). Their base populations are much smaller of course—the sum total of people living in these three countries today is half that of Germany—but there is no disputing their extraordinary rate of growth.

The model projections tell us that by 2050 human populations will be larger in all of the NORC countries except one. The glaring exception is Russia, where falling births, rising deaths, and an aging population promise a precipitous decline of nearly one in five people. Of the NORCs, Russia alone joins Japan, Germany, South Korea, and Italy as a population loser by 2050. But even with 24 million fewer Russians, the total population of the eight NORC countries is still projected to rise by 76 million people

Some Population Densities and Trajectories 2010–2050

Country	Density (people/km²)	2010	2050	Change (%)
India	369	1,214,464,000	1,613,800,000	33
Canada	3	33,890,000	44,414,000	31
United States	33	317,641,000	403,932,000	27
Iceland	3	329,000	407,000	24
Norway	13	4,855,000	5,947,000	22
United Kingdom	255	61,899,000	72,365,000	17
Mexico	57	110,645,000	128,964,000	17
Sweden	21	9,293,000	10,571,000	14
Spain	90	45,317,000	51,260,000	13
Brazil	23	195,423,000	218,512,000	12
China	141	1,354,146,000	1,417,045,000	5
Netherlands	401	16,653,000	17,399,000	4
Finland	16	5,346,000	5,445,000	2
Denmark	127	5,481,000	5,551,000	1
Italy	199	60,098,000	57,066,000	-5
South Korea	487	48,501,000	44,077,000	-9
Germany	230	82,057,000	70,504,000	-14
Russia	8	140,367,000	116,097,000	-17
Japan	336	126,995,000	101,659,000	-20

(*Source:* United Nations Population Division)

(+15%). Most of this will be driven by growth in the United States (+86 million, with perhaps +15 million in northern states[389]) and Canada (+11 million), with nearly +3 million more arriving in Sweden, Norway, Finland, Denmark, and Iceland.

Where will all these new people live? Outside of Europe, the NORCs control most of the land areas lying north of the forty-fifth parallel. Excluding the Greenland Ice Sheet, this is over forty million square kilometers of land, more than quadruple the area of the lower forty-eight U.S. states. By my calculations[390] roughly fourteen million square kilometers—about one and one-half times the size of the United States or China—are quite livable. Might these be the lands into which new settlements will spread?

The First Waves

Actually, they already have. The forty-fifth parallel does miss Toronto, Canada's largest city, but captures virtually all of the rest of Canada, plus a row of northern U.S. states from Minnesota to Washington. The cities of Portland, Seattle, Vancouver, Edmonton, Calgary, Winnipeg, Minneapolis–St. Paul, Ottawa, and Montreal are all contained within the planet's northern quarter of latitude. Tracing the forty-fifth parallel farther east, we see it snares all of Germany and the United Kingdom, and indeed much of Europe, including the cities of Paris, Brussels, and Budapest. Looking still farther east, it swallows Russia, most of Mongolia, and a good chunk of northeast China, including the city of Harbin.

To the north, we find that even the harshest Arctic hinterlands have long been occupied (albeit thinly). The first people to see the Arctic Ocean were probably Mongolic, reaching the northern coast of what is now Russia by thirty to forty thousand years ago, if not sooner.[391] By at least fourteen thousand years ago, their descendants had crossed the Bering Strait into Alaska. From there, groups spread south and east across North America, some reaching eastern Canada and Greenland by about forty-five hundred years ago. A later wave of Mongolic invaders again swept across Arctic Canada to Greenland, supplanting the first. The ancestors of today's Aleut, Yupik, Inuit, Chipewyan, Dogrib, Gwich'in, Slavey, Cree, Nenets, Khanty, Komi,

Dolgan, Evenk, Yakut, Chukchi, Tlingit, and many others migrated and grew. Our circumpolar colonization was nearly complete.

Northern Europe got a later start because it was buried under an ice sheet. But after the glaciers retreated it was invaded and reinvaded many times, beginning about twelve thousand years ago. From genetic studies it appears that its most ancient occupants today are the Sámi and Karelians of northern Scandinavia and northwestern Russia.[392] A second clue comes from linguistics: Today's Sámi and Karelians (and Finns and Estonians) speak derivatives of Finno-Ugric, predating the arrival of Germanic (Swedish and Norwegian), Baltic (Latvian and Lithuanian), and Slavic (Russian) Indo-European languages in the region. This is why Swedes, Norwegians, and Icelanders today can sort of understand each other whereas Sámi and Finnish sound like pure gibberish to them and also to Russians. The last bits of undiscovered land—Iceland and the Faroe Islands—weren't colonized until the Vikings found them in the ninth century A.D.

Next came more waves of expansion and rediscovery. French and British trappers and traders arrived in the New World; Russian Cossacks surged east through Siberia all the way to the Pacific Ocean. In the nineteenth and twentieth centuries almost three million Scandinavians emigrated to the American Midwest and rural Canada. Today, there are Nigerians moving to Fort McMurray, Iraqis to Stockholm, Filipinos to Yellowknife, and Azerbaijanis to Noril'sk. There are growing cities, guest-worker programs, and multinational corporations. As I drove across the Arctic Circle in my rental car, just a few hours north of Fairbanks, it was with a Starbucks Venti latte still clutched in my hand. The latest invasions have begun.

So, unlike the Arctic Ocean seafloor, even our northernmost landmasses are hardly a vacant frontier. Siberia has thirty-five million people, most living in million-plus cities. Canada and Alaska share thirty-four million, the Nordic countries twenty-five million. However, we are still talking about some of the lowest population densities on Earth, especially in Canada and Russia with only three and eight people per square kilometer, respectively (see preceding table). If all Canadians could be airlifted from their cities and sprinkled uniformly across the country, every man, woman, and child would get their own eighty-two-acre spread. The same exercise in China would yield less than two acres per person; in India less than one.[393] But no landscape on Earth is settled uniformly like that. We concentrate in specific places for specific reasons—for arable soil, at strategic trading crossroads, along rivers, and so on. Physical limitations have always influenced human settlement patterns in the past, and they will continue to do so in the future. Obviously, one of the biggest limitations on human settlement in these northern areas has always been the cold.

The Uneven Cold

As a general rule the higher the latitude the more severe the cold (and seasonality, of course)—and the fewer the people. However, being near an ocean does change things. Thanks to the geography of continents and the sluggish, heat-carrying thermal properties of water, air temperatures do not simply vary from south to north, or from low elevation to high, but also with distance from a westerly ocean.[394]

Take, for example, the line of 45° N latitude defined earlier. On the Pacific coast of Oregon, the average January daytime temperature along this line is 52°F. Moving east through the Montana-Wyoming border, South Dakota, and Minneapolis it tumbles to 22°F. Temperatures persist in the low twenties through Green Bay, Wisconsin (home of the Packers), Ottawa (20°F), and Montreal (22°F) but leap abruptly for ship captains on the Atlantic Ocean, thanks to the Gulf Stream current and its north-flowing extensions that carry warm water north all the way from the tropics. Their heat warms 45° N's landfall on a beach in southern France (49°F) and lingers for a while over western Europe. But by Milan (40°F) the warm touch is fading again, and by Stavropol, Russia (25°F), it is gone. Tracing the January averages along this single line of latitude, we found temperature swings of over thirty degrees!

This is the so-called continental effect, in which the interiors of continents experience colder winters and hotter summers with distance from a major ocean, especially on their eastern halves.[395] The continental effect helps create the numbing cold of the "Siberian Curse" described in Chapter 5, and the southerly dip of permafrost in eastern Canada and eastern Russia. It is what forces people living in Ottawa to bundle into parkas in winter, while due east in Milan they get by with light jackets and fashionable scarves. It is an important reason why the northern penetration of human settlements has been greater in western Canada than eastern Canada, and in western Russia than eastern Russia. Together with heat from the Gulf Stream and North Atlantic Current, it explains why most of the Eurasian population north of 45° N is piled onto the western end of the continent, and thus the historical agrarian settlement pattern of Europe.

The Coastal and Lowland Imperative

Another important consideration for human settlement patterns, especially in cold places, is terrain.[396] Even prehistoric nomadic hunters, who worried little about permafrost or crop yields, preferred low-lying valleys and coasts.

The reason again is temperature. High elevations are colder than low elevations, and usually more rugged too. As a general rule of thumb, air temperatures fall roughly 6.5°C for each kilometer of increased elevation (18.8°F per mile). Thus, high-elevation ground is colder. It allows permafrost to exist farther south than it otherwise would in the mountains of Norway, the mountain cordillera of western Canada, and on the Tibetan Plateau. In Russia east of the Yenisei River, high elevation compounds the continental effect, making these lands among the coldest on Earth. They are deeply frozen in permafrost, useless for agriculture, and frighteningly cold in winter. In North America, temperatures grow colder mainly from south to north, but in Russia, it's from west to east.

For these and other reasons the northern high latitudes have never been a strong draw for southern settlers. Their extreme seasonality makes for a short (if intense) growing season. Abundant water and hot summers create a moist haven for hordes of mosquitoes. The freshly scoured landscape, exposed only since the last ice age, has poorly developed soils. Biological richness is low and essentially still colonizing since the glaciers' retreat. It's not surprising, therefore, that our past historical expansions have left vast northern land areas only lightly touched.

In Canada, most French and British colonial settlements hugged southern coastlines and rivers. Farms would later spread across her low, flat prairies, bracketed by rugged mountains to the west and rocky Precambrian crystalline shield to the east. All of Alaska's major settlements are either in low-elevation terrain, along the coast, or both. Norway's long-axis mountain spine crowded its settlements along its shores where grew societies of fisher-men, explorers, and (now) offshore oil and gas drillers. Sweden, Finland, and northwestern Russia, in contrast, are low-elevation and permafrost-free. They have been widely settled since prehistory and their reindeer-herding, dairy, and cool-weather agricultural societies count among the oldest in Europe.

Given all this, it took prospects of financial gain to attract nonnative settlers to remote northern areas. In the ninth century seafaring Vikings—ancestors of today's Norwegians, Swedes, and Danes—variously plundered or settled Russia, Greenland, Canada, Iceland, and the Faroe Islands. The (re)discovery of North America attracted French and British trappers and traders who penetrated across Canada in search of beaver. From Siberia came the call of sable fur. After defeating the Khan near modern-day Tobol'sk, Russian Cossacks swept three thousand miles east from the Urals all the way to the Pacific Ocean in 1697, completing the Russian version of "manifest destiny" a full century and a half before the United States did. Their legacy was a system of remote outposts where Russian fur traders and missionar-ies interacted with dozens of aboriginal groups. It took gold discoveries to bring new rushes of people to the Yukon and Alaska. Some remained after, commingling with the existing aboriginal population to grub out frontier lives as miners, trappers, and small farmers. That was pretty much how the situation remained, until the Second Wave.

If expansions of early settlement were shaped by climate, terrain, and gold, in the twentieth century they were shaped by politics and war. Two major transformations happened that altered huge areas of the Northern Rim forever. The first was Joseph Stalin's decision to grow the Gulag, a vast network of thousands of forced-labor work camps and exile towns across Russia between 1929 and 1953. The second was the decision of the U.S. Army to invade western Canada during the depths of World War II.

The Second Wave: Stalin's Plan and the U.S. Occupation of Canada

Even before Japan's December 7, 1941, attack on Pearl Harbor, the United States was worrying about how to defend Alaska. It was impossibly remote, reachable only by ships or air, with no road connecting it to the rest of the country. Meanwhile, Hitler's armies were devouring Europe, and Japan's advance across Southeast Asia and the Pacific Islands seemed unstoppable. From Washington, the view of the entire northwestern corner of North America—not just Alaska but western Canada as well—was of a broad soft flank, completely vulnerable to an overland invasion by Japan.

Bases were thrown up in Anchorage, Fairbanks, and the Aleutian Islands and several thousand troops rushed to them. After Japan bombed Pearl Harbor, American fears went into overdrive and a deal was struck between Washington and Ottawa: Canada would allow the U.S. Army to develop her frontier and connect it to Alaska, so long as everything was turned over to her after the war. The U.S. military-industrial machine swung into gear and selected its beachhead, a sleepy Canadian farm town called Dawson Creek, at the end of a minor western spur line for Northern Alberta Railways.

In March 1942 the residents of Dawson Creek got the shock of their lives. The train arrived, but instead of bringing the expected dry goods and furniture, it was loaded with heavy equipment and work crews of the U.S. Army. They had come to carve a fifteen-hundred-mile-long emergency road through uncharted wilderness—through British Columbia and the Yukon, connecting Dawson Creek all the way to Fairbanks—in under a year.

Canada's government watched from Ottawa as the U.S. Army opened up her western frontier. Forty thousand American soldiers and civilian contractors poured into a vast wilderness of forest and bog, a place with no roads and hardly any settlements. It was home to fewer than five thousand Canadians, mostly aboriginal hunter-gatherers.

Dawson Creek became the gateway of what would eventually be called the Alaska Highway. To ferry supplies, dozens of new airfields were cut into the wilderness to form the Northwest Staging Route, later used to shuttle some ten thousand American-built airplanes— painted with the Soviet red star—to Alaska, where they were handed over to Russian pilots.[397] Another six-hundred-mile road and pipeline were built to bring crude oil south from fields at Norman Wells. Yet another road was built to link the new highway to the Alaskan port of Haines. The old gold rush town of Whitehorse had a new population explosion and sprouted pipelines running north and south. A telephone network was built, together with new shipping facilities along the Mackenzie River. Through immense manpower and treasure, the United States had opened up another country's wilderness and connected Alaska by road to the rest of the continent.[398]

The same thing was going on elsewhere in the Northern Rim. A major airport and base were built in Keflavík, Iceland, and more than thirty thousand troops kept there during and after the war. That facility is now Iceland's international airport. Another built at Sondre Stromfjord is now Greenland's international airport; the American-built road there is now the longest in the country. Another airport in northern Greenland (Thule Air Base) is still retained by the U.S. military and is now the northernmost air base for the United States.

The close of World War II changed only the enemy, not the construction projects. Three chains of remote "distant early warning" (DEW) radar stations were strung through Alaska, Canada, and Greenland to deter Soviet bombers. A joint U.S.-Canada base was built at Fort Churchill, Manitoba, and another at Frobisher Bay (now Iqaluit). More than sixty thousand troops were stationed in Alaskan bases that are maintained there to this day. By the end of the Cold War, the American military had built a first skeleton of roads, airports, and outposts throughout the northern high latitudes, leaving an indelible template still shaping the region today.

Stalin's Gulag

America's northern investments during World War II and the Cold War were purely military.[399] But Joseph Stalin's underlying purpose for building the Gulag ran far deeper. It was more than just a convenient way to punish criminals and silence political dissidents—it was a deliberate decision to industrialize the Soviet Union using the slave labor of his own people. It intended to advance certain socialist ideologies, like asserting man's triumph over nature and *Engels' dictum*—the notion that industry should be evenly distributed geographically across a country. It was nothing less than a forced settlement of his country's barely habitable Siberian territories—then thinly occupied by aboriginals and a sprinkle of outposts—with ethnic Russians.

The use of prison camps in Russia dates to the tsars, but Stalin took it to a whole new level. By the 1930s he had established camps throughout all twelve Russian time zones. By the program's peak in the early 1950s, the total camp population had swelled to 2.5 million prisoners. Nearly five hundred complexes were built, comprising thousands of individual camps, each holding a few hundred to thousands of people.[400] Many were condemned for minor crimes. Over the course of the program perhaps eighteen million are believed to have passed through the camps; another six million were banished to live near them in exile.

Stalin's Gulag is one of the darkest chapters in Russian history.[401] Its atrocities include uncounted deaths from starvation, exposure, exhaustion, and even outright murder. Thousands of projects were blazed into the wilderness without supplies, a plan, or competence. Many were ultimately abandoned. But as a blunt tool to force massive industrialization and resettlement of Siberia, the program was a resounding success.

A large fraction of this giant captive labor force was aimed directly at the heart of the frozen frontier. Prisoners blasted mines and cut forests. They built roads, bridges, railroads, and factories. The Soviet Union proceeded to industrialize on the backs of these workers and the iron, coal, and timber they produced. If they survived their sentences, many prisoners were forbidden to return home. Millions of exiles and prisoners' family members instead moved into the growing towns and cities near the camps. Factory towns grew huge, attracting further subsidies and emigration programs from Moscow. Even after Stalin's death and the dismantling of the Gulag system in 1953, these subsidies continued. By the 1980s there were gigantic industrial cities scattered across some of the coldest terrain on Earth: Novosibirsk, Omsk, Yekaterinburg, Khabarovsk, Chelyabinsk, Krasnoyarsk, Noril'sk, Irkutsk, Bratsk, Tomsk, Vorkuta, Magadan. . . . At staggering cost in blood and treasure, Mother Russia had urbanized Siberia.

Contrasting Patterns of Settlement

The decision of Soviet planners to relocate millions of people and force the growth of giant cities across her coldest, most remote terrain has created one of the most fascinating contrasts in human settlement found on Earth.

On a world map or globe Norway, Sweden, and Finland *look* coldest of all. Their settlements and infrastructure are arranged in a north-south direction and extend even farther north than most Siberian cities. But don't be fooled. They are bathed in heat from the North Atlantic Current and enjoy much warmer winters than Russia, even high above the Arctic Circle. I once visited Norway's lovely city of Tromsø, at 70° N latitude, in January, its coldest month of the year. Its residents were out in force, romping in the snow and chatting amiably in their front yards. Even in January the average daytime temperature is +25°F in Tromsø, warmer than Minneapolis. Reykjavík, the capital of Iceland, which is plunked squarely in the warm current, averages a balmy +35°F. But in Russia's Novosibirsk, way down south at 55° N latitude, temperatures are subzero, averaging −2°F. Worry not for the Nordics.

Returning to our world map or globe, Canada and Russia at first look rather similar. Both are enormous countries with long east-west coastlines fronting the Arctic Ocean. Both have vast emptiness in the northern parts of their bulk. Both have a band of cities also running in an east-west direction, just north of and roughly parallel to their long southern borders.

But upon closer inspection some differences emerge. Canadian cities hug the U.S. border like a long spotted eel, whereas Russian cities are arranged more like a shotgun blast. Owing to the peculiar orientation of Russia's climatic gradient (recall that in Canada, temperatures grow coldest moving from south to north, but in Siberia from west to east), Russian cities, unlike Canada's, push deeply into the coldest parts of the country. It is roughly analogous to Canada establishing her population centers in a band of huge cities running from south to north, from the U.S. border all the way to the Arctic Ocean.

Under the Soviet planned economy, metropolises were grown in places that don't make sense: in harsh cold, separated from one another and from potential international trading partners by great distances. They are precariously linked by absurdly stretched infrastructure, if even linked at all. Their subsidization so saddled the Soviet economy that some researchers believe they helped bring about the collapse of the USSR in 1991.[402] Afterward, of course, the subsidies evaporated and in the 1990s the giant Siberian cities depopulated faster than Detroit in a bad layoff year. In eastern Siberia the population fell by *half*—from about twelve million to six—a free fall that is only now stabilizing as the region's population downsizes to some semblance of a free-market equilibrium economy. But even after this depopulation, the

Russian Federation is unique among the NORCs in having so many big cities in its coldest, most remote territories.

Perhaps one day, having far-flung urban cores and disseminated infrastructure throughout the New North will pay off. But for now, Russia continues to pay the price for the inefficient layout and bitterly cold temperatures of her Siberian cities. The economic geographer Tatiana Mikhailova estimates that remote distance and cold temperatures cost the country at least 1.2% GDP annually in extra energy and construction costs alone.[403] That is nearly half the GDP contraction suffered by the United States during the recession of 2008–09 (2.5%).

An extreme example of this kind of geographic inefficiency is Yakutsk, the capital of the Sakha Republic in eastern Siberia. Despite having a population of over two hundred thousand—more than ten times larger than Canada's Yellowknife—Yakutsk is essentially a fly-in city. Getting there otherwise requires either a thousand-mile-long boat ride down the Lena River during its short shipping season, or braving the "Road of Bones," a twelve-hundred-mile-long rutted track from Magadan, built by Gulag prisoners, that is only really drivable in winter. The road even ends on the wrong side of the bridgeless Lena River from Yakutsk. Finishing the trip thus requires driving over the river ice in winter or a ferryboat in summer. During the Lena's violent, ice-jammed spring flood, Sakha Republic's capital city is completely cut off from the world except by airplane.

In contrast to Russia, the vast share of Canada's population and infrastructure closely hugs its border along the warmest and most accessible parts of the country. Its big population centers are close to U.S. population centers. This proximity, together with NAFTA and a historically friendly border, encourages massive volumes of trade and traffic between the two countries. Many Americans are surprised to learn that the United States' largest trading partner is not China but Canada. However, there is a penalty (or benefit, depending on one's point of view) that Canadians pay for clustering along the American border. With its small population and economy concentrated in the south, the vast share of their country is inaccessible except by airplane or temporary winter road servicing a sprinkle of little villages. The same is true, on a smaller scale, for Alaska. To this day, much of Canada and Alaska remain stunning wilderness.

If Canada and Alaska are wild, and Russia colonized, then the Nordic countries are downright civilized. On my first visit to Iceland I was amazed to learn that I could lease a sleek sedan from my choice of several multinational rental-car companies, then drive it comfortably at high speed around the entire island. My accustomed experience with driving through thinly populated Arctic countries was crawling with clenched teeth over a potholed gravel road in a battered 4WD truck, either hired from a local driver or borrowed from the government, praying that the one decrepit gas station two hundred miles away would actually have some gas and a working telephone. But Iceland, Norway, Sweden, and Finland all have beautiful paved thoroughfares extending to their remotest extremities. They are dotted with gleaming service centers, stores, and restaurants. Cell phones work everywhere. Reaching the Arctic Circle in Alaska, Canada, or Russia can quickly turn into a bushwhacking expedition. In the Nordic countries, it's just a pampered weekend getaway.

Soviet planning didn't turn out exclusively bad results. The next time you pay your natural gas utility bill or fill up your fuel tank, you might nod your head to some ghosts of Soviet planners past: If not for their uneconomic, market-forces-be-damned decision to develop a remote Arctic swamp half a continent away from Moscow, you'd surely be paying a lot more than you just did.

The West Siberian Lowland is a vast, soggy plain bounded by the Ural Mountains in the west and the Yenisei River in the east and from 52° to 73° N latitude. It spans nearly one thousand miles in every direction, is one-third the area of the continental United States, and nearly six times larger than Germany. Weather alternates between a long subzero polar night in winter and dank mosquito heaven in summer. It is blanketed in wet, semifrozen peat and covered with lakes; its northern half is permanently frozen in permafrost. The fate of this frozen, carbon-rich peat, which is relatively fresh (<12,000 years old), is discussed in Chapter 9. But trapped in the rocks below the peat, thousands of feet down, we find another form of carbon that is considerably older. It is the cooked remains of twenty-three trillion tons of organic-rich muck that settled to the bottom of a long-gone sea between 152 and 146 million years ago. That muck is now called the Bazhenov Shale,[404] and it has changed not only Russia but the entire world.

New Hydrocarbon Cities

In 1960 West Siberia was empty except for mosquitoes and aboriginal rein-deer herders. But when four supergiant oil fields were discovered there between 1962 and 1965, Soviet planners back in Moscow made an extraor-dinary decision. The USSR would massively develop the West Siberian Lowland, no matter how stunningly remote it was. Never mind that there was no good way to get there, that it was loaded with permafrost, frozen solid in winter, and a flooded swamp in spring. Never mind that everything would have to be built from scratch—ports, roads, railroads, drilling pads, and pipelines—by sending boats up the Ob' River. Never mind that the magnitude of the initial investment would mean the venture could not be profitable for decades. It was a historically crazy decision that could never be done by private energy companies today.[405]

For years Moscow poured money into a place few Russians had even heard of. It was east of the Urals, for starters, so might as well have been on the moon. But three decades later the supergiant oil fields Samotlar, Fedo-rovskoye, and Mamontovskoye were household names. From the region's tens of thousands of wells flowed one-fifth of the world's oil and natural gas. The once-empty boggy plain was dotted with cities—Surgut, Nizhne-vartovsk, Noyabr'sk, Novy Urengoy, and others—and a population of over three million people.

I have been to these cities and driven across thousands of miles of West Siberia.[406] Much of it is still empty and indescribably beautiful in the way that only endless mossy bogs, tea-colored rivers, and scraggly forests can be, silently existing under a surreally glowing northern sky. But entrenched across this Pleistocene still-life is the spoor of the oil and gas industry: wide paved highways, belching bulldozers and diesel trucks with tires taller than your head, thousands of wells, and a labyrinth of pipelines streaming west. Piled high are stacks of rusty pipe, mountains of sand, and jumbled iron graveyards of cannibalized trucks. Our grubby faces were stared at in disbelief over surrendered American passports:[407] To West Siberians, a clump of dirty, sunburned field scientists wearing fleece and rubber boots hardly fit the profile. To them, Americans are smiling oil company executives, with briefcases and business suits.

Because of the West Siberian Lowland, the Russian Federation is now the world's largest producer of natural gas and second-largest producer of oil. It is home to her major oil companies and Gazprom, the state-owned natural gas monopoly. After almost five decades of operations in the region, Russia's energy industry has amassed enormous political, brainpower, and economic presence there. West Siberia is to the Russian energy industry what Silicon Valley is to technology, New York is to finance, or Los Angeles is to entertainment in the United States.

The Third Wave

The Third Wave of human expansion in the Northern Rim thus stems from our relentless search for fossil hydrocarbons. It began in the 1960s with major discoveries in Alaska, Canada, and the West Siberian Lowland and shows no signs of abating. World interest in the Arctic, in particular, is fueled either by environmental concern for its threatened ecosystems, or excitement over perceived new bonanzas in oil and gas.

The newest frontier is the Arctic seabed. The previous chapter discussed the geopolitical commotion this dawning realization has spawned, and the critical importance of the United Nations Convention on the Law of the Sea (UNCLOS). A 2008 auction offered by the U.S. Minerals Management Service sold a whopping $2.8 billion worth of Arctic offshore leases; the Canadian government similarly won record-breaking bids for leases in the Beaufort Sea.[408]

In 2009 a first comprehensive assessment of the Arctic Ocean's oil and gas potential was published in *Science* by the U.S. Geological Survey, and the associated data files were released to the public.[409] This assessment, which is still incomplete and ongoing, suggests that nearly a third of the world's undiscovered natural gas and 13% of its undiscovered oil lies north of the Arctic Circle (see maps pp. viii–xi). All that in a place covering barely 4% of the globe.

Two huge winners are revealed by the USGS data: northern Alaska for oil, and Russia for natural gas. Of forty-nine geological provinces analyzed so far, these two places tower above all others. The Alaska Platform, covering the North Slope and extending a roughly similar area offshore, is thought to hold between 15 and 45 billion barrels of oil with a best guess of about 28 billion. This number approaches the proved reserves of Nigeria and is about one-fourth those of Iraq. Russia's South Kara Sea alone is thought to hold between 200 and 1,400 trillion cubic feet of natural gas, with a best guess of 607 trillion. That number, if correct, is more than twice the proved reserves of the United States and Canada combined.

There are other promising geological provinces besides these two. For oil, they are Canada's Mackenzie Delta, the north Barents Sea, the West Siberian Lowland, and three provinces off the east and west coasts of northern Greenland. For natural gas they are the south Barents Sea, the Alaska Platform, and the north Barents Sea.[410]

If recent offshore lease sales are any indication, many or all of these places will experience increasing levels of interest, exploration, and investment in the coming decades. But will drilling platforms and thousands of new offshore wells mushroom in these Arctic waters by 2050?

Possibly, but don't bet on it. Offshore energy development is more likely to grow cautiously and incrementally. Even in ice-free oceans—which the Arctic Ocean most certainly is not—offshore drilling is complicated and expensive. Northern environments are environmentally delicate, so they demand above-normal protections. Existing ports and other maritime facilities, as discussed in Chapter 5, are scarce. Ice-resistant platforms and other new technologies still need to be invented. Outside of the Alaska Platform, the vast share of hydrocarbon in the Arctic is not oil but natural gas, which is harder to capture and transport. Oil can be simply pumped from the ground and poured into a tanker. Natural gas needs either pipelines or an expensive LNG or gas-to-liquids facility—essentially a refinery—to liquefy it. Even by 2050 these are formidable obstacles in a place that is environmentally sensitive, remote, and still inaccessible for much of the year.

Something Old Is New Again

What *is* assured over the next forty years is intensification of oil and gas exploitation in and around the places they already exist today. Offshore, these include the giant Shtokman and other gas fields of the Barents Sea, and Sakhalin Island in the Russian Far East. On land—despite some added engineering complications from thawing permafrost—they include West Siberia, the North Slope of Alaska, the Mackenzie Delta, and Alberta. Norway's gas and oil activities are mostly offshore, but by 2050 large areas of north-central Russia, Alaska, and Canada will look quite different than they do today.

Russia in particular will continue to aggressively develop her Siberian gas fields. When I cornered Alexei Varlomov, deputy minister for the government agency overseeing all natural resources of the Russian Federation, he told me, "The most important factor is the needs of industry," and that absolutely nothing should get in the way of energy exploration.[411] His view is understandable, given the prominence of his country as a world energy supplier. Russia produced 3.6 billion barrels of oil in 2008, second only to Saudi Arabia.[412] It produced 603 billion cubic meters of natural gas—and held 43.3 *trillion* more in proved reserves, both second to none in the world.

Two out of every three barrels of this oil and 85% of this gas comes from West Siberia. However, as is the case with all oil provinces, the size distribution of its petroleum fields is log-normal and the region's oil production has entered decline.[413] Russian production peaked in 1987–88. Samotlar, one of the largest oil fields in the world, peaked at 3.4 million barrels per day in 1980. It has since dropped over 90%, producing just 300,000 barrels per day from its approximately five thousand wells.[414] The region's three major gas fields have also peaked, and their production is expected to fall 75% by 2030.[415] There will always be more pockets to be found, but as discussed in Chapter 3, like any other hydrocarbon province on Earth they are growing exponentially smaller and thus less economical to develop.

West Siberian exploration is therefore shifting away from the middle reaches of the Ob' River—where most of the basin's crude oil is found—to the immense concentrations of natural gas found farther north. The largest known natural gas reserves on Earth are found in approximately sixty to one hundred fields in this area. Just offshore is the South Kara Sea, now thought to hold perhaps 1,400 trillion cubic feet more. The Yamal Peninsula, stuffed with natural gas, condensate, and oil, lies at the heart of this bonanza and will doubtless be developed.

While less glamorous than the prospect of proliferating offshore platforms in the Arctic Ocean, 2050 will likely see a great degassing of the Yamal Peninsula, feeding thousands of miles of pipeline heading west to Europe and east to China. It is unclear whether a port can be built on its shallow west coast, or that environmental damages will be avoided, but pipelines will spread across the Yamal. Already, at least two are planned and the first one has just broken ground.[416]

Pay Dirt

Even from several miles away, through the fogged window of the little airplane and a dreary splatter of rain, I could see the heavy curtain of smoke and glowing spots of orange flame. It was Tolkien's Mordor brought to life, the soil ripped away to expose pitted blackness beneath. Giant trucks dug away at the spoils like orcs. Near the fuming smokestacks were yellow mountains of pressed sulfur blocks, waste excrement from the transformation of low-grade bitumen into synthetic, crude oil. It was a depressing and evil-looking landscape, at least to anyone who finds boreal wetlands and green pine forest attractive.

It was northern Alberta, not Noril'sk. Beneath me sprawled the open sores of the Athabasca Tar Sands, economic engine of Fort McMurray and almost one-half of the Canadian oil industry. Though they are more commonly called "oil sands," what they hold is nothing like conventional oil. The pure, light, sweet crude pumped with ease from Saudi oil fields is a dream compared to this stuff. It is tarlike *bitumen,* a low-grade, sulfur-rich, hydrogen-poor hydrocarbon that has soaked into vast expanses of Alberta sandstone.

Extracting liquid oils from this mess is an extraordinarily invasive, consumptive, and environmentally damaging process. At present, the most common way to do it is strip mining, with about two tons of tar sand needed to obtain a single barrel of oil. Gigantic trucks and shovels scrape the stuff off the surface. Then it is crushed and dumped onto conveyor belts heading to swirling tubs of water. The resulting slurry is piped to an extraction facility, where it is churned in a heated witches' brew of steam, water, and caustic soda. This splits the bitumen from the sand and clay, which sink to the bottom. The bitumen floats off to an "upgrader" (a sort of refinery) to remove sulfur and add hydrogen (from natural gas), creating synthetic crude oil. The waste liquid and dirt are sent to tailings ponds; the yellow blocks of sulfur are simply stacked up.[417]

Tar sands are an environmentalist's nightmare. The extraction process gobbles enormous quantities of energy and water. Migratory birds land in the tailings ponds and die.[418] Sulfur dioxide, nitrogen oxides, and particulates are released into the air alongside up to three times more greenhouse gas than released by conventional oil drilling. Depending on the technology used, it takes 2–4 cubic meters of water, and 125–214 cubic meters of natural gas, to produce a single cubic meter of synthetic oil.

The water is pumped from groundwater or diverted from the Athabasca River, reducing inflow to the Peace-Athabasca Delta, a UNESCO World Heritage Site and Ramsar wetland, about 150 kilometers downstream.[419] Most mines will operate for about forty years and excavate about a hundred square kilometers of land. No tailings ponds have ever been fully reclaimed, and putting the overburden back afterward mitigates the damage but does not really restore the original ecosystem. Since 1967, when the first mining began, only one square kilometer has been certifiably restored and returned to the public.[420] These and other problems have environmental organizations yowling against any further increase in tar sands production.

They face a difficult battle. Short of being outlawed, it's hard to imagine how the growth of this industry will ever stop. The oil reserves the tar sands contain are estimated at an astonishing 175 billion barrels which, if correct and recoverable, is the second-largest oil endowment on Earth after Saudi Arabia (estimated at 264 billion barrels). That means Alberta holds more oil than Iraq (115 billion barrels), Kuwait (102 billion), Venezuela (99 billion), Russia (79 billion), or Norway (7.5 billion). The cost to produce it has dropped from thirty-five dollars per barrel in 1980 to twenty dollars per barrel in recent years, making even fifty-dollars-per-barrel oil prices very profitable.[421] Huge new supplies of natural gas, needed for energy and hydrogen, will come online with construction of the Mackenzie Gas Project, a long-anticipated 1,220-kilometer pipeline that will carry Arctic gas from the Mackenzie Delta area to the tar sands and other North American markets.[422] History tells us that Canada's adherence to international climate-change treaties crumbles before market forces like these: The tar sands are the biggest reason why Canada not only failed to meet her pledged reduction in carbon dioxide emissions under the Kyoto Protocol (to -6% below 1990 levels), but actually *grew* them +27% instead.[423]

So far, about 530 square kilometers have been strip-mined, an area not much greater than the city of Edmonton. But tar sands underlie a staggering *140,000* square kilometers of Alberta, nearly one-fourth of the province and about the size of Bangladesh. Of this large area, about 20%—sixty more Edmontons—are shallow enough for strip mining. The rest can be exploited using underground extraction, which involves injecting 450°F pressurized steam underground for several years to heat the ground, eventually fluidizing the tar enough to pump some of it out.[424] This type of underground extraction has the potential to spread across nearly all of northern Alberta. If it does, new pipelines, roads, and towns must follow.

This future springs not simply from my fertile imagination but from cold, hard cash. Those 175 billion barrels of grubby bitumen lie right next door to the world's largest and friendliest customer, whose other suppliers have either entered decline or soon will. Energy companies are no fools. By early 2009 the government of Canada had already leased more than seventy-nine thousand square kilometers in tar sands contracts. Future production is anticipated to rise from 1.3 million barrels per day today, to 3.5 million by 2018, to *6 million barrels* per day by 2040.[425] If that black torrent of tar becomes reality, its flow will be nearly ten times greater than the amount of conventional oil flowing south from Alaska's North Slope today.

America is ready and waiting.

The visual image of Canadian oil flowing from north to south is exactly the right one to hold in mind. Under the North American Free Trade Agreement (NAFTA), it passes over U.S. borders unencumbered by tariff.

And compared to the world's other geopolitical relationships, America and Canada remain two countries in a happy marriage.

Their embrace far transcends the energy industry. It is just one part of a bigger cross-border dependency that has long existed, thanks to friendly borders and the geographic proximity of their neighboring population cores, as described earlier. But this lovers' gaze has not always been so transfixed. Throughout much of the twentieth century, Canada was focused more on domestic integration than the cross-border sort.

A serious schism was fractious Québec, the French-Canadian province with a long history of separatist movements and terrorism. A wave of bombings by terrorist cells of the *Front de Libération du Québec* culminated in 1970 with the kidnapping of two government officials, one of whom—Minister of Labor Pierre Laporte—was found strangled and dumped in the trunk of a car. The 1970s also marked the emergence of aboriginal rights movements and rising economic and political power in Canada's energy-rich western provinces. Debate was raging over a national bilingualism policy. During this period of history most Canadians were focused on bridging their country's internal cultural divisions, not furthering its integration with the United States.

The New Cascadians

But the passage of NAFTA in 1994 marked the beginning of a stunning reorientation in Canada's political and economic geography. It quickly began to integrate in a north-south direction with parts of the United States, rather than in the old east-west orientation across Canada. Very recent studies of this phenomenon are discovering it runs far deeper than simply increased cross-border trade and traffic; there is an actual melding of cross-border economies under way.[426] This is not being steered by Ottawa and Washington but rather by a proliferation of cross-border networks of business groups, chambers of commerce, NGOs, mayors' councils, and other forms of grassroots enterprise.

The end result of this north-south reorientation is the emergence of new cross-border "super-regions" with distinct economic footprints and cultural auras of their own. Names are even being floated for two of them. "Cascadia" refers to the melding economies of the Pacific Northwest and western

Canada, centered on the Vancouver-Seattle-Portland corridor. "Atlantica" links upstate New York, Vermont, New Hampshire, and Maine with Nova Scotia, New Brunswick, and Prince Edward Island.[427] A key super-region is the Toronto-Hamilton-Detroit corridor integrating southern Ontario—the industrial heart of Canada—with Michigan's automotive industry and manufacturing sectors in Indiana, Ohio, and other Midwestern states.

For each of these emerging super-regions, the two respective halves across the U.S.-Canada border are also knitting culturally. New surveys reveal that the social values of Atlantic Canada now resemble those of the U.S. East Coast, whereas those of Alberta and British Columbia now resemble those of the western United States.[428] Apparently, proximal Canadians and Americans identify better with each other than with their own countrymen living farther away. In North America big doors are opening wide along this long border, with the widest hallways running north and south.

The Friendly Globalizers

The happily knitting border between Canada and the United States is not unique in the North. Unlike the Arctic Ocean seabed, territorial boundaries on land are long settled and calm among the eight NORC countries.[429] Borders between Norway, Sweden, and Finland are among the friendliest in the world and their citizens (like Cascadians) identify more closely with each other than with the rest of Europe. The closest thing to a troubled border, if there is one, zigzags through more than seven hundred miles of forest to disentangle Finland from Russia.

Throughout history the Finns were subjugated, first by Sweden and then by Russia, before capitalizing on the disarray of the Bolshevik Revolution to win peaceful independence from Russia in 1917. Finland has been grappling with how to coexist with her giant and occasionally unruly eastern neighbor ever since. The countries fought twice during World War II, and Finland was forced to cede substantial territories to the Soviet Union. One of these, Finnish Karelia, contains the beautiful port city of Viipuri (now Vyborg) and remains a source of great bitterness to Finns today. From time to time Finnish politicians make noises about seeking its return. Less noticed was the loss of Petsamo (now Pechenga), a small corridor that once connected Finland to the Arctic Ocean.

Its loss shuts Finland out of any UNCLOS claim there. It is reasonable to expect that Finnish regret over this region will rise in the coming decades.

But none of this renders the Fenno-Russia border militarized or tense. A regional cross-border economy in roundwood (unmilled lumber) is emerging between the two countries, not unlike the one between Canada and the United States.[430] Many Russians now own vacation homes in Finland—to the delight of local merchants and consternation of old-timers—and Finnish tourists pour into Karelia. In fact, the only reason this border even warrants mention is because all the other borders around the Northern Rim are so placid. Compared with other neighboring countries around the world, the NORCs are an extraordinarily peaceful bunch.

They also rank among the most rapidly globalizing, business-friendly countries on Earth. Compiled on the following page are index performance scores for fifteen countries, representing the six largest national economies, the BRICs, and the NORCs.[431] These respected indices ingest a wide range of econometric and other data to derive country performance rankings in things like openness to trade, tendency to make war, treatment of citizens, and so on. Rather than dissect the merits or agendas of each index, I simply provide rank-based scores from all of them.[432] Each uses a different scoring system, so they are presented as percentiles for easy comparison. A score of 86, for example, means a country ranked higher than 86% of all of the countries in the world that are measured by that particular index. Also shown is a single composite score for each country, averaged across the five numeric indices.

A remarkable story leaps from these numbers. With the exception of Russia, the NORC countries are the most stable, trade-liberal, rapidly globalizing players on the planet. Who knew that Denmark and Canada are even more open to free trade than Japan, Germany, or the United States? Of particular relevance to energy production is that this openness also pervades the oil and gas industry, in contrast to the worldwide trend toward nationalization described in Chapter 3.[433] Civil and political freedoms run remarkably high except in Russia. Six are among the most peaceful nations in the world. Viewed collectively, the NORCs appear particularly well-positioned to succeed in our rapidly integrating world.

Aside from cold winters, NORC cities also count among the world's happiest places to live. According to the London-based Economist Intelligence

Some Common Measures of Economic Globalization, Peacefulness, and Civil Liberties, Relative to the World

	Economically globalizing?			Peaceful?		Political freedoms?	Average score
	WSJ/Heritage	*EFW*	*KOF Globalization*	*GPI*	*EIUDI*	*Freedom House*	
Denmark	96	91	97	99	97	free	96
Canada	96	95	96	94	93	free	95
Finland	91	90	94	94	96	free	93
Iceland	92	91	83	97	98	free	92
Norway	84	84	90	99	99	free	91
Sweden	85	77	97	96	99	free	91
United States	97	94	82	42	89	free	81
Russia	18	28	79	6	36	not free	34
Germany	86	88	89	89	92	free	89
United Kingdom	94	96	87	76	87	free	88
Japan	89	81	66	95	90	free	84
France	64	68	92	79	86	free	78
Brazil	41	32	62	41	75	free	50
India	31	45	41	15	79	free	42
China	26	34	56	49	19	not free	37

[expressed as percentile of all sampled countries in the world]

(*Sources:* 2009 Index of Economic Freedom, Heritage Foundation, and *Wall Street Journal* (179 countries); 2008 Economic Freedom of the World Index (141 countries); 2009 KOF Index of Globalization (208 countries); 2009 Global Peace Index (144 countries); 2008 Economist Intelligence Unit Democracy Index (167 countries); 2009 Freedom in the World Country Rankings (193 countries))

Unit, four of them rank among the world's top ten most livable cities (with Vancouver in first place), citing low crime, little threat from political instability or terrorism, and excellence in education, health care, infrastructure, and culture.[434] Remember Lagos, Dhaka, and Karachi, three megacities of 2025 presented in Chapter 2? They scored in the bottom ten.

Acceptance of Global Immigrants

But it takes more than just natural resources in the ground, ameliorating climate, stable governance, and pleasant cities for a civilization to expand. It also takes people.

Like the rest of the developed world, all eight NORC countries are graying and fertility rates dropping. The Russian Federation also faces a sharp population contraction (projected to fall -17% by 2050, see table on p. 173). However, the other seven are expected to grow anywhere from +1% to +31% by 2050. Much of this growth will come from international immigration.[435] Thus, global flows of *people* are already changing the face of the Northern Rim and are critically important to how its future will unfold.

The specific rules and quotas of future immigration policies are impossible to divine here. However, an examination of current laws and trends reveals some surprisingly different attitudes toward foreigners among the NORC countries. National policies differ on the number, origin, and skill sets of foreign immigrants admitted. And culturally, some places are more welcoming than others.

The Russian Federation faces the bleakest prospect. Its demographics are in free fall, with sixteen people dying for every ten new babies being born.[436] Its total population is now dropping by nearly eigtht hundred thousand people per year. After the collapse of the Soviet Union, some three million ethnic Russians moved from former satellites into the new Russian Federation, but by 2003 that wave of return had largely ended. In an effort to repatriate more, the Putin administration created a national program to recruit twenty million Russian expatriates to "return home" in 2006. However, it now appears impossible to attract more than two and a half million in total, even counting migrants from the Baltic States.[437]

Russia's labor pool for construction, agriculture, and other seasonal work thus depends heavily on migrants from Kazakhstan, Ukraine, Uzbekistan, Kyrgyzstan, Moldova, Tajikistan, and—increasingly—China in the Far East. Many are "irregular migrants" and would be called "undocumented workers" or "illegal aliens" in the United States. Perhaps ten million may be living inside Russia. Up to a million Tajiks—almost half of the entire workforce of Tajikistan—migrate to Russia in search of seasonal work each year.

Russian leaders have long realized they need to raise legal immigration into the country, but policies to do so are unpopular. Before the spring 2008 elections the Putin administration slashed the quota for foreign labor migrants from six million to two, and several years earlier abolished laws allowing multinational firms to easily hire skilled foreign workers. The reason for such moves is purely political, as Russia suffers from widespread xenophobia. Resentment of foreign migrants runs deep, especially in large cities where they tend to concentrate. In 2008 alone, at least 525 migrants suffered hate-crime attacks, with 97 of them killed.[438]

The United States is similar to the Russian Federation in that its economy also draws heavily from undocumented migrant labor. Throughout history it, too, has suffered from bouts of xenophobia, presently directed at Hispanics. However, by any global measure, the culture of the United States is immigrant-friendly. Its population, fueled greatly by foreign immigration, is growing smartly by over 2.6 million people per year. Each year approximately 1 million new immigrants are admitted as legal permanent residents (LPRs), another 1 million become citizens, and another 1 million are apprehended at the border trying to enter illegally.[439] Nearly 4 million more are admitted as temporary residents. The number of undocumented migrants is difficult to know but is probably around 10–12 million people, roughly comparable to the Russian figure.

The first and foremost goal of stated U.S. immigration policy is family reunification. Applicants who already have relatives living in the United States enjoy highest priority for legal permanent residency, and over 65% of all LPRs are admitted for this reason. The other stated U.S. objectives, in decreasing priority, are admitting skilled workers, protecting refugees from political, racial, or religious persecution at home, and ensuring cultural diversity. Competition is fierce, especially in the last category with 6–10 million applicants per year vying for just 50,000 slots. Even family reunification applicants face processing backlogs of five to ten years. In a world of aging population and falling births, the United States is remarkably advantaged among developed OECD countries, still with no lack of willing settlers ready to move to the United States from all over the world.

Canada enjoys a similar situation, but with some important differences. Like the United States, its immigration policy objectives are to reunite families, to attract skilled workers, and to protect refugees. However, the priority of the first two is reversed. The first and foremost goal of Canada's immigration policy is to admit people with economically valuable work skills.

Out of the quarter-million legal immigrants admitted to Canada in 2008, skilled workers outnumbered family members by nearly three to one.[440] Since 1967 an intricate point system has been used to score an applicant's value for the workforce, e.g., up to twenty-five points for education level, up to twenty-four points for language skill, up to ten points for suitable working age, and so on. Put simply, Canada has sharpened its immigration policy to attract educated, multilingual, skilled workers above all else.[441] While less emotive than U.S. policy of prioritizing family reunification first, it clearly makes Canada's workforce globally competitive despite its much smaller population.

Canadian policies have also suffered from ugliness, such as exclusion of non-Europeans until 1976. Since then, however, the country's culture has become unusually welcoming of immigrants from all over the world. Nearly one in five Canadians today is foreign-born. Not long ago I watched thousands of Tamil protestors flood the streets of downtown Ottawa, badly snarling traffic on Parliament Hill. Entrapped drivers just calmly waited it out, some politely tooting their horns in support. A popular television show in Canada is *Little Mosque on the Prairie,* a situation comedy about Muslim immigrants trying to adjust to small-town life in Saskatchewan. My favorite example comes from the CBC television network, which recently introduced sports commentators Parminder Singh and Harnarayan Singh to broadcast *Hockey Night in Canada* (equivalent to *Monday Night Football* in the United States) in Punjabi, now poised to become the country's fourth most-spoken language. A photograph of these two gentlemen preparing to call out a game of the Toronto Maple Leafs appears in the pictorial section of this book.

In the Nordic countries, public sentiment and national immigration policies tend to fall somewhere between the dysfunctional xenophobia of Russia and the fast-growing ethnic cauldrons of Canada and the United States. Viewed collectively they are morally sympathetic to the plight of refugees and appreciate the need for immigrant labor, but are also wary of diluting their ethnic makeup and (especially) languages and culture. Compared with North America, Russia, and larger countries of Europe, their populations are small and quite homogenous. Other than Sweden, none has a long history of absorbing foreigners. Xenophobia is present and most people, if asked, are more worried about preserving things as they are, rather than population decline or finding enough construction workers.

In principle all of the Nordic countries have adopted policies of allowing free inflows of workers from any country in the European Union, even though Norway and Iceland are not members of the EU.[442] This is more welcoming than Russia, which demands worker permits even from its fellow members of the Commonwealth of Independent States (CIS). However, winning citizenship in a Nordic country is much harder, often with a language-test requirement. Immigrants from outside the EU are unwelcomed and restricted mainly to a small number of refugees.

To be sure, some subtle differences exist among the Nordic countries. The stereotype Swede is blond and blue-eyed, but in fact there are many dark-skinned immigrants in Sweden. About 12% of Sweden's population is now foreign-born, similar to the proportion in the United States and Germany. Iceland has also become quite dependent on immigrant labor. Its foreign-born population rose as high as 10% just before its 2008 banking collapse. From there the numbers decline for Norway (7.3%), Denmark (6.8%), and Finland (2.5%).[443] Finland, despite belonging to the EU and thus technically open to migrants from all EU countries, is the least welcoming Nordic country, in part due to the difficulty of the language but also owing to a lack of concerted recruitment programs. Not surprisingly, population growth in this country is projected to be among the lowest of the NORCs, pegged at just +2% by 2050 (see table on p. 173). Forced to choose, many Finns prefer less immigration over more, even at the cost of their country's population and economic growth.

Imagining 2050

Our thought experiment has gained human texture. Against a global backdrop of rising material wealth, environmental stress, and total human population, we find the likelihood of smaller, flourishing cultures growing amid the milder winters and abundant natural resources packed into the northern quarter of the planet. From all indications these resources can and will be divided peacefully between nations, and global market forces allowed to exploit them. While Russia's population is contracting, she reigns supreme in the economic potential of her enormous northern holdings of natural gas. In all other NORC countries populations are growing, led especially by the United States and immigrant-friendly Canada, with a growth rate very near that of India.

Key settlements and physical infrastructure exist already, but their geography and quality vary widely. North America is efficient but condensed, Russia remote but far-reaching. Best developed are the Nordic countries: Perpetually warmed by the North Atlantic Current, they have extensive high-quality roads and rail, stable governance systems, and towns, ports, companies, and universities already in place, stretching from their southern capitals all the way north to the remote Arctic.

Global immigration explains most of the projected population growth around the Northern Rim. But it is flowing into the larger cities, to places like Stockholm and Toronto, Fort McMurray and Anchorage. These are urban outposts in the midst of beautiful, expansive wilderness. Who will rule the rest?

CHAPTER 8

Good-bye Harpoon, Hello Briefcase

"The foundation of our culture is on the ice, the cold, the snow."
—Sheila Watt-Cloutier (1953–)

"Inuvialuit are a proud and adaptable people. We wouldn't have lasted for so many generations . . . if we weren't."
—Nellie J. Cournoyea (1940–)

"M*EIDÄN ELÄMÄ ON AINA VAIHTUNUT,"* said my host, rapping the rustic wooden corral fence with gnarled hands for emphasis.

I eagerly returned my eyes to my new Finnish translator—perhaps too eagerly. She was gorgeous and something was definitely in the air. I didn't know it yet, but just six weeks later we would agree to get married.

"She says, 'We're always changing.'"

"Hm? Oh, yes. Ask her to elaborate."

In my defense, I might have been distracted from the interview no matter who was translating. What I was hearing from my subject, a fiftyish Sámi reindeer herder in Lapland, was quickly turning into what I'd already heard in many other interviews around the Northern Rim. It was fast becoming clear to me that the perspective I'd carried into this project would need to be broadened considerably.

I had come up here—I thought—to write a book about climate change. My plan was to document not only the physical realities of thawing ice and soil, but their corresponding impacts upon traditional aboriginal societies. I'd wanted to find the faces and tragedies hiding inside the pixels of my satellite images and climate models. I'd envisioned being welcomed with gratitude, after traveling thousands of miles to record personal accounts of meatless hunts, starving wildlife, and perilously thinning ice. In my year-plus vacation from number crunching, I would become the Anna Polit-kovskaya of Arctic climate change.

In retrospect it's a bit embarrassing. Instead of gratitude I got a resigned look and the tired recitation of stories told once too often. Often I was the third, fourth, or tenth outsider interrupting someone's busy summer, demanding to know how climate change was destroying their life. In air-planes and hotels I bumped into camera crews and book authors, all asking for leads to a stricken hunter to interview, a melting lump of ice to film.

I got all of those stories of woe. My notebooks are overflowing with them. Our Sámi reindeer herder is now spending a bundle on hay, because bizarre winter rains have made her animals unable to scrape through the ice-crusted snow to eat.[444] There is no question that climate change is wreaking havoc upon northern peoples, as described in earlier chapters. These problems will only get worse in the future. But to isolate climate change, and portray it as the sole concern facing northern societies is disingenuous. It is but one part of a much bigger story.

Across a vast chunk of Canada's bitterly frozen extreme north, a place with no permanent roads and too cold even to grow wood, a remarkable politi-cal experiment is unfolding.

The new Nunavut Territory—the first redrawing of Canada's map since 1949—has just celebrated its first decade. With 1.9 million square kilome-ters, approximately the size of Mexico, Nunavut is geographically large enough to be a good-sized country. But if it were, with barely thirty thou-sand people it would have the lowest population density on Earth.

Its residents are hard at work changing that. Nunavut has the fastest pop-ulation growth rate of anywhere in Canada, and it isn't relying on foreign immigrants to do it. It is birthing twenty-five babies per thousand people versus the national average of eleven. With a median age of just twenty-three years (Canada's average is forty), Nunavut is extraordinarily youthful. More than a third of its population is under the age of fifteen.[445]

As of Canada's last census in 2006, Nunavut's population had leapt more than 10% in just five years. Iqaluit—its new capital sprouting from the site of an old Cold War U.S. Air Force base—jumped nearly 20%. With vacancy rates near zero, new housing can't be built fast enough in Iqaluit to keep up with demand. Apartments go for two to three thousand dollars per month, and the city vies with Fort McMurray for the dubious distinction of being the most expensive rental market in Canada.

I first met Elisapee Sheutiapik, Iqaluit's mayor, in 2007. She bubbles with enthusiasm about Nunavut's potential. It is a very exciting time for northern aboriginal people, she explains. We are regaining control of our homeland. There are more jobs and new opportunities. The whole world is watching.

She'll also describe its problems—soaring food prices, the housing shortage, substance abuse, and climate change. Nunavut's main travel platform—sea ice—is becoming unreliable. Various other problems commence if temperatures go above 21°C in summer. With a contagious laugh she explains that Iqaluit's new buildings are being built with air-conditioning, something never seen before by the Inuit. Then, getting serious, she'll talk about plans to convince the Canadian government to build a deepwater port for its newest capital city.[446]

She just might get one. With two giant military neighbors and virtually no presence in the region, Canada suffers deep insecurities about Arctic sovereignty and knows that her aboriginal settlements are key to shoring it up—even to the point of past abuses like relocating Inuit families to bleak High Arctic outposts in the 1950s. While Canada's Inuit are a tiny people—only fifty thousand in 2006 (up from forty thousand in 1996), mostly in isolated villages scattered across the Arctic coast—they are the dominant human presence in such a vast empty place. In the Arctic, small numbers of people gain outsized importance. A village of two hundred becomes a major destination, two thousand a metropolis.

From all current appearances Canada's sovereignty anxiety is sharper than ever. The world is staring hard at the Arctic in general and the Northwest

Passage—which actually contains several possible routes—in particular. Like everywhere else Canada's rural areas are depopulating; her fast population growth is fueled mainly by foreign migrants flocking to southern cities. Canada knows that the remote Inuit towns are her essential outposts, and that without them her entire northern front would be empty. But after decades of ham-fisted treatment, like discouraging native languages and yanking kids off to residency schools to be assimilated, the relationship between Canada's central government and her northern aboriginal citizens is finally on the mend, an improvement that seems unlikely to reverse course.

One big example is Nunavut. With a population that is 85% Inuit, its creation marks the first time in history that an aboriginal minority has formed a standard governance unit—in this case a territory[447]—within a modern western country. Imagine creating a new U.S. state seven times bigger than Nevada, with the small aboriginal population of Nevada building its entire new state government from scratch. That is the scope of Nunavut.

It is a process wracked with false starts and growing pains. The Inuit people have milled across this tundra for millennia, but today's permanent towns and institutions are very new. Nunavut's evolving government is being invented and used at the same time, rather like assembling a truck while driving it. It is challenged by far-flung settlements unconnected by roads, high suicide rates, not enough educated workers to fill the new jobs, and an increasingly risky winter travel platform. But optimism abounds. A brand-new northern society is being built from scratch, and the Inuit are in charge. They know this is a grand opportunity—not simply to re-create the old ways, but to build the new.

Aboriginal Demographics

The NORCs hold between 6 and 20 million aboriginal people, depending on how the Russian population is counted.[448] The Russian Federation probably has about 20 million, but only about 250,000 are legally recognized as such, so officially that's 0.2% of Russia's total population (unofficially 14%). The United States has 4.9 million (1.6% of total population), Canada 1.2

million (3.8%), Denmark 50,000 (0.9%), Norway 40,000 (0.9%), Sweden maybe 20,000 (0.2%), and Finland 7,500 (0.1%).[449] Iceland, discovered empty by the Vikings in the ninth century A.D., has none.

Clearly, aboriginal population percentages in the NORC countries are small. Why, then, an entire chapter dedicated to their status and trajectories? Because aboriginal people are a key component of our northern future.

First, the national statistics above mask the importance of geographic distribution. In the coldest, most remote territories of the NORC countries— the same places where many of the more extreme phenomena described in this book are happening—aboriginal populations are disproportionately large, capturing *large minorities or even a majority* of the population. Alaska is 16% aboriginal. In Canada, aboriginal people capture 15% of the populations of Saskatchewan and Manitoba, 25% of Yukon Territory, 50% of NWT, and a whopping 85% of Nunavut. In certain northern areas of Sweden, Norway, and Finland, they have 11%, 34%, and 40% population shares, respectively. Denmark's Greenland is 88% aboriginal. In northern Russia, even the officially recognized population share is 2%—ten times the national average—and that number ignores almost four hundred thousand aboriginal Yakut people comprising one-third of the population in Sakha Republic.[450]

Second, in North America aboriginal populations are growing very quickly. As of Canada's last census it had ballooned 45% in just ten years—a growth rate nearly six times faster than the country's population as a whole. U.S. aboriginals, currently totaling 4.9 million, are projected to rise to 8.6 million by 2050.[451]

So we see that the fast population growth of Iqaluit is not unusual but simply reflects a much broader demographic trend. Yet, a serious attitude contrast exists between the people of Iqaluit and the far larger numbers of aboriginal groups scattered in hundreds of impoverished reservations throughout southern Canada and the conterminous United States. Why are the people of Iqaluit bustling while those living on reservations are depressed? What are the implications for the future of the Northern Rim? The answers start across the border to the west and invoke a theme that is by now, I hope, familiar.

The state of Alaska was barely eight years old—even younger than Nunavut is now—when the largest oil field in North America was discovered at Prudhoe Bay on its northern coast. What followed was land-grab pandemonium.

It was 1968 and the fledgling state hadn't even finished negotiating its land transfers from the U.S. federal government yet. Oil companies grasped immediately that the strike was huge but the waters too icy to reach by tanker ship. Instead, a very long pipeline over public lands was needed to decant it to southern markets, either to a year-round port in the Gulf of Alaska, or through Canada. Modern environmentalists, freshly inspired by Rachel Carson's 1962 book *Silent Spring,* readied themselves for an epic battle.

Meanwhile, another group had also galvanized to win closure of a long-suffering wound: Who owns the land upon which aboriginal people have always lived? Even before the United States purchased Alaska from Russia in 1867, aboriginal Alaskans had long asked when and how the tsar had come to acquire title to their homelands.[452] But no one seemed to care much about this issue. It had simmered, neglected and out of public consciousness, for over a century.

By the time oil was found in Prudhoe Bay, times had changed. America's civil rights movement had taught a new generation the power of organized protests and lawsuits. The Alaska Federation of Natives and other groups had been litigating Washington to block transfers of federal land to the new state of Alaska until their ancestral claims were adjudicated. Many of the claims overlapped and, when added up, covered a total land area larger than that of the new state. It was a mess, and in 1966, Secretary of the Interior Stewart Udall (father of the current senator from New Mexico Tom Udall) declared a "land freeze," effectively stopping all transfers of land to the new state until the mess was cleaned up. When oil was struck and talk of a pipeline began, the legal implications of the aboriginal claims blew sky-high. Who, exactly, owned this land? Suddenly, Alaska—a place that was about as conspicuous as Nunavut is today—mattered to everyone. No pipeline could be built until the issue was resolved.

State legislators and oil companies began lobbying for quick congressional action on an arcane issue ignored since the Alaska Purchase of 1867. After three years of lively politics on Capitol Hill, the final result was the Alaska Native Claims Settlement Act (ANCSA), signed into law by President Richard Nixon in 1971.[453]

ANCSA's grand bargain was this: Aboriginal Alaskans would forever relinquish all of their ancestral claims to land within the state of Alaska, as well as their traditional rights to hunt and fish without regulation. Also, their old reservation treaties would be nullified. In return, they won fee simple property title and mineral rights to forty million acres of land—about one-ninth of the state of Alaska—nearly $1 billion in cash, and a business plan.

The U.S. government had just made Alaska Natives (aboriginals) the largest private landowners in the state of Alaska.[454] The land was geographically divided among twelve "Regional Corporations" to manage the new property and cash holdings, and oversee further incorporation of more than two hundred village corporations within their boundaries. All of the new companies were then free to pursue whatever profits they could from their new assets, which were then returned as dividends to their shareholders. To become a shareholder, one had to possess one-quarter Alaska Native blood, be a U.S. citizen, and enroll with a regional or village corporation. A special landless corporation was even set up for eligible shareholders living outside of the state.[455]

ANCSA differed from all previous aboriginal treaties in at least two important ways. First, an enormous amount of land was granted, more than the area of all historical Indian reservations in the United States combined. Some grumbled that even forty million acres was a pittance compared to what had been stolen in the first place, but there is no questioning it was colossal compared with past treaties. Second, ANCSA did not create permanent sanctuaries for an everlasting traditional subsistence life. Instead, it incentivized use of the granted land not simply for hunting and fishing but for capitalist enterprise, with aboriginal-owned companies and shareholders running the show, to spur development and economic growth. ANCSA had blown up the traditional model of Aboriginal Reservation and replaced it with a new one of Aboriginal Business.

Today, Alaska's aboriginal-owned regional corporations and their sub-sidiaries are worth billions. They've spawned hundreds of companies in construction, oil and gas field support, transportation, engineering, facili-ties management, land development, telecommunications, and tourism, to name a few. They publish shareholder reports, elect boards, and write five-year management plans. In common with other corporations some have done well and others not. Some have been mismanaged into bank-ruptcy. Others have squandered their cash endowments, clear-cut their forests, and sold off land or deeded it to their shareholders. But the suc-cessful ones, especially in remote areas, have become a dominating force in Alaskan politics and society. They create jobs and attract other businesses by offering logistics services. They pay thousands of dollars per year to their shareholders.

ANCSA was really just the beginning of aboriginal empowerment in Alaska. It also set the stage for home rule governments like the North Slope Borough, an enormous success story, which has built schools, sewer sys-tems, and water treatment facilities, and brought many other quality-of-life improvements to the North Slope by taxing oil field activities. Much of its success can be traced back to the ANCSA model. Not surprisingly, aborigi-nal Alaskans today are far more supportive of oil and gas exploration, of land development, and of business in general, than any prior generation.

Out of Alaska

What happened in Alaska inspired aboriginal groups around the world and propelled an era of comprehensive modern land claims agreements across Canada. By 1973 the Inuit, Cree, and others also had legal teams pressing their land claims and, following Alaska's example, thwarting outside natural resource development projects until they were settled.[456] Just four years after ANCSA, aboriginal resistance to a series of new hydropower dams led to the James Bay and Northern Quebec Agreement, Canada's first modern land claims settlement. In 1974 the Dene, Métis, and Inuit people stunned the world by blocking the Mackenzie Gas Project, a long-planned pipeline to bring Arctic natural gas to southern markets and a cornerstone of Canada's northern development plan. Their negotiations took even longer, but today, with their land claims agreements and businesses in place, most are now avid supporters of the pipeline.[457] Like ANCSA their aboriginal-owned corporations and companies will benefit greatly from the project, which could begin as soon as 2018.[458]

Canada's modern land claims agreements have evolved well beyond the simple business corporations of ANCSA. From the outset their aboriginal negotiators insisted that the new agreements affirm not only property rights but political, social, and cultural ones as well. Many settlements also set up political self-governance. They collect royalties from the extraction of subsurface minerals, oil, and natural gas, not only from the granted property but from surrounding public lands.[459] Aboriginal corporations and the Canadian government now make joint decisions on development, wildlife management, and environmental protections on these public lands. Outside companies must hire prescribed numbers of aboriginal workers and companies. Numerous protections of native language and culture reverberate throughout these documents. Such complex agreements take years to negotiate, run hundreds of pages long, and often contain provisions for still more negotiations in the future.[460]

After nearly four decades the era of modern, geographically large land claims agreements in North America is drawing to a close. Over half of Canada is now under jurisdiction of one settlement or another, most recently in 2008 and 2009.[461] The final push will be a wave of smaller agreements across Canada over the next decade or two.[462] Then it will all be done.

Greenland Rules!

The third place where northern aboriginal people have clawed back political power from distant southern capitals is in Greenland. For almost three centuries this enormous, glacier-buried island just four hundred miles east of Iqaluit was a colony of Denmark, but its population and language—currently around fifty-seven thousand people—is overwhelmingly Greenlandic Inuit ("Greenlanders") with a fair mixture of Danish blood.

As in Canada, the Alaska Native Claims Settlement Act of 1971 did not go

unnoticed in this icy Danish province. In the year of its passage Greenlanders voted into their provincial council[463] some radical youth, including an
unknown twenty-four-year-old schoolteacher, Lars-Emil Johansen (whom
I would meet years later as the former prime minister of Greenland), and
the young firebrand Moses Olsen. These two began stridently objecting
to Denmark's sovereignty of Greenland, and for the first time in memory,
Greenlanders began thinking seriously about disentangling themselves from
Copenhagen's colonial rule.

One year later, Greenlanders heartily rejected Denmark's referendum to
join the European Community (predecessor to today's EU) with 70% of
the vote. Alongside their growing nationalism, natural resources were again
a root cause, but this time going the other way: Danish membership in the
EC would impose fishing restrictions and a sealskin ban on Greenland, both
dear to her small aboriginal economies. The referendum passed anyway,
but the vote was a wake-up call to Copenhagen. Within months the Danish Parliament was cooperating with Greenland's minister and provincial
council to explore the possibility of political self-rule. Greenlanders then
overwhelmingly passed a public referendum on whether or not to advance
the idea. In 1979 the Greenland Home Rule Act was passed, and Greenland
became a politically autonomous country within the Kingdom of Denmark.[464] In 1982 she withdrew from the European Community.

Greenland Home Rule wasn't a "land claim" in the property title sense,
as there is no private land in Greenland (while privately owned structures
may be built, all land title is held for the public good). But the end result
was the same. For the next thirty years Greenlanders controlled the use of
their land and set about building up an autonomous government, services,
and political apparatus, just as Nunavut is doing today.

This continued for three decades. Then, in 2008, Greenlanders headed
back to the polls. A new Greenland referendum was proposing even further
divorce proceedings from Denmark. Its sweeping reforms would include
taking over the police force, courts, and the coast guard. Greenland's official
language would be changed from Danish to Greenlandic. Revenues from
future oil and gas development would be shared between the two countries,
so that the Danish subsidies needed for Greenland's survival could be phased

out. Greenland would conduct its own foreign affairs with other countries. The referendum passed overwhelmingly and entered effect in 2009. This island—bursting with offshore gas along both flanks—is now on a political path to full and complete independence.[465]

The Unfair Geography of Aboriginal Power

The modern land claims agreements in North America, and Home Rule in Greenland, are big deals. Politically, they portend a fundamental shift of power from central to aboriginal governments. Economically, they portend abolishment of a culture of paternalism and welfare in favor of engaging aboriginal people in the modern global economy. These new commitments are here to stay. In Canada, for example, the new land claims agreements are even protected by a constitutional amendment.[466] While not perfect, this devolutionary trend is a giant step forward relative to abuses of the past. It signals a profound return of autonomy and dignity to many northern aboriginal peoples.

Notice I said "northern." Nearly all of this new control lies north of the sixtieth parallel. It is an assemblage of people living way up there—the Alaska Natives, Inuit, Yukon Indians, Dene Nation, Greenlanders, and others—who have most cause to celebrate.

Geography and luck greatly explain this uneven spatial pattern of new aboriginal clout. Many remote northern groups escaped being cajoled into old-fashioned treaties during the eighteenth, nineteenth, and early twentieth centuries because their homelands were distant and undesired. Permafrost tundra and spruce bogs held little appeal to white homesteaders. While northern aboriginals were infected, harassed, and resettled, they were not forced to sign away their claims to the land. With no historic treaty signed, their ancestral claims to the land were never extinguished.[467] Legally, this left them in a strong bargaining position by the time a more progressive interpretation of their legal and civil rights arrived in the 1970s.

Most importantly, their remote geography meant there was actually still something left to negotiate *for*. In North America, the perfect trifecta of fossil fuels, hydropower, and civil rights converged upon empty federal and

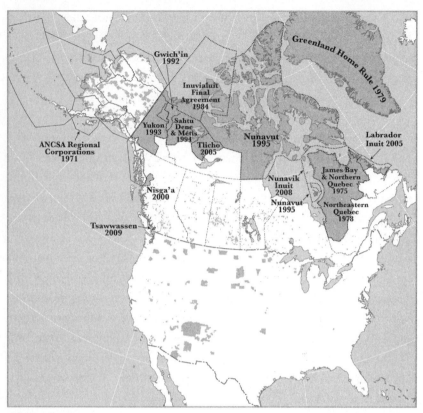

The unfair geography of aboriginal power. Shaded areas indicate lands where aboriginal groups have total or partial control, either through reserves, deeded property, or joint management through modern land claims or home-rule agreements. Alaska boundaries delineate the jurisdictional borders of the twelve Regional Corporations established by ANCSA. Aboriginal control is greatest in the far north and Greenland; in southern Canada and the lower 48 U.S. states it is constrained to much smaller reserves and reservations set by historical treaties. (Map data assembled from multiple sources.[468])

Crown lands, controlled only by Washington and Ottawa. Until the new land claims agreements were instituted, virtually none of this land had ever been privatized. These new aboriginal corporations are thus the first and only major private land owners in northernmost North America.

The situation is totally different in southern Canada and the lower forty-eight American states. While British colonial law did hold a modicum of respect for aboriginal land rights,[469] you know already the centuries-long tale of death and displacement following European colonization of the New World. Between 1492, when Christopher Columbus found Haiti (and misnamed its inhabitants "Indians"), and 1923, when the last Indian Reserve Treaty was signed in Canada, North America was a place where white settlers shot, diseased, and connived aboriginal people off their land. Millions of deaths and four centuries of lopsided treaties later, their descendants are boxed into tiny scattered fragments of land, often enveloped by private property. These treaties, no matter how unfair by today's standards, extinguished their land claims. They have no hope of another. Reservations can grow, but only by buying out their neighbors, if they'll sell, at market value. Even if surrounded by public lands, with their claims extinguished they have no legal standing to sue for a new treaty.[470]

ANCSA deeded about forty million acres of property to Alaska Natives. The modern land claims agreements in Canada ceded joint or total control of just over one billion more,[471] with dozens of smaller claims still pending. In contrast, the sum total of all aboriginal reservations in the lower United States is about seventy million acres, which, if you could sweep all the bits together, might add up to the area of Colorado. Reservation populations may be growing, but their borders are not. There will not be another Nunavut.

President Keskitalo's Argument

I was in Tromsø sitting with Aili Keskitalo, president of the Norwegian Sámi Parliament. She was describing the plight of her Sámi people (Lapps[472]), the aboriginal occupants of northern Europe. The petite thirty-eight-year-old mother leaned forward in her chair, speaking quietly but blue eyes blazing.

"Our language. Our symbols. Our traditional knowledge. They are threatened. In some areas, to a very large extent. We need to have a say in how the natural resources are exploited!"

I nodded. Once again my climate-change project was going down the tubes. When was she going to talk about crusty snow and starving reindeer? But then, while explaining how her parliament was very busy yet politically toothless, with no vote in Oslo, she unknowingly connected the dots for me.

"The climate change, it makes the oil, the gas, the mineral resources in the North more accessible. So the need to get control over the resource management is even more important, because of the climate change." She sat back in exasperation. "If you have no representation, how can you have an influence on resource management?"

If there was ever a moment when my perspective suddenly broadened on the future of the northern countries I was traveling, that was probably it. We talked some more, so I could assemble in my own head what was already so obvious in hers. Everything is linked. Shrinking ice, natural resource demand, and political power were all tugging on each other. My scientist's training had wrongly led me down the path of dissect, isolate, and rank. This works well for a focused problem, but is not always best for gaining a synoptic understanding of the world.

Northern aboriginal people don't like being portrayed as hapless victims of climate change. Nor are they waiting for their central governments to come in and solve their problems. Quite the opposite. After numerous interviews with aboriginal leaders,[473] the resounding message I've heard is a desire for more autonomy, more control, more say over what happens or does not happen on these lands. The damages inflicted by climate change—already coming into view—only intensify their sense of urgency. More control affords more resilience, more adaptability, to deal with the consequences. The people I've met are not hoping that outside task forces will be dispatched to save them from climate change. They want the power—and yes, the resource revenues—to save themselves.

With this new understanding in view, I could see why president Keskitalo was pissed off. In the three countries discussed thus far—the United States, Canada, and Denmark—northern aboriginal people are becoming politically powerful. In the Nordic countries and Russia, they are not.

The Sámi Situation

Europeans are fascinated by their Sámi. Long after great cities had spread across Germany and France, the Sámi were still living in tents, migrating with their reindeer, living off the land by fishing, trapping, and hunting.

Their mystic, highly spiritual culture is permeated with ties to the natural world, expressed in beautiful chanted songs called *joiks*. Furthermore, they are white. Unlike most northern aboriginals, they have a European rather than Mongol origin. Many Sámi have fair skin, blue eyes, and blond hair. This is partly due to centuries of mingling with Nordics, but genetically, the Sámi are much closer relatives of Basques than of Inuit.

Today, about seventy thousand Sámi live in "Sápmi," their ancestral homeland stretching across northern Fennoscandia (see map on pp. x–xi). But Sápmi today is chopped up into four bits owned by Norway, Sweden, Finland, and Russia. It is dismembered.

The Sámi population can never form a single collective political unit within a country, as has happened in Canada and Greenland. Traditional reindeer herding, which moved animals all around Sápmi, is difficult or impossible. Also ethnic Norwegians, Swedes, Finns, and Russians have moved in, bringing industrial development, land privatization, and the loss of grazing and hunting grounds. With four different court systems to navigate, the collective ability of the Sámi to mount legal challenges to such encroachment is dissipated and constrained. And unlike what happened in North America and Greenland, none of the four governments are signaling any possibility of a sweeping land claims agreement, or a new Sápmi state, or individual home rules for each fragment.

However, there are differences among the four countries. Since 1989 Norway, Sweden, and Finland have introduced elected Sámi parliaments, whereas Russia has not. These parliaments are politically weak, serving mainly as forums and advisors to their central governments, but they do provide a voice for the Sámi. Norway's parliament, being the oldest and largest, is most consequential of the three.

Also, when it comes to sticking up for aboriginal rights under international law, Norway is one supportive NORC. It was the first country in the world to ratify International Labour Organization Convention 169, thus committing the Norwegian government to preserving its aboriginal people, cultures, and languages through deliberate action (later, Denmark also ratified this ILO treaty). Norway was also one of five NORCs to adopt the United Nations Declaration on the Rights of Indigenous Peoples in

2007.[474] In part to meet its obligations under these laws, Norway passed a sort of pseudo land claims law, called the Finnmark Act, in 2005. While not specific to Sámi people, it did transfer land ownership from the Norwegian government to its largest and northernmost county, where the population is about 34% Sámi.[475]

While weak by North American standards the Finnmark Act is about as good as it gets in this part of the world. Similar trends are not apparent in Sweden, Finland, or Russia. None of these countries have ratified ILO Convention 169, nor is there any talk of land claims settlements. In Lapland, Sámi complain of imposters stealing their culture, wearing fake clothes, and butchering their language for tourists.[476] The Sámi situation is most depressing in Russia, where a small population of two thousand has little to look forward to.

Trapped on the Kola Peninsula—the militarized, industrialized heart of the Russian North—they are mostly unemployed with no parliament. What few reindeer herders remain complain of grazing lands privatized and closed, and horrid environmental pollution from mining, smelting, and leaking radiation from old nuclear reactors. Russian soldiers sometimes shoot their animals to eat or for fun.[477] Snared in poverty, lacking land tenure, and with no political voice, they are quickly losing their aboriginal language. Of Sápmi's four fragmented pieces, Russia's has the most uncertain future.

The Mi-8 Time Machine

We thudded over the taiga in an orange Soviet-era Mi-8 helicopter, crammed against one of its little porthole windows. Below us was an endless plain of mossy lakes, cottongrass sedge, and hunched conifers stretching to infinity. My doctoral student Karen Frey murmured from behind a video camera while I wrote notes and GPS coordinates into a pad. Faint reindeer trails splayed here and there across the tundra, but the landscape was motionless. We'd been at it for over half an hour with no sign of life.

Suddenly the Mi-8's rotors whined and we were hovering. There were scraping noises up front and men speaking in Russian. The ponderous helicopter slowly eased its bulk onto the ground and a door clanged open. From its cavernous interior white Russian hands produced a burlap sack full of potatoes. From outside, dark, weathered hands reached up to take it.

We had dropped by the campsite of a Nenets family, one of the largest of several aboriginal reindeer peoples of the Russian North. Their *chum,* a circular tent halfway between a teepee and yurt, was made of lashed wooden poles and reindeer hides. There were corrals and long sleds with curved wooden runners. Grubby, cute kids were peeking at us. Freshly flayed reindeer skins were drying. The whole place hung with smoke from burning smudge fires. Our Mi-8 wasn't a helicopter, it was a time machine: The Nenets are one of the last people on Earth still following the ancient practice of moving around with their reindeer.

Anthropologists, even Russian ones, have long romanticized Siberian scenes like this. But most of Russia's northern aboriginal people do not lead nostalgic frontier lives out on the land. Instead, they live in gritty, impoverished, multiethnic villages rife with unemployment, alcoholism, and suicide.[478] Life expectancies are low. Aboriginal control over outside resource exploitation is virtually nil, as is the amount of royalty they receive when resources are developed. There is no prospect of winning private land title as has transpired in North America,[479] and even if there were, under Russian law all subsurface mineral and energy rights still remain with the state. Vastly outnumbered as they are by ethnic Russians, there is no hope for sizable aboriginal political majorities except in small *okruga* (regions) and *raiony* (districts). Exceptions, like a tiny pod of Yukagir people who won self-governance in Sakha Republic,[480] are rare. With so little political power, even their wild food is constantly under threat by commercial interests. In one recent case, aboriginals of Kamchatka beseeched President Medvedev and Prime Minister Putin to halt auction lease sales of their salmon rivers so they wouldn't starve.[481]

Russia's northern aboriginals don't have time to debate political governance models or resource revenue-sharing schemes. Their priority is simply retaining access to wildlife and land, and keeping at bay the encroaching industries that would damage them. The Russian anthropologist Aleksandr Pika, who devoted his life to studying northern aboriginals before drowning in a 1995 Bering Sea boat accident, alongside five Eskimos and three Americans, once wrote:

The numerically small [aboriginal] peoples of the North live on lands rich with oil, natural gas, uranium, tin, timber, and other resources. Society has not yet learned to take these resources without damaging nature. Society cannot live, in fact, without touching these resources. The peoples of the North are often guilty simply in that they live on these lands and their very existence poses problems for the state. Indeed, many feel that without these peoples, there would be no such problems, and that the peoples of the North should understand this, and not complain too loudly or too often.[482]

This does not mean that the Russian government, or Russians more generally, care nothing for their aboriginal people. My student and I were sternly admonished to respect the Nenets family's privacy by not photographing them, and thirteen of the aforementioned Kamchatka commercial salmon leases were, in fact, retracted to protect traditional aboriginal fishing rights. Under old Soviet law, aboriginals had no legal claim to land or its resources, but that has changed somewhat under the Russian Federation. Its 1993 Constitution now mandates that both be protected "as the basis of life and activities of the peoples" who live on them, and holds central and regional governments responsible for protecting "traditional ways of life." To flesh out these general constitutional requirements, three meatier federal laws specifically addressing aboriginal land rights were adopted in Moscow by 2001.[483] Chief among the new reforms is a revival of *obshchiny*, small group-owned plots of land to which families, clans, or villages can request exclusive use for traditional subsistence.

It is a well-known adage in Russia that obedience to federal laws is inversely proportional to geographic distance from Moscow. However, these new ones, at least on paper, are a significant advance for aboriginal Russians. While Russia has not yet ratified ILO Convention 169, it is clear that these new laws were written to conform with many of its guidelines. Interestingly, the country's recent recentralization of power, begun under Vladimir Putin and reviled by the western press, is good news for Russia's forty-five officially recognized aboriginal groups: If Moscow demands that the far-off regional governments implement and enforce the new federal laws, these people will be better protected.

Imagining 2050

A final and keenly important distinction must be drawn between the emerging new aboriginal policies of North America and Greenland versus those of northern Europe and Russia. While the former do accord value and protections to the traditional cultures of the past, they also seat chairs at the table of the future by devolving political power, land management decisions, and natural resource revenues including oil and gas royalties. But in the Nordic countries and Russia, emerging policies seek to preserve "traditional" cultures and ways of life above all else. Indeed in Russia, demonstrable proof of such activity—raising reindeer, for example, or subsisting by hunting and fishing—is a key requirement for winning aboriginal protections and privileges, including *obshchiny*. Also, the old Soviet tradition of limiting legal recognition of aboriginal status to populations having fifty thousand or fewer persons has been retained, such that small, scattered aboriginal groups can win these privileges but not large ones. At first blush, such policies sound noble—what's wrong with trying to protect vanishing ancestral cultures from going extinct? But, as put by the recent *Arctic Human Development Report,* "one must question the tendency to consider change as a threat to some immemorial 'tradition' in discussing indigenous societies, when it is called progress in western societies."[484]

Put bluntly, the Nordic and Russian aboriginal policies encourage the mummification of aboriginal people and their historical practices into bits of living folklore. By not going far enough, the new legal protections—well intentioned and keenly desired by their subjects as they may be—lapse into paternalism, pure and simple. Aboriginals win permission to carry on their ancient ways—to the gratitude of village elders and future anthropologists—but are denied forms of empowerment that matter most for the future: political power, a say over land use and development, a say over environmental protection, and the right to receive royalties from all the natural gas, oil, and minerals that will be plucked from beneath their feet. Their cultures are denied the right to evolve. Instead, they are pickled under a glass bell jar.

When I try to imagine the role of NORC aboriginals in 2050, I sense two very different scenes unfolding. In the eastern hemisphere, I see fascinating historical enclaves, where people can still carry out ancestral subsistence traditions on the land. Their lives are not so different from today except they have become living museum displays, beset by anthropologists and a global tourist trade. In the western hemisphere, I imagine unprecedented new societies taking hold. They are a unique blend of the old and the new, choosing some parts of traditional culture to retain and others to abandon. People run their business corporations in the morning and go hunting in the afternoon (the ringed seals and polar bears are now protected but harbor seals and salmon are moving in). Pipelines and ports are spreading, natural gas is flowing south, and royalties are flowing north. In Canada, the first university above the sixtieth parallel has been founded.[485] The global fleet bristles offshore but the land belongs to them. I see the original stewards of this land taking it back again.

PART THREE

ALTERNATE ENDINGS

CHAPTER 9

The Pentagon Report

Our thought experiment so far has been propelled by big drivers, the four global forces of demography, natural resource demand, globalization, and climate change. A fifth—enduring legal frameworks—cropped up in discussions of sovereignty over the Arctic seafloor and the political power of aboriginal peoples. Throughout the book we have stayed within the confines of the following ground rules as stated in the opening chapter:

No Silver Bullets (incremental and foreseeable advances in technology),

No World War III (no radical reshuffling of our geopolitics and laws),

No Hidden Genies (like a global depression, a killer pandemic, a sudden climate change),

and

The Models Are Good Enough.

These overarching drivers and ground rules have served the 2050 thought experiment well to this point. I hope it has kept the book from being shelved in the science fiction sections of bookstores and libraries. The described outcomes are deduced from big trends and tangible evidence already apparent today, rather than political ideology or my wonderful imagination. They favor the likely over the unlikely. I honestly expect, should I live long enough, to see many or all of them materialize within my lifetime.

In this chapter and the next, let's step out of the comfort zone a bit. What are some other outcomes these trends could provoke? Are the four forces robust, our ground rules reasonable? If not, how might they surprise us? This chapter explores six less assured, but plausible, developments that could affect some of the big trends presented thus far. Five of them originate in the North, but have global or far-reaching consequences. Let's begin with climate change, by breaking the ground rules on hidden genies and computer models.

The Evolution of Climate Models

The motivation for running climate models is nothing like the motivation for making weather forecasts on the nightly news. Those seek to identify specific events, like a storm front, and are meaningful only a few days into the future. But climate models forecast *average* climate variables, like mean January temperature, and are meaningful many decades into the future. They do this by taking account of certain things—like deep ocean circulation and increasing greenhouse gas concentrations—that simply don't matter for short-term weather. It's not possible to know what the exact temperature will be in Chicago next August 14 or January 2 at three o'clock in the afternoon, but it's very possible to know what the average August or January temperatures will be. One is weather, the other is climate.

Climate models are also amazing tools for figuring out how our complex world actually works. Suppose that it is an observed fact that summer rainfall is declining in Georgia, but this phenomenon simply won't show up in a climate model's simulations no matter how many times it is run. Puzzled, its programmers realize that something is missing and wonder what it might be. Into the model goes a hypothesis—say, loss of forest (trees pump enormous volumes of water vapor back to the atmosphere), because many trees have been removed to build Atlanta suburbs. Does the model now correctly simulate the measured rainfall decline? If so, congratulations—new scientific understanding has been won about how rainfall works in Georgia, and the climate model has been made more realistic. If not, on to test the next hypothesis down the list. Eventually the missing bit of physics is discovered, the model is improved, and its creators move on to ponder its next little failure.

At their core, climate modelers seek to understand how the atmosphere functions, and how it responds to changing drivers. By studying when and where the models break down, we improve scientific understanding of how the real world works, and our models become more accurate. After more than fifty years of trial and error, they have now evolved far beyond their primitive ancestors of the 1960s. We've learned a great deal about how Earth's climate system actually operates. In today's generation of models, complicated things like El Niño and the Hadley Circulation emerge organically without programmers having to "add" them at all. That is very encouraging, because it tells us the models' assumptions and physics[486] are realistic and working correctly.

The big push now is to hone down climate model spatial resolutions (i.e., the "pixel size" of their simulations) from hundreds of kilometers, useful for broad-scale projections like the ones presented in this book, to kilometers, which is what local planners need. But even at the coarser spatial scale of today's generation of models, many important conclusions about our future are now well vetted and uncontroversial. All of the megatrends discussed so far—rising global average temperature, the amplified warming in the Arctic, rising winter precipitation around the northern high latitudes—fall within this uncontroversial category.

More troublesome are the short-sellers and inside traders of natural climatic variability. Volcanoes, wildfires, and sunspot cycles are just a few of many phenomena imprinting their own natural variations over the underlying greenhouse gas signal. But now these volatile (and fairly common) phenomena, too, are being added to climate models and tested.

Where climate models suffer most is in capturing rare events lying totally outside of our modern experience. Most weather stations are less than a century old; the satellite data era began only in the 1960s and '70s. These records are far too short to illuminate the full range of our Earth's twitchy behavior. Shifting oceans and ice sheets are key drivers of climate yet contain toggles and circuits with longer patience than our short instrumental records. They add boosts, buffers, and dips to the overall greenhouse effect, so we must understand them as well.

Unfortunately, a naturally twitchy climate makes the steady, predictable push from anthropogenic greenhouse gases more dangerous, not less. From the geological past we know the Earth's climate has not always been so quiet as it is now. Therefore, through greenhouse loading we are applying a persistent pressure to a system prone to sudden jumps in ways we don't fully understand. Imagine a wildcat quietly sleeping on your porch—it looks peaceful but is by nature an ill-tempered, unpredictable beast that might spring into a flurry of teeth and claws in an instant. Greenhouse gases are your knuckles pressing inexorably into its soft slumbering belly; the global ecosystem is your exposed hand and arm.

Rare or threshold behaviors—like a permanent reorganization of rainfall patterns, accelerated sea-level rise, or a giant burp of greenhouse gas from the ground—all pose legitimate threats to the world. We know they are plausible but, unlike greenhouse gas forcing, don't know yet how probable. But their behaviors, too, must be added to climate models somehow. Just because something seems unlikely doesn't mean it won't happen, or that its impacts are not potentially enormous if it does. These are the climate genies, and we are just beginning to discern the outline of their various sleeping forms. To find them at all, we must turn to the prehistoric past.

The Flickering Switch

One of my personal heroes in science is Richard B. Alley, an outstandingly accomplished glaciologist and professor of geosciences at Penn State University. Not only has he cranked out one landmark idea after another, published nearly forty times in *Science* and *Nature,* been elected to the National Academy of Sciences, and written a wonderful popular book explaining it all for the rest of us,[487] he is also about the nicest and most enthusiastic guy one could ever hope to meet.

In 1994, Alley came to deliver a guest lecture at Cornell University, where I was a lowly second-year graduate student. Everyone was abuzz that Richard Alley was coming, because he had just published a pair of back-to-back articles in *Nature* that had stunned the climate-science community.[488] Even my thesis advisor—who was pretty famous himself, having written the paper putting together the theory of plate tectonics[489]—was talking about them. But a great thing about academia is that it is on open, democratic affair even when it comes to its pop icons. Visiting celebrities will hang out for a day or two happily chatting with whomever, even lowly second-year graduate students. Landing a meeting with one is largely a matter of getting to the sign-up sheet first, which of course I did.

When my time slot arrived I went to meet Alley, armed with a list of questions about his *Nature* papers so I could hear more from the great man himself. That lasted about forty-five seconds, before he insisted on hearing all about *my* work. I couldn't believe it. It was a dumb little side project of my research, but Alley's enthusiasm was totally contagious. We relocated to my lab hole, where he huddled alongside me, giving all manner of helpful advice and inspiration. By the time he ran off late to his next appointment, I was so excited about my project I barely remembered I'd forgotten to ask more about his. That's just the kind of guy he is.[490]

What had everyone gabbling was what Alley and his colleagues had dug out of the Greenland Ice Sheet. The U.S. National Science Foundation had funded construction of a drilling and laboratory camp on top of it to extract a two-mile-long ice core called GISP2, an enormous task taking about four years.[491] Preserved in the upper sections of ice cores are annual layers, like the rings of a tree. Each one contains the compressed equivalent of a full year's worth of snow accumulation falling on the ice sheet surface (cores are drilled from deep ice sheet interiors where it never melts). By counting the layers down-core and measuring their thickness and chemistry, a very long reconstruction of past climate variations is obtained. We even get tiny samples of the ancient atmosphere, by cracking into air bubbles trapped in the ice. From these high-resolution annual measurements in Greenland, Alley and his colleagues had discovered that around twelve thousand years ago, just when we were pulling out of the last ice age, the climate began shuddering wildly.

The shudders happened faster than anyone had dreamed possible. Our climatic emergence from the last ice age, it seems, was neither gradual nor smooth. Instead it underwent rapid flip-flops, seesawing back and forth between glacial and interglacial (warm) temperatures several times before finally settling down into a warmer state. These large temperature swings happened in less than a decade and as quickly as three years. Precipitation doubled in as little as a single year. Around Greenland, at least, there was no gradual, smooth transition from a cold ice age to the balmy interglacial period of today. Alley's team had shown that climate could some-times teeter as well, like a "flickering switch," between two very different states. Furthermore, it had happened other times in earlier millennia, so this was not a totally isolated event. The extreme rapidity of these changes, concluded Alley, implied "some kind of threshold or trigger in the North Atlantic climate system."[492]

Thus was born a brand-new subfield of climate science known today as "abrupt climate change." Twenty years ago anyone who hypothesized a sudden, showstopping event—a century-long drought, a rapid temperature climb, or the fast die-off of forests—would have been laughed off. But today a growing body of evidence from ice cores, tree rings, ocean sediments, and other natural archives tells that such things have happened in the past. We've long known the Earth's climate has experienced big changes before but assumed they only occurred slowly over geological time, like the gradual turning of a dial. Now we know they can sometimes happen abruptly as well, like flipping a switch. The implications of this are global, as we shall see next.

The Pentagon Report

From a societal perspective, an abrupt unexpected climate change is more destabilizing than one that is gradual and anticipated. Military analysts concede that the expected gradual climate changes pose national security threats, and by late 2009 the U.S. Central Intelligence Agency had opened a new center specifically dedicated to assessing them.[493] A recent study, for example, projects a more than 50% increase in armed conflict and nearly four hundred thousand more battle deaths in Africa by 2030.[494] But one of the few attempts to assess the societal impact of an *abrupt* climate change was commissioned by the U.S. Department of Defense in 2003.

This document, titled "An Abrupt Climate Change Scenario and Its Implications for United States National Security," is not based on climate model projections, but instead on a known prehistoric event seen in ice cores, sediments, and fossils. About 8,200 years ago, several thousand years after the really big swings that Alley had studied, temperatures near Greenland suddenly tumbled by about 6°–7°C. Cold, dry, windy conditions spread across northern Europe and into Asia; certain African and Asian monsoon rains faltered, and temperatures probably rose slightly around the southern hemisphere. These conditions persisted for about 160 years before reversing again.

This event was not unique but simply the last and smallest of several climate shudders seen in Greenland ice cores as the last ice age wound down. It was less severe, shorter-lived, and less geographically extensive than its predecessors (especially the Younger Dryas event, the monster cold snap studied by Alley that abruptly kicked in about 12,700 years ago, then persisted for nearly 1,300 years).[495] That said, let's hope that it never happens again. The Pentagon's report, which outlines possible social scenarios if what occurred 8,200 years ago were to happen again today, is quite scary.

It describes wars, starvation, disease, refugee flows, a human population crash, civil war in China, and the defensive fortification of the United States and Australia. "While the U.S. itself will be relatively better off and with more adaptive capacity," the authors conclude, "it will find itself in a world where Europe will be struggling internally, large numbers of refugees washing up on its shores, and Asia in serious crisis over food and water. Disruption and conflict will be endemic features of life."[496] The report's authors insist that their assessment, while extreme, is plausible.

Could this really happen? Nobody knows for certain, but the good news is that the physical mechanism underlying these North Atlantic cold shudders is now fairly well understood, and its behavior successfully replicated by climate models, so we can at least test the probability. The culprit appears to be a slowdown of the global thermohaline circulation—the long, ribbon-like "heat conveyor belt" of ocean currents, one arm of which carries warm tropical water from the Indian Ocean all the way to the Nordic seas, bathing western Europe and Scandinavia in all that heat so undeserved for its latitude as described in Chapter 7. The North Atlantic region is a critical pivot for this global circulation pattern. It is where the warm, salty north-flowing surface current finally cools sufficiently so that it becomes heavier than the surrounding colder (but less saline) water, sinks down to the ocean floor, and begins its millennia-long return south, crawling along the dark bottom of the abyss.

All of this is driven by density contrasts. If sufficiently large, a local freshening of the North Atlantic can slow or even halt the sinking, thus killing this entire overturning arm of the global heat conveyor belt. This has immediate implications for the Earth's climate. Heat becomes less mixed around the planet. Cold temperatures (especially winters) and drought descend upon Europe. The southern latitudes warm; the Asian and African monsoons weaken or drift. It's rather like adding hot water to a cold bath, in which stirring the water around helps to even out the temperature contrasts. But with no water circulation, one's back grows cold but feet are scalded.

The most likely source of water for the sudden freshening of the North Atlantic was one or more massive floods released from the North American continent at the end of the last ice age, as its giant glacial ice sheet melted away. As the sheet retreated north into Canada, huge freshwater lakes, some even larger than the Great Lakes today, pooled against its shrinking edge. Then, when a pathway to the sea emerged from beneath the rotting ice, out the water went. The deluge that tore out through Hudson Bay must have been biblically awesome in scale.[497] I wonder if any aboriginal version of Noah witnessed and survived it, creating a legend for generations of the Great Flood that drained the Earth's water to the sea, bringing seemingly endless winter upon the land.

Figuring out hidden genies takes time and a lot of work. The above hydrologic explanation for the North Atlantic climate shudders was first proposed by Columbia University's Wallace Broecker back in 1985.[498] Its finer details are still being tinkered with today. But now that we understand this genie rather well, and its physics are reproducible in climate models, we can assess the likelihood of another such shudder happening again in the future.

So far, most simulations agree that a complete collapse of the thermohaline circulation is unlikely anytime soon, for the simple reason that it's hard to find a big enough freshwater source with which to sufficiently hose down the North Atlantic. The Laurentide ice sheet that once covered Canada and much of the American Midwest is long gone. The projected increases in high-latitude precipitation and river runoff appear sufficient to weaken the circulation, but not enough to kill it outright.[499] This weakening shows up in most future climate model projections as a little bull's-eye of below-average warming centered over the North Atlantic. It's not enough to create outright cooling, but it does reduce the magnitude of warming locally over this area. Let's hope these simulations are correct—because if they're wrong, losing even part of the Asian monsoon would be really, really bad.

There is, of course, another big source of potential freshwater—one that happens to be plunked right in the middle of the North Atlantic. No serious scientist thinks the Greenland Ice Sheet will melt away anytime soon, and if it ever does we'll be dealing with even bigger worldwide problems than a cold, dry Europe and faltering monsoonal rains. But this genie, we're nowhere near to understanding well enough to model yet.

Genie in the Ice

Two smelly straight guys sharing a tent sized for one is bad enough. But waking up covered in yellow dust, with no hot water for days, is the pits. It was impossible to keep the stuff out, even barricaded inside the lone wind-rated tent we had thought to bring with us.

The Greenland Ice Sheet was in charge, not me and not Ohio State geography professor Jason Box. We were camped next to its southwestern edge, where one of its many outlet glaciers finally succumbs to a grinding wet death, killed by the sun among the tundra grasses, caribou, and musk oxen. Every night, we squeezed head-to-toe in the little tent and buttoned up tight. Every night a fierce katabatic wind would pour off the ice sheet, lift tons of grit from its gravelly outwash plain, and fling it against our shuddering tent. The silt pushed through closed zippers and tiny mesh slits. It entered our nostrils and encrusted our hands as they gripped the tent's violently shaking poles.

But by morning the winds would die down and we went to work. Jason installed time-lapse cameras to track the speed of the glacier's sliding snout; I submerged electronic sensors in its outgoing torrent of meltwater to monitor how much was flowing off to the sea. We were studying these things to help answer a burning scientific question that should worry us all. Chapter 4 showed that we are facing decimeters of sea-level rise by century's end. Many scientists wonder if even these estimates might be too low. Could climate warming cause the Greenland and West Antarctic ice sheets to accelerate their dumpage of ice and water into the sea, thus cranking up its rise even faster than is happening already? Could the world's oceans go even higher, say a couple of meters by the end of this century?

The short answer is maybe. The geological record tells us sea levels are certainly capable of responding quickly to shrinking glaciers. And over the long haul—meaning several thousands of years—it looks like the Greenland Ice Sheet is in trouble and could well disappear completely.[500] Glaciers and ice sheets are nourished on their tops by snow. They are removed at their margins by melting and—if they float out into an ocean or lake—by calving off icebergs into the water. When nourishment exceeds removal, glaciers grow, storing water up on land, so sea level falls. When removal exceeds nourishment, glaciers retreat and their stored water returns to the ocean. In this way sea levels have danced in a tight waltz with glaciers, falling and rising anywhere from about 130 meters lower to 4–6 meters higher than today over the past few ice ages. Other things—especially thermal expansion of ocean water as it warms—also drive sea level, but the waxing and waning of land ice is a huge driver.

As the last ice age unraveled, sea levels commonly rose 1 meter per century, and sometimes as fast as 4 meters per century during intervals of very rapid glacier melting.[501] Looking forward, if average air temperatures over Greenland rise by another +3°C or so, its huge ice sheet, too, must eventually disappear. Depending on how hot we allow the greenhouse effect to become, this will take anywhere from one thousand to several thousand years, raising global average sea level by another 7 meters or so.

Based on the emissions scenarios currently being bandied about by policy makers, the temperature threshold to begin this process will indeed be crossed in this century, and the long, slow decline of Greenland's ice sheet will begin.[502] It is already something of a stubborn relic of the last ice age; if it magically disappeared off the island tomorrow, it's doubtful this ice sheet could grow back.[503] One thousand years from now, eighteen of the twenty-seven megacities of 2025 listed in Chapter 2 will lie partially or wholly beneath ocean water that might once have been blue ice in Greenland.[504]

But over the shorter term, meaning between now and the next century or two, the scary genie of Greenland and Antarctica isn't from their ice sheets melting per se (indeed, it will never become warm enough at the South

Pole for widespread melting to occur there) but from their giant frozen rumbling ribbons of ice that slide over hundreds of miles of land to dump icebergs into the sea. Already, there are many such ice streams in Antarctica and Greenland moving tens of meters to more than ten thousand meters per year. They empty out the deep frozen hearts of these ice sheets, where temperatures are so cold the surface never melts at all.

Of grave concern is collapse of the West Antarctic Ice Sheet. This vast area is like a miniature continent of ice towering out of the ocean, much of it frozen to bedrock lying *below* sea level. If it became unstuck, a great many Antarctic glaciers would start lumbering toward the water, eventually raising average global sea level by around five meters. There is geological evidence that this has happened before,[505] and if it happens again it would hit the United States especially hard. For various reasons a rise in global average sea level does not translate to the same increase everywhere—water will rise by more than the average amount in some places and less than the average in others.[506] Such a collapse would produce above-average inundation of the Gulf Coast and eastern seaboard, putting Miami, Washington, D.C., New Orleans, and much of the Gulf Coast underwater. When it comes to climate genies, the West Antarctic Ice Sheet is an ugly-looking lamp.

Frankly, we don't understand the physics of sliding glaciers and ice sheet collapses well enough yet to model the futures of Greenland and Antarctica with confidence. Many things affect the speed and dynamics of that long slide that are hard to measure or see. They include the interplay between the sliding ice and its bed, the heat and lubrication added by meltwater percolating to the bed from the surface, the importance of buttressing ice shelves (which help dam ice up on the land), the ocean water temperature at the ice edge, and others.[507] Computer models and field studies—like the one Jason and I were conducting in Greenland—are in their infancy. Scientists are still discovering new things and debating what may or may not be important. This is why the likelihood of accelerated sea-level rise was kept out of the last IPCC assessment, and may be kept out of the next one as well. Might the ice sheets start slipping faster, with higher sea levels right behind? Perhaps—but without well-constrained models, we don't yet know how likely that is.

Genie in the Ground

Digging into a permafrost landscape usually goes something like this: After cutting through a thick living mat of vegetation, the spade turns over a dark, organic-rich soil, almost like the mulch that one buys to spread in a garden. Usually there are bits and pieces of old dead plants poking out of it. Then, anywhere from several to tens of inches down, the blade goes *chunk* and will bite no farther. But it's not a stone. At the bottom of the hole, there is just more of the same organic-rich goop but it is frozen hard as cement, often with a little black ice peeking through. Going any deeper is a major job, requiring a big drill and lots of manpower.

Why on Earth would anybody go all the way to the Arctic to drill holes into frozen black muck? The reason is organic carbon, and we now know that frozen northern soils hold more of it than any other landscape on Earth. In fact, the more we study these soils the more carbon we find. As of 2010 the latest estimate is 1,672 billion tons (gigatons) of pure organic carbon frozen in the ground. [508] That's roughly half of the world's total soil carbon crammed into just 12% of its land area.

The reason there's so much carbon there is because this is a place too cold and damp for living things to fully rot away when they die. Live plants draw down fresh carbon from the atmosphere and store it in their tissues. When they die, decomposing microbes chow down, pumping the carbon back to the atmosphere in the form of carbon dioxide (CO_2) or methane (CH_4) greenhouse gases. But while plants and trees can still grow in cold places, even on top of permafrost, the microbes are hard pressed to finish off their remains because their metabolisms are strongly temperature-dependent (just as stored food decomposes more slowly in a refrigerator than at room temperature). Very often a mulch-like layer of peat will accumulate, building up the ground elevation over time as successive generations of plants root into the semirotted remains of their ancestors. Some decomposition continues underground, but once permafrost sets in, even that halts, and the stuff becomes cryogenically preserved. Since the end of the last ice age, this excess of plant production over plant decomposition has slowly accumulated one of the biggest stockpiles of organic carbon on Earth.

To put that earlier 1,672 gigatons (Gt) of carbon estimate into greater perspective, all of the world's living plants hold about 650 Gt. The atmosphere now holds about 730 Gt of carbon, up from 360 Gt during the last ice age and 560 Gt before industrialization. The world's remaining proven reserves of conventional oil hold about 145 Gt of carbon and coal about 632 Gt. Each year we release around 6.5 Gt of carbon from burning fossil fuels and making cement. The total target reduction for "Annex 1" (developed world) signatory countries to the Kyoto Protocol was 0.2 Gt per year.

Put bluntly, there is an absolutely gigantic pile of carbon-rich organic material just sitting up there in a freezer locker, lying at or very near the surface of the ground. The big question is, what will happen to that carbon as it thaws out? Will it stay put, perhaps even offsetting the greenhouse effect thanks to faster-growing plants, thus storing more carbon even faster than before? Or will the microbes wake up and chow down, feasting on thousands of years of accumulated compost and farting voluminous quantities of methane and carbon dioxide back into the air? I'm not suggesting that sixteen hundred gigatons of deeply frozen soil carbon could all be returned to the atmosphere at once, but even 5% or 10% of it would be enormous.

This possibility is another one of those climate genies that we are only just beginning to assess. Compared with the previous two, relatively little work has been done on it. Most permafrost research has traditionally focused on engineering, i.e., how to build structures without thawing the ground, thus slumping it and destroying what was built. Hardly anyone cared much about permafrost carbon until recently.

We don't know how quickly or deeply permafrost will thaw or how quickly and deeply the microbes will get to work. The microbes themselves generate heat, and we're not sure how much this will further enhance the permafrost thawing process. The net outcome—net carbon storage versus net carbon release—hinges on a small difference between two far larger and opposed numbers (i.e., the rates of plant primary production versus microbial decomposition). Both numbers are difficult to measure and have large uncertainties associated with them.

Much also depends on hydrology. The millions of lakes sprinkled across permafrost landscapes are themselves heavy greenhouse gas emitters and even bubble forth with pure methane, so their fate, too, is intimately tied to our climate future. Also, if thawed permafrost soils become dry and aerated (as might be expected if deep permafrost goes away), then microbes will release stored carbon in the form of carbon dioxide. If soils stay wet (as might be expected from climate model predictions of increased northern precipitation), then microbes will release it as methane, which is twenty-five times more potent a greenhouse gas than carbon dioxide. Given all these uncertainties, our current generation of computer models contain significant knowledge gaps. I'd wager we have twenty years' work ahead of us before a solid scientific consensus can be reached on what will happen to this big mess of carbon as it defrosts.[509]

We do know this very same landscape switched on to become a major source of greenhouse gas once before—at the end of the last ice age, when northern peatlands first began to form. About 11,700 years ago, as temperatures rose at the end of the Younger Dryas cold shudder, a threshold was crossed, plants began growing, and peatlands sprang up all around the Arctic, pumping out enormous volumes of methane.[510] We also know, from a single study in Sweden, that rising air temperatures penetrate permafrost soils more quickly and deeply than we thought. From two other studies in West Siberia, we know that although thawed soils ooze up to six times more dissolved carbon into rivers and lakes than frozen soils, they also store carbon faster—or at least they did for the past 2,000 years. This is at odds with a different study in Alaska, which suggests that faster-growing plants will not be able to outpace the faster-decomposing microbes once the permafrost disappears. Finally, we know some simple math: If even 2% of this frozen carbon stock somehow returns to the atmosphere between now and 2050, it will cancel out the Kyoto Protocol Annex 1 target reductions more than four times over. Like the West Antarctic Ice Sheet, this is one genie with global repercussions that we should all hope stays asleep.

Globalization Reversal

Might any of the four global forces of demography, natural resource pressure, globalization, and climate change screech to a halt between now and 2050, thus ruining all of our best projections?

Three of these have tremendous inertia. Demographic trends are a slow-moving ship, taking a generation—fifteen to twenty years'—before even major course corrections will be felt. Population momentum ensures that our fastest-growing countries will keep growing for decades, even if their fertility rates fall to 2.1 tomorrow (replacement level), because their age structures are so youthful.[511] And with a projected population increase to around 9.2 billion by 2050—especially a modernized, urban, consumptive one—it's hard to envision how our demand for water, energy, and minerals will decrease from what it is today, even with great strides in conservation and recycling. Greenhouse physics dictates that we are locked in to at least some climate change and higher global sea level no matter what; the big uncertainties are how far we will allow greenhouse loading to go, what the impacts on global rainfall patterns and hurricanes will be, and lurking climate genies.

That leaves globalization. In today's world of Walmart and iPhones, it's easy to take our continued economic integration for granted. But as discussed in Chapter 1, the current globalization megatrend did not simply happen by itself. It was set into motion by the United States and Britain very deliberately, with a long string of new policies dating to the Bretton Woods summit in 1944. While the Internet and other information technology have enhanced globalization, they did not create it. Global social and information networks surely seem here to stay, but unlike population momentum or greenhouse gas physics, there is no natural law commanding that current policies favoring our global economic integration must continue.

History tells us of past balloons of economic integration and technological advance followed by puncture. In 221 B.C. the Qin armies first unified northeastern China out of a bedlam of warring fiefdoms. Successive Han, Sui, T'ang, Yuan, and Ming dynasties then expanded the world's biggest trade empire into central and southeast Asia, India, the Middle East, and the Mediterranean. By the fifteenth century, China had trade outposts in Africa and led the world in medicine, printing, explosives, banking, and centralized government. But then, its rulers lost interest in a global empire. They began a series of fateful political decisions that shut down China's overseas trade while discouraging scientific advances at home. Its nascent industrialization cut short, China stood frozen in time, and the much smaller European states commenced to take over the world.

Europe wasted little time ramping up the next round of globalization.

By the 1600s colonialist governments were working hand in hand with private corporations like the Dutch and British East India companies—the equivalent of today's multinational corporations—setting up remote trading posts and shipping routes. Merchant capitalism flourished, fueled by furs, timber, gold, spices, and coal imported from overseas. Guided by multinational banks, by the 1870s goods and capital were flowing across national borders as freely as they do today. Steamships, the telegraph, and railroads were opening up the world just as standardized shipping containers, jet aircraft, and the Internet would do again a century later. Many countries decided to peg their paper currencies to a gold standard, creating fluid international currency markets and huge flows of cross-border capital. The British pound became the dominant circulating world currency much as the U.S. dollar is now. Remarkably, by 1913 the industrialized national economies were enjoying even greater levels of foreign investment than today.[512] It was a golden age of economic globalization.

It unraveled surprisingly fast. The June 28, 1914, assassination of Archduke Franz Ferdinand in Sarajevo initiated a chain of events setting off a world war, the suspension of gold-backed currencies, and a near-total collapse in global investment and trade. Even after hostilities ended, former trading partners remained bitterly divided, a collection of protectionist states heaping tariffs upon one another. Only after a second world war, followed by the United States and Britain's deliberate reboot of the global economic order at Bretton Woods, did things start to recover. It took sixty years for merchandise exports to regain the levels of 1914.[513] The rapidity of this collapse proves that unlike the three other global forces, it is possible for globalization to come to a fast halt. It is also a sobering reminder that national leaders can, on rare occasions, take their countries to war with trade partners even if it means gutting their own economies in the process.

Besides another world war, at least two things could plausibly weaken or halt the global economic integration of today. The first is obvious: Central governments could decide to abandon proglobalization policies in favor of a return to economic protectionism. A variant of this would be a shift from "globalization" to "regionalization," with separate economic blocs emerging in North America, Europe, and East Asia.[514] Some economists have argued that the 2008–09 global financial crisis will mark the end of an era for twentieth-century globalization and neoliberal policies. It is even conceivable that well-meaning carbon-reduction policies, by penalizing emissions by different amounts in different countries, could trigger tariff wars if countries respond by imposing border taxes to recoup their losses.[515]

A second possibility is the rising cost of oil. Global trade is fueled by cheap energy, and container ships and long-haul cargo trucks cannot readily be electrified like passenger cars as described in Chapter 3. And as environmental damages, too, are increasingly priced into production costs in manufacturing countries like China, the apparent profit margin of a global versus local trade network will narrow.

A deglobalized world with extremely high energy prices might be an oddly familiar one, with local farmers feeding compact walking cities, a return to domestic manufacturing, and airplane travel afforded only by rich elites. One could even imagine a reversal of the urbanization trend as farming returns to being a labor-intensive industry, no longer propped by cheap hydrocarbon for fuel, fertilizers, and pesticides. Overseas tourism would fade, perhaps to be replaced by virtual experiences or even uninterest and disengagement from foreign affairs.

Political genies are even harder to anticipate than permafrost genies. In my mind's eye I imagine an even more integrated world in 2050 than 2010. But no one really knows if our globalization megatrend will accelerate, slow, or reverse over the next forty years. Of the four global forces, this one is the hardest to foresee.

Dragon Swallows Bear

At the smaller, more regional scale, the future of the Russian Far East is similarly murky.

This region is Russia's gateway to eastern Asia. By any measure it is vast, resource-laden, and practically empty of people. It covers some 6.2 million square kilometers, about two-thirds the size of the United States and triple the area of Britain, France, and Germany combined. It is rich in oil and natural gas (especially Sakhalin Island and the Sea of Okhotsk), minerals, fish, timber, and a surprising amount of farmland. It holds one-third of

Russia's landmass but, with just 6.6 million people and falling, less than 5% of its population. Averaging barely one person per square kilometer, the Russian Far East has one of the lowest population densities on Earth.

Except for a tiny 20-kilometer border with North Korea, its main southern neighbor, following a 3,000-kilometer border along the Amur River, is the People's Republic of China. Its three bordering provinces of Heilongjiang, Jilin, and Liaoning hold more than 100 million people. On the Chinese side of the Amur, population densities average fifteen to thirty times higher than on the Russian side. The city of Harbin alone contains more people than the entire Russian Far East.

This stark contrast does not go unnoticed by Russians. They have long feared the "yellow peril," a perception that millions of Chinese are poised to flood across the border and swallow up this region. The fear has fomented an intense xenophobia toward Chinese immigrants, something Russian politicians and media often stoke by asserting that millions are illegally entering the country. One individual even suggested that forty million Chinese would sneak into Russia by the year 2020.[516]

Most migration experts estimate illegal Chinese immigration to be in the hundreds of thousands, not millions. Nor do Russians let their fearmongering get in the way of putting undocumented Chinese migrants to work, for example in the farm fields of the Amur Oblast breadbasket.[517] However, the fact remains that this "yellow peril" fear is deeply ingrained in the Russian psyche, something that is perhaps unsurprising when one considers the history of this region.

Much of what is now the Russian Far East actually belonged to China until 1860. Ethnic Russians began arriving in significant numbers only in the 1930s, after Soviet planners closed the border and set about turning the region into a deeply subsidized supplier of raw materials for the centralized Soviet economy and a protective military fortress to the outside world. The Soviet arms buildup there deeply troubled China, Japan, and South Korea. Tensions with China scraped bottom in the 1960s with a series of border skirmishes, including a bloody clash for Damansky Island on the Ussuri River, in 1969.[518]

Attempts to link the economies of European Russia with Asian Russia never made much sense. The only real transportation link between them was (and is) the Trans-Siberian Railroad, with 9,300 kilometers separating Vladivostok from Moscow. By the 1980s the Soviet Union was ready to abandon the fortress resource colony model for the more sensible idea of opening up the Russian Far East to Asian Pacific trade. Mikhail Gorbachev gave a famous speech in Vladivostok in 1986 that called for the region's deep subsidies from Moscow to be scrapped and Russia's eastern flank opened up. When the Soviet Union collapsed in 1991, those subsidies did indeed go away. So also did much of the military defense spending that supported up to 40% of the jobs in this region. The place descended into deep economic malaise and people began to leave.

At its peak population in 1991, the Russian Far East contained a hair over eight million people. Today its population is 20% smaller and will likely shrink further. More detached than ever from distant European Russia, this region struggles to reconcile its dire and obvious need to glom economically on to China, South Korea, and Japan with its deep xenophobic fear of being swallowed up by China. It is the poorest, least healthy, most economically strapped region in all of Russia. Despite its oil and gas riches, electricity is spotty and expensive. A corrupt bureaucracy and perverse tax system dissuade foreign investment. Its resource-hungry neighbors China, Japan, and South Korea, while more than happy to buy raw materials from the region, hesitate to pour badly needed capital into it. Repeated plans from Moscow to develop and improve the region's quality of life have failed. However, Russia is Heilongjiang's largest trading partner, and as of 2008 the province had concluded more than two thousand collaborative projects there worth about USD $2.9 billion. Trade between China and Russia's Primorsky Territory was over USD $4.1 billion in 2009.[519]

What does the future hold for the Russian Far East? Politically, the relationship between Beijing and Moscow is better than it has ever been; and all of the old border disputes are now settled (Damansky Island is now Zhenbao Island). Even the huge demographic contrast does not predicate a territorial takeover, a political act. But over the long run, given its geographic remoteness and thinning economic ties to the west, the pressures for the Russian Far East to integrate with eastern Asia are obvious. Its 3,000-kilometer border with China is roughly triple its physical distance from Moscow. This region has a huge natural resource base, shrinking labor pool, and dire need for capital investment. Neighboring China has a huge resource demand, bottomless labor pool, and is well on track to become the world's biggest economy in 2050. Somehow, over the long run, these two things must converge.

A NAFTA-like free economic zone in this part of the world seems the most obvious outcome. Indeed, there are plenty of signs that the Russian government strongly desires this direction, for example, through consistently strengthening its ties with the Association of Southeast Asian Nations (ASEAN) trade bloc including regular ASEAN-Russia summits since 2005, and a pending petition for membership in the East Asia Summit. In 2012 Russia will host the Asia-Pacific Economic Cooperation (APEC) summit in Vladivostok. However the far-out possibility of military seizure or outright sale—as Russia did long ago with nearby Alaska—cannot be ruled out. Just as I once learned in school about the U.S. Alaska Purchase of 1867, perhaps one day schoolchildren in Beijing and Moscow will be reading about the Yuandong Purchase of 2044. If either of these things happens, the economic opening of the Russian Far East, spurred by the demand of Asian markets for its abundant natural resources, would not be far behind.

Blue Oil

Demographic models tell us that billions of new people are coming around the hot, dry southern latitudes of our planet, places water-stressed today that will be even more stressed in the future. With a few notable exceptions the water-rich North, in contrast, is expected to become even wetter. Given this obvious mismatch, might northern countries one day sell their water to southern ones?

The idea is not crazy. International bulk water sales have been popping up elsewhere, for example from Lesotho to South Africa and from Turkey to Israel. Indeed, Turkey built a $150-million water export facility at the mouth of the Manavgat River to sell water to regional buyers by tanker.[520] A French company is considering an underground canal to send Rhône River water from France to Spain.

The most ambitious example of all is in China, where a massive, decades-long reengineering of its river networks to shunt water from its wet south to the parched north is now under way. This "South-to-North Water Diversion" megaproject will link together four major drainage basins and build three long canals running through the eastern, central, and western parts of the country. Its costs will include at least USD $62 billion—more than three times the cost of China's Three Gorges Dam—the relocation of three hundred thousand people, and many negative environmental impacts. When finished, the amount of water artificially transferred from south to north each year will total more than half of all water consumption in California.[521]

Might another megaproject emerge to redirect water from north to south, say from Canada to the United States, or from Russia to the dry steppes of central Asia? There are certainly some precedents, and not just the one going on now in China. The last century saw the construction of many major engineering projects in the Soviet Union and North America, including two huge schemes to transfer water from one drainage basin to another: Canada's James Bay Project for hydropower, and California's State Water Project, a massive system of canals, reservoirs, and pumping stations to divert water from the northern to southern ends of the state.

Most audacious of all were two megaprojects designed in the 1960s but never built. Both proposed the massive use of dams, canals, and pumping stations to replumb the hydrology of the North American continent and shunt its water from north to south. They were the North American Water and Power Alliance (NAWAPA), proposed by the Ralph M. Parsons engineering company in Pasadena, California (now Parsons Corporation); and the Great Recycling and Northern Development (GRAND) Canal, proposed by a Canadian engineer named Tom Kierans.

NAWAPA was colossal in scale. It proposed redirecting north-flowing rivers headed to Alaska and northern Canada into the Rocky Mountain Trench—thus forming a giant inland sea—then pumping the water south through connections linking all of the major drainage basins of western North America and the Great Lakes. Flows in the Yukon, Peace, and other distant northern rivers could end up in the Great Lakes, California, or Mexico.

NAWAPA's price tag and ecological damages were immense. Reviled by environmental groups and most Canadians, and with an estimated cost of $100 to $300 billion in 1960s dollars,[522] this grandiose plan did better at attracting media attention than financial backing. But NAWAPA firmly planted the idea of massive north-south water transfers in the minds of generations of engineers and politicians. A half-century later, it continues to inspire revulsion, awe, and smaller spin-off project concepts.

The second immense north-south water scheme of the 1960s, the GRAND Canal, continues to have its advocates today. Its idea is to build a dike across James Bay (the large cove at the southern end of Hudson Bay, see map on p. ix), thus retaining runoff from this lowland's many north-flowing rivers prior to their entering the ocean. The enclosed part of James Bay would become a giant freshwater lake, and its water then pumped back south again toward Lake Huron.

The GRAND Canal plan's inventor, Tom Kierans, now in his nineties, remains its tireless proponent. He points out that the only place the project would deprive of water is Hudson Bay, a brackish sea overwhelmed by jellyfish. Every now and then the plan is resurrected by Canadian politicians.[523] But with a current estimated cost of USD $175 billion—not to mention many environmental impacts and a local climate change over the region[524]— its revival seems distant, at least for now.

Smaller projects in the same area, like the very recently proposed "Northern Waters Complex" concept[525] (see maps on pp. viii–xi), could realistically win support sooner. This particular plan is to impound seasonal floodwater from three north-flowing rivers, temporarily inundating about eleven hundred square kilometers of land before pumping it south again. According to its proponents, the Northern Waters Complex would cost only USD $15 billion, could be finished by 2022, and would generate $2 billion annually in hydropower and perhaps another $20 billion annually in water sales. With economic incentives like these, the great sucking sound from the United States could start sounding better to many Canadians.

Giant water projects cause significant environmental damage and are no longer popular in either the United States or Canada. In fact, the American trend today is to remove dams, not build them. But smaller-scale water exports can happen with pipelines, tanker ships, and bottling plants. The Great Lakes, fronted and shared by both countries, can be replenished at one end and decanted from another, for example at the Chicago Diversion. In his book *The Great Lakes Water Wars,* author Peter Annin describes how

Great Lakes governors and premiers—fearing the specter of long, greedy straws coming at them from the American Southwest—are engaged in a flurry of cooperative lawmaking, hoping to barricade themselves against future water diversions out of the region.

An open question—one feared by many environmentalists—is whether Canada could in fact become obligated to sell bulk water to the United States and Mexico under NAFTA, the North American Free Trade Agreement. Unlike oil, the much more controversial issue of water was deliberately left unaddressed during the writing and ratification of this treaty. Legal scholars point out that if even one province, say Quebec, were to start selling bulk water to the United States, it could establish legal precedent, thus committing Canadian water providers to sell to U.S. or Mexican customers as well as their own. In such a world, North America would grow accustomed to buying not only oil, but also water, from its northernmost country.

Most Canadians oppose the idea of becoming water purveyors to the United States, although their provincial governments are generally more open to the idea. Alongside environmental concerns, Canada suffers water shortages of its own. A water-rich country on paper, most of its uncommitted surplus lies in the far north, flowing over thinly populated permafrost to the Arctic Ocean or Hudson Bay. The south-central prairies are prone to drought, with spotty rainfall and heavy reliance on a few long, oversubscribed rivers fed by distant melting snow and glaciers. If any future megaprojects arise to divert water from northern Canada to the United States, a cut will likely go to southern Canada.

One place where we could well see the resurrection of a massive twentieth-century water-transfer idea by 2050 is in Russia. Siberia's mighty rivers, flowing untouched to the Arctic Ocean, have long been contemplated as a potential water source for the dry steppes and deserts of central Asia. In the 1870s, tsarist engineers noted the favorable, if long, topographic gateway linking wet western Siberia with the Aral-Caspian lowland, in what is now Kazakhstan and Uzbekistan. By the 1940s the Soviet engineer M. M. Davydov had drawn up a grand plan for north-south water transfers out of western Siberia, complete with canals, pumping stations, and the creation of a giant inland lake that would have inundated the same area that is plastered in oil and gas wells today.

From the late 1960s to the early 1980s the USSR studied, revised, and finalized a scaled-down version of Davydov's plan. The idea was to tap Siberia's mighty Ob', Irtysh, and Yenisei rivers using a 2,544-kilometer-long canal to irrigate cotton fields around the Aral Sea (see map on page xii). Diversion of the Aral's feeder rivers was already careening the region toward the desiccated disaster it is today. By 1985 the canal's route had been surveyed and the first work crews arrived in Siberia to commence the "project of the century," known by then as "Sibaral" (short for Siberia to Aral Sea Canal).[526] But then, the new Soviet leader, Mikhail Gorbachev, abruptly halted the project in 1986, citing a need for further study of Sibaral's environmental and economic impacts. Nothing further happened and when the Soviet Union collapsed, the project, after decades of planning, was abandoned.

Today, Sibaral continues to rear up from the grave with surprising regularity. The megaproject is more politically awkward than before because six sovereign countries—Russia, Kazakhstan, Turkmenistan, Uzbekistan, Kyrgyzstan, and Tajikistan—are now involved instead of one. However, all five former Soviet republics want Sibaral to happen and continue to clamor for it.

Support for this in Russia is mixed. In 2002 Moscow's mayor, Yuri Luzhkov, wrote a letter to President Vladimir Putin urging the plan's revival, citing destabilization of Central Asia over water shortages and the specter of refugees pouring across the Russian border. Russia's deputy minister of natural resources also wrote support for the plan.[527] By 2004 Luzhkov was stumping the project in Kazakhstan; and the director of Soyuzvodproject, a government water agency, said they were assembling archived project materials from more than three hundred institutes in order to revisit and revise the old plans. Most Russian scientists are opposed to Sibaral but some note that reducing river runoff to the Arctic Ocean could slightly mitigate the anticipated weakening of the North Atlantic thermohaline circulation described earlier in this chapter.[528] Modeling studies are needed to confirm or disprove this hypothesis, but if correct, Sibaral could conceivably win the support of environmental groups worried more about global climate change than ecological damages in Siberia.

It remains to be seen if China's ongoing South-to-North Water Diversion will rekindle humanity's past passion for massive water projects. Given the enormous obstacles—financial, environmental, and political—I am skeptical that any of these north-to-south water transfer megaprojects will materialize by 2050. But of the ones described here, Sibaral is the most developed. Central Asia is getting very, very dry, and its population is growing. Unlike the North American schemes, something about this project refuses to die. Despite serious likely environmental damages, it really could happen one day.

Regardless of whatever water engineering schemes are or are not undertaken by 2050, one thing remains clear. When it comes to water, the NORCs will be the envy of the world.

CHAPTER 10

The New North

Within hours after the CCGS *Amundsen* docked at Churchill, my life had changed completely. After months of railroads through desolate boreal forest, long empty coastlines, and the cold salt air of Hudson Bay, I sank back into the smog-choked din of my sweating desert megacity. It was familiar but surreal, exhilarating but perturbing, all at once—in short, the typical reaction most Arctic scientists have at summer's end when they migrate from north to south, like overeducated birds, to reintroduce themselves to society.

What makes coming home so jarring, compared to other returns from other exotic places—isn't simply culture shock. It's *human* shock, seeing so many people again after dwelling in a place so empty of them. Even Iowa farmlands seem crowded after one has steamed for days along the Labrador coast or flown hundreds of miles overland, seeing virtually no trace of humans. To experience true northern solitude is both spooky and thrilling, like being time-warped to another planet without us. The question is how many more years things will remain like this.

The number of people wishing to visit, exploit, or simply become informed about the Arctic grows larger every year. The count of prospective students contacting me to pursue graduate degrees has leapt from none to dozens per year. At annual conventions of the American Geophysical Union, research presentations about the Arctic now overflow giant convention halls where before there was a tiny room of lifers talking only to each other. Some ten thousand scientists and fifty thousand participants from sixty-three countries participated in the 2007–2009 International Polar Year.

Research and development spending is rising too. The U.S. National

Science Foundation alone now funnels nearly a half-billion dollars annually toward polar research, more than double what it did in the 1990s. I wish that this trend meant winning a research grant was half as hard, but with so many new young scientists around, they are more competitive than ever. Global investments in the International Polar Year totaled some USD $1.2 billion. NASA and the European Space Agency are developing new satellites to map and comprehend the polar regions as never seen before. NASA's investment alone will likely reach USD $2 billion by the middle of this decade.[529]

Thanks to heavy media coverage, images of drowned polar bears, bewildered Inuit hunters, and satellite maps of shrinking sea ice are now commonplace in people's minds. In a remarkably short span of years these phenomena have changed the world's perception of the Arctic from unconquerable ice fortress, to militarized zone buffering two nuclear superpowers, to frail ecosystem on the verge of collapse (or business bonanza, depending on one's point of view). A place perceived as a maritime graveyard and killer of men even into the 1980s is now perceived as dissolving into a frontier ocean, laden with natural-resource riches for the taking. With so few actual Arctic residents around to protest these frames, all of them have been freely cemented into public consciousness by the words and images of their times.

On the following page, the image on the left reflects the height of the Arctic exploration craze in the late nineteenth and early twentieth centuries. The one on the right is a popular stock photo currently circulating on many climate-change Web sites and blogs. Both portray the same geographic location—the Arctic Ocean—but to very different effect. At left (*Abandonment of the Jeannette*, circa 1894) it is a darkly foreboding place, deadly and impregnable. At right (*The Last Polar Bear*, circa 2009) it is a place of sunny skies, an alluring glass-calm sea, and a magnificent animal doomed to extinction.

Both are stylized, of course. The craggy spires ensnaring the *Jeannette* more closely resemble alpine mountains than sea ice; upon magnification, shadow angles and other subtle details in the photo reveal that the polar bear has almost surely been digitally inserted. Each has its own message it is trying to advance. But stylized or not, it is images like these that powerfully

Frames of the Arctic: (Left) impregnable killer of ships and brave men, ca. 1894; (Right) imperiled ecosystem or business bonanza, depending on one's point of view, ca. 2009.

reflect—and shape—the perceptions of their times. And, as any good advertising executive knows, when it comes to spending money, perception is everything.

In the nineteenth and early twentieth centuries, explorer accounts of glorious adversity shadowed by death persuaded urban donors around the world to loosen their wallets and fund expeditions to the Northwest Passage and North Pole. During World War II and the Cold War, fears of Japanese invasion, atomic bombs, and communist ideology loosened enormous national expenditures of blood and treasure to essentially open up the North for the first time. Today, scientists, through USGS oil and gas assessments and climate model projections, are convincing governments and investors that the region is a place of rising strategic value that is opening for business. And history tells us, when it comes to human decisions about spending money, this growing perception is as equally important—perhaps even *more* important—as the climate changes themselves.

Viewed in this light, disappearing sea ice in the Arctic Ocean is profound—but so also are the decisions of NORC governments to begin military exercises there, to start buying frigates and F-35 fighter jets, or to commit to the long and costly process of filing UNCLOS claims to its seafloor. Thawing permafrost is profound—but so also are the business decisions of private capital to snap up Canada's northernmost railroad and port of Churchill, to buy USD $2.8 billion in Arctic offshore energy leases, and to begin developing specialized LNG tanker ships and platforms for offshore drilling in icy environments.[530]

Environmental groups around the world, horrified at the prospect of an entire ecosystem going extinct, are raising money for, and awareness of, the Arctic. And unlike most other fields of geoscience, when yet another polar ice shelf crumbles the news media actually reports on it. My colleagues and I routinely field reporters' questions on subjects, like soil carbon storage in permafrost, once relegated to the dusty bin of academic obscurity.

All of this publicity has spurred a massive increase in tourism to the area. In 2004 more than 1.2 million passengers traveled to Arctic destinations on cruise ships. Just three years later the number more than doubled; by 2008 there were nearly four hundred cruise ship arrivals in Greenland alone.[531] Many passengers cite the desire to "see the Arctic before it's gone" as motivation for the pricey tickets. And while a liquid Arctic won't arrive anytime soon, the new tourism companies, port-of-call businesses, and other new stakeholders springing up to meet this demand will.

This tide of interest in the Arctic is being spurred by the dramatic climate-change impacts that are happening there. They are recasting the world's perception of what the place is. By transforming its frame from empty fortress to ecological catastrophe, from military theater to business opportunity, climate change is triggering yet another powerful feedback loop in the region, a distinctly human one, that will transform it in very tangible ways. For the region's development, its perceived strategic value, and its ties and economic linkages to the rest of the world, this may prove the most profound feedback of all.

But does a thawing Arctic deserve all of this hype? I myself travel often to this remarkable area to study the torrid pace of climate change there. But as we've seen, climate is but one of four global forces driving this story of change. Furthermore, the Arctic proper (northward of the Arctic Circle, approximately 66°33' N latitude) is actually tiny relative to the outsized attention it enjoys with news media, science funding agencies, commonly used map projections, and the public imagination. Only 4.2% of the planet's surface and 4.6% of its ice-free land (meaning not buried under glacial ice) lies north of this parallel, nearly all of it treeless, deeply frozen in permafrost, and plunged into polar darkness for much of the year. North of the 45° N parallel, however, we find 15% of the planet's surface area and a whopping 29% of its ice-free land.[532] While the Arctic is unique, extreme, and home to remarkable people, it also drags the spotlight away from the vaster NORC areas to the south. With their greater land area, population, biological productivity, and economic clout, it is these much larger regions—together with their Arctic hinterlands—that form the heart of a New North, a place of rising world interest and human activity in the twenty-first century.

By a more generous definition,[533] the "Arctic" contains some twelve million square kilometers, four million people, and an economy of USD $230 billion per year. Those numbers are surprisingly large to most people. However, that entire GDP is less than one-half of the annual U.S.-Canada trade figure alone. Relative to the total NORC geographic area, population, and economy its numbers are dwarfed, even after restricting U.S. participation to the northernmost states.[534] Even using this narrower geographic definition of NORC members, they collectively control some thirty-two million square kilometers, a quarter-billion people, and a $7 trillion economy. Such a bloc, if so measured, would represent the world's fourth-largest economy behind the BRICs (Brazil, India, Russia, and China, $16.4 trillion), the European Union ($14.5 trillion), and the United States in its entirety ($14.3 trillion). Its population would approach that of the entire United States, its land area more than triple that of China. Viewed in this way, the NORCs are an impressive collective (see table on following page).

Unlike the European Union or United States of America the NORCs are not, of course, a formal alliance or free-trade bloc. However, the previous chapters reveal numerous connections among these countries that go well beyond the obvious geographic ones. Nearly two decades after NAFTA the

Gross Domestic Product, Land Area, and Population of NORCs versus Other Major World Economies

	GDP (billions $USD)	GDP world rank	Land area (km²)	Population
USA North[a]	2,693	[6][b]	1,471,053	55,039,000
Russia	2,103	9	17,098,242	140,367,000
Canada	1,287	15	9,984,670	33,890,000
Nordics	1,006	16	3,424,422	25,304,000
total NORCs	**7,089**	**[4]**	**31,978,387**	**254,600,000**
BRICs	16,442	[1]	38,441,883	2,834,617,000
EU	14,520	1	4,324,782	491,583,000
USA (all)	14,260	2	9,826,675	307,641,000
China	8,767	3	9,596,961	1,338,613,000
India	3,548	5	3,287,263	1,156,898,000
Germany	2,812	6	357,022	82,330,000
U.K.	2,165	7	243,610	61,113,000
Brazil	2,024	10	8,459,417	198,739,000

Sources: CIA World Factbook, 2009; Bureau of Economic Analysis, U.S. Department of Commerce.

[a] See note 530.

[b] Note: Brackets [ ] indicate relative world GDP rank for economic blocs not included in the 2009 *CIA World Factbook* rankings.

economic and cultural embrace between Canada and the United States is arguably stronger now than at any other time in their history. It will clench even tighter if oil production from the Alberta Tar Sands (and possibly water exports someday) rises as projected. Despite memberships in the EU, Sweden and Finland feel greater cultural and economic kinship to Iceland and Norway than to Italy or Greece. Since the 1990s, even cantankerous Russia has been forging ties with its NORC neighbors, including participation in the Arctic Council, healthy trade with Finland, the Ilulissat Declaration, an expressed desire to open a shipping lane to Canada's port of Churchill, and an orderly filing of seafloor claims under UNCLOs Article 76. The NORCs collaborate constantly on issues of fisheries, environmental protection, search-and-rescue, and science. They share peaceful, stable borders that count among the friendliest in the world. Aboriginal groups like the Inuit and Sámi both identify and organize across national borders. Among these eight NORC countries, I see many more ties and similarities than say, among the BRICs, or even many countries of the European Union.

The foundations of this New North run far south of the Arctic Ocean, to global immigrant destinations in Toronto and natural gas markets in Western Europe. They are laid by the global forces of demographics, natural resource demand, globalization, and climate change—together with lesser actors like the shipping industry, UNCLOS, and aboriginal land claims agreements. A broad set of "push" and "pull" forces—physical, ecological, and societal—are set in motion. The changes will unfold along the preexisting bones of geography and history, the legacy of past political decisions, the birth rates and migrations of people. They will be constrained by physical realities like the continental effect, sea ice, thawing ground, and an uneven distribution of natural resources. In many ways, this coming-of-age for a new geographic region is just business as usual in the history of the world. But unlike past human expansions it will probably be orderly, bearing little resemblance to the violence and genocides so common in the past.

———

In many ways, the New North is thus well positioned for the coming century even as its unique ecosystem is threatened by the linked pressures of hydrocarbon development and amplified climate change. But in a globally integrated 2050 world of over nine billion people, with mounting megatrends of water stress, heat waves, and coastal flooding, what might this mean for motivating renewed human settlement of the region? Extending the thought experiment further, to what extent might a wet, underpopulated, resource-rich, less bitterly cold North promise refuge from some of the bigger pressures described in the first four chapters of this book?

If Florida coasts become uninsurable and California enters a Perfect Drought, might people consider moving to Minnesota or Alberta? Will Spaniards eye Sweden? Might Russia one day, its population falling and needful of immigrants, decide a smarter alternative to a 2,500-kilometer-long Sibaral canal is to simply invite former Kazakh and Uzbek cotton farmers to abandon their dusty fields and resettle in Siberia, to work in the gas fields?

Such questions demand consideration of what makes civilizations work in the first place. In his book *Collapse,* my UCLA colleague Jared Diamond scours human history to ask the question of why civilizations fail. By studying past collapses like Easter Island and Rwanda, and close calls, like eighteenth-century Japan, he identifies five key dangers that can threaten an existing society. In no particular order, they are self-inflicted environmental and ecosystem damage, loss of trading partners, hostile neighbors, adverse climate change, and how a society chooses to respond to its environmental problems. Any one of these, Diamond argues, will stress an existing settlement. Several or all combined will tilt it toward extinction.

Turning the question around, what causes *new* civilizations to *grow*? My approach suggests that first and foremost will be economic incentive, followed by willing settlers, stable rule of law, viable trading partners, friendly neighbors, and beneficial climate change. No one of these alone is enough to spawn a major new settlement, but several or all combined might tilt one into existence, or encourage existing outposts to grow.

At first blush all eight NORC countries fulfill these requirements to some degree. Save Russia, they rank among the most trade-friendly, economically globalized, law-abiding countries in the world. Whether a boon or a curse,[335] they control a valuable array of coveted natural resources. Already, they enjoy more petitions from prospective migrants than they can or will absorb. Media hype about Arctic scrambles notwithstanding, they are friendly neighbors. Their winters will always be frigid, but less bitterly so than today. Biomass will press north, including some increased agricultural production in contrast to the more uncertain futures facing much larger agricultural areas to the south.

Already the NORCs possess a sprinkle of sizable settlements from which to grow. Their biggest hubs, like Toronto, Montreal, Vancouver, Seattle, Calgary, Edmonton, Minneapolis–St. Paul, Ottawa, Reykjavík, Copenhagen, Oslo, Stockholm, Helsinki, St. Petersburg, and Moscow are growing fast and attract many foreign immigrants today. Smaller destination cities include Anchorage, Winnipeg, Saskatoon, Quebec City, Hamilton,

Göteborg, Trondheim, Oulu, Novosibirsk, Vladivostok, and others. Some truly northern towns that might grow in a New North include Fairbanks, Whitehorse, Yellowknife, Fort McMurray, Iqaluit, Tromsø, Rovaniemi, Murmansk, Surgut, Novy Urengoy, Noyabr'sk, Yakutsk, and others. The ports of Archangel'sk, Churchill, Dudinka, Hammerfest, Kirkenes, Nuuk, Prudhoe Bay, and others are poised to benefit from increased exploration and shipping activity in the Arctic Ocean.

Fueled by West Siberian hydrocarbons, Noyabr'sk and Novy Urengoy—brand-new cities that did not even exist until the early 1980s—are now up to a hundred thousand people apiece. Canada's Fort McMurray is the fat tick of the Alberta Tar Sands, feeding on bitumen and water like Las Vegas feeds on gamblers. Its population boom, closing in on a hundred thousand within the decade, is probably just the beginning. Covering an area roughly the size of Bangladesh, this vast plain of tar-soaked dirt is thought to hold 175 billion barrels of oil, second only to Saudi Arabia and 50% more than Iraq. Despite devastating environmental damages, tar sands development is fast proceeding and by 2040 is projected to produce ten times more oil than Alaska's North Slope does today.

Cities are key to the New North because the NORCs—like everywhere else—are rapidly urbanizing. Even in the remote Arctic and sub-Arctic, people are abandoning small villages or a life in the bush to flock to places like Fairbanks and Fort McMurray and Yakutsk. Tiny Barrow, Alaska—a metropolis by Arctic standards—is absorbing an influx of people from remote hamlets across the North Slope. Paired with reduced winter road access and ground disruptions from thawing permafrost, this urbanization trend suggests abandonment of large tracts of remote continental interiors. These lands will remain wild even as the oceans become busy. It is not unreasonable to suppose that one day people will visit them not to hunt or live on, but as global tourists wishing to see the last great wilderness parks left on Earth.

Ultimately, this question of future population expansion boils down to economic opportunity, demographics, and willing settlers. All of the NORC urban cores offer diverse global economies and attract large numbers of immigrants, offsetting their aging populations and falling domestic fertility rates. However, the Russian Federation faces sharply falling population, low aboriginal birth rates, and a generally hostile attitude toward foreigners. The Nordic countries are growing but slowly, have tiny aboriginal populations, and while generous to foreign immigrants are culturally resistant to the notion of throwing open their doors to millions more. Only Canada and the United States absorb large numbers of immigrants while also having substantial, fast-growing domestic aboriginal populations. Canadian policies favor admitting qualified workers above all else, benefiting her skilled labor force especially in southern cities. Her rising aboriginal population is fueling growth in remote northern towns as well. Canada continues to integrate economically and culturally with the United States, where nearly one hundred million more people will be living by 2050. These powerful trends are but three reasons why I have begun socking away Canada-region mutual funds in my retirement plan. After all, I need to be proactive: With a graying planet, the probability that a comfortable taxpayer-funded pension will be waiting for me is slim.

But outside the cities and towns it's hard to attract new settlers, especially in the NORC countries' Arctic hinterlands. With four million people and a gross domestic product slightly larger than Hong Kong's[536] the circumpolar Arctic holds a bigger population and economy than most people realize, but both are still fleetingly small. For example, with just fifty-seven thousand people and $2 billion GDP per year, Greenland's population and economy are 1% of Denmark's. Furthermore, the mainstay of the Arctic economy is simply exporting raw commodities like metals, fossil fuel, diamonds, fish, and timber. Public services comprise the second-largest sector, followed by transportation. Tourism and retail are significant only in a few places. Universities are rare, and manufacturing extremely limited except for a robust electronics industry in northern Finland around the city of Oulu (Nokia is one of the better-known companies operating there). Thus, unlike the southern NORC cities, the Arctic economy is a restrictive blend of resource-extraction industries and government dollars, with an underskilled and undereducated workforce.

With few exceptions most of these natural resource profits leave the far North, creating an apparent "welfare state" situation in which NORC central governments prefer to deeply subsidize public services rather than surrender these profits to local taxation. Career choices are limited and although salaries are high, so also is the cost of living. One can expect to pay $250 per night for a cheap hotel room and $15 for a cheeseburger in an Arctic town. Gas pipelines and diamond mines generate enormous wealth but most of this revenue flows south (or west, in Russia), controlled by an array of private, multinational, and state-owned actors and central governments. In North America, much of what's left is now controlled by aboriginal-owned business corporations and/or regulated through comprehensive land claims agreements. Northern Transportation Company Limited, Canada's oldest Arctic marine operator, whose vice president, John Marshall, so kindly showed me around the shipyard in Chapter 6, was bought by two such corporations in 1985 and is now 100% aboriginal-owned.[537]

Put simply, the Arctic is not an easy place for fresh arrivals and business start-ups outside of a narrow range of activities. Add to all this the infernal cold and darkness of the polar winter, followed by the steaming heat and billions of mosquitoes of the polar summer, and we see the Arctic is not and will never be a big draw for southern settlers. Even the sub-Arctic boom cities of Fort McMurray, Noyabr'sk, and Novy Urengoy must recruit heavily to attract enough foreign workers. While Arctic settlements will grow with the region's rising energy, mining, and shipping base, its fast-growing aboriginal population (in North America), and the ongoing urbanization trend, it's hard to imagine big new cities spreading across it by 2050 or even 2100.

Instead, a better envisioning of the New North today might be something like America in 1803, just after the Louisiana Purchase from France. It, too, possessed major cities fueled by foreign immigration, with a vast, inhospitable frontier distant from the major urban cores. Its deserts, like Arctic tundra, were harsh, dangerous, and ecologically fragile. It, too, had rich resource endowments of metals and hydrocarbons. It, too, was not really an empty frontier but already occupied by aboriginal peoples who had been living there for millennia.

Like the New North, the American West presented a strong geographic gradient in terms of attractiveness for settlement, varying roughly with longitude instead of latitude. East of the Rocky Mountains, across the Great Plains and into Texas (then part of Mexico), there was sufficient rain to have a go at dryland farming, but not farther west in the harsher landscapes of what are now Arizona, Nevada, Utah, New Mexico, and California. What drew settlers there were gold and silver, culminating in rushes to California in 1849 and to Nevada a decade later. These metal rushes populated the American West just as tar sands and natural gas are doing today in Alberta and West Siberia, and as offshore finds might one day populate port towns along the shores of the Arctic Ocean.

Just as Mexico once ceded what is now all or partly Arizona, California, Colorado, Nevada, New Mexico, Texas, and Wyoming to the United States in the 1848 Treaty of Guadalupe Hidalgo, perhaps one day the Russian Federation will cede its Far East to the People's Republic of China. One shining difference is that we are unlikely to reexperience brutality toward northern aboriginals, unlike the forced displacement and genocide of American Indians throughout the settlement and expansion of the United States. Indeed, in Alaska, northern Canada, and Greenland, aboriginals are poised to lead the way.

Flying over the American West today, one still sees landscapes that are barren and sparsely populated, looking not much different now than they did then. Its towns and cities are relatively few, scattered across miles of empty desert. Yet its population is growing, its cities like Phoenix and Salt Lake and Las Vegas humming economic forces with cultural and political significance. This is how I imagine the coming human expansion in the New North. We're not all going to move there, but it will become integrated into our world in some very important ways.

I imagine the high Arctic, in particular, will be rather like Nevada—a landscape nearly empty but with fast-growing towns fueled by a narrow range of industries. Its prime socioeconomic role in the twenty-first century will not be homestead haven but economic engine, shoveling gas, oil, minerals, and fish into the gaping global maw. These resources will help to supply and grow cities around the world, as described in Chapter 2. Its second important role is innovative social experimenter with aboriginal home rule, through still-evolving power devolutions in northern North America and Greenland. These new societies will inspire other marginalized groups around the world, even as their ecosystems and traditions are decimated by some of the most extreme climate changes on Earth.

Many of the transformations I've presented in this book are negative, and most that are positive exact a toll someplace else. And as painfully demonstrated by the 2008–09 economic contraction, in a globally integrated world, even "winners" suffer pain from the losers. More hydrocarbon development risks not just local damages to northern ecosystems, but global damages through still more greenhouse gases released. For most NORC residents, the downside of milder winters is more rain instead of snow, making them dark, wet, and depressing; while farther north it means conversion of land that is barely livable to land that is hardly livable. The 23.5° tilt in the Earth's axis of rotation commands that there will always be darkness and cold at high latitudes, even if climate warming causes Februaries in Churchill to warm up to Februaries in Minneapolis.

The identified trends have strong inertia, but none are inevitable. The projections of computer models are not edicts, but bent by social choices. Africa's violent cities can be changed. Even the four global forces of demographics, resource demand, globalization, and climate change, being human-generated, must—by definition—lie within human control. And through personal choices, everyone has the ability to shape the perceptions and choices of others. Recent studies, using public data posted on Facebook, have shown that individual actions disseminate unexpected influence over strangers by blazing quickly and deeply through extended social networks. Put simply, a surprising number of one's personal decisions are swayed not deliberately by an advertising billboard, but unintentionally by an unknown friend of a friend of a friend. So each day, by choosing the red pill or blue, we also shape the actions of others. And in turn, the course of history.

To me, the old debates of Malthus and Marx, of Ehrlich and Simon, miss the point. The question is not how many people there are versus barrels of oil remaining, or acres of arable land, or drops of water churning through the hydrologic cycle. The question is not how much resource consumption the global ecosystem can or cannot absorb. It's moot to wonder whether the world should optimally hold nine billion people or nine million, colonize the sea, or all move to Yakutsk. No doubt we humans will survive anything, even if polar bears and Arctic cod do not. Perhaps we could support nine hundred billion if we choose a world with no large animals, pod apartments, genetically engineered algae to eat, and desalinized toilet water to drink. Or perhaps nine hundred million if we choose a wilder planet, generously restocked with the creatures of our design. To me, the more important question is not of capacity, but of desire: *What kind of world do we want?*

NOTES

1 The October 2008 median home price in Los Angeles County, California, was $355,000. *Los Angeles Times,* November 19, 2008.

2 Personal communication with Marsha Branigan, Manager, Wildlife Management Environment and Natural Resources, Inuvik, NWT, December 4, 2007.

3 "Hairy Hybrid: Half Grizzly, Half Polar Bear," *MSNBC World Environment,* May 11, 2006.

4 Of particular relevance to the pizzly story is the recent discovery that transient grizzly bears are now regular visitors to Canada's Arctic Archipelago, and a small but viable population may be establishing itself in or around Melville Island. See J. P. Doupé, J. H. England, M. Furze, D. Paetkau, "Most Northerly Observation of a Grizzly Bear *(Ursus arctos)* in Canada: Photographic and DNA Evidence from Melville Island, Northwest Territories," *Arctic* 60, no. 3 (September 2007): 271–276. The second hybrid animal was shot April 8, 2010, near the Canadian town of Ulukhaktok. Genetic tests confirmed it was the offspring of a polar-grizzly mother and a grizzly father. "Bear shot in N.W.T. was grizzly-polar hyprid," CBC News, April 30, 2010, http://www.cbc.ca/canada/north/story/2010/04/30/nwt-grolar-bear.html?ref-rsss; also "Grizzy-polar bear cross confirmed," *Vancouver Sun,* May 3, 2010; "Tests confirm offspring of hybrid polar-grizzly bear;" CTV News, May 2, 2010.

5 6.1 km/yr average range shift from a quantitative assessment examining historical data for >1,046 species. C. Parmesan, G. Yohe, "A Globally Coherent Fingerprint of Climate Change Impacts across Natural Systems," *Nature* 421 (2003): 37–42. Springtime phenological shifts averaged 4.2 days earlier per decade between 32° and 49° N latitude, and 5.5 days earlier per decade from 50° to 72° N latitude. T. L. Root et al., "Fingerprints of Global Warming on Wild Animals and Plants," *Nature* 42 (2003): 57–60.

6 In February 2010 successive blizzards buried Washington, D.C., and were followed by snowstorms that closed schools from Texas to the Florida Panhandle to the coasts of Georgia and South Carolina, whitening places that hadn't seen snow in a decade or more. Classes were canceled in Florida, Alabama, Georgia, Louisiana, and Mississippi. M. Nelson, "Rare snowflakes start falling from Miss. to Fla.," Associated Press, February 12, 2010, http://www.google.com/hostednews/ap/article/ALeqM5glTiXzNo68z_xAn_fl4DY8L-fpnQD9DQT2Joo. The collection of storms was dubbed "Snowpocalypse" and "Snowmageddon" by pundits, e.g., S. Bezrob, "Covering the Snowpocalypse," FoxNews.com, February 10, 2010, http://liveshots.blogs.foxnews.com/2010/02/10/covering-the-snowpocalypse/?test=latestnews. Meanwhile, snow sport events at the Vancouver Winter Olympics were mired in rain, e.g., S. Almasy, "4,000 to miss out on snowboard cross because of rain," CNN.com, February 15, 2010, http://www.cnn.com/2010/SPORT/02/15/snowcross.refund/?hpt=T3.

7 This is an actual supply chain. For an in-depth examination of the globalization of the tomato, see Bill Pritchard, David Burch, *Agri-Food Globalization in Perspective: International Restructuring in the Processing Tomato Industry* (Burlington, Vt.: Ashgate Publishing, 2003), 308 pp.

8   G. A. Strobel et al., "The Production of Myco-diesel Hydrocarbons and Their Deriva-tives by the Endophytic Fungus *Gliocladium roseum*," *Microbiology* 154 (2008): 3319–3328, DOI:10.1099/mic.0.2008/022186-0.

9   S. Pinker, "A History of Violence," *The New Republic* 236 (March 19, 2007): 18–21; D. Jones, "Human Behaviour: Killer Instincts," *Nature* 451, no. 7178 (2008): 512–515.

10 To name just two examples, economic growth models seldom consider political changes to immigration policy; climate model projections depend strongly on their assumptions about cloud physics.

11 "The Fox knows many things, but the Hedgehog knows one big thing." This phenomenon has been statistically studied by Philip Tetlock at UC Berkeley, who discovered predictions made by economic and political pundits often fare little better than flipping a coin. But by casting a wide net for subject matter, the probability that an important factor will be missed is reduced. P. E. Tetlock, *Expert Political Judgment: How Good Is It? How Can We Know?* (Princeton, N.J.: Princeton University Press, 2006), 352 pp.

12 The following global population estimates are taken from the U.S. Census Bureau Inter-national Data Base (updated June 18, 2008), http://www.census.gov/ipc/www/idb/world pop.html (accessed September 26, 2008).

13 We will return to Thomas Malthus and his 1798 *An Essay on the Principle of Population* in Chapter 3.

14 Paul R. Ehrlich, *The Population Bomb* (New York: Ballantine Books, 1968).

15 The term *death rate* usually refers to the crude death rate, measured as the number of deaths per thousand people in a population. There are different measures of population fertility; this book uses the total fertility rate (TFR), which is the average number of children for a woman within that population. Because it is a statistical average, it is pos-sible to have noninteger values of TFR, for example 1.7 children per woman, a real-world impossibility. I also use the term *birth rate* to refer to TFR, not to be confused with crude birth rate, the raw number of births per thousand people. For a good introduction to population demography, including its definitions, the demographic balancing equation, and data collection issues, see J. A. McFalls Jr., "Population: A Lively Introduction," 5th ed., *Population Bulletin* 62, no. 1 (March 2007).

16 W. Thompson, "Population," American *Journal of Sociology* 34 (1929): 959-975. See also M. L. Bacci, *A Concise History of World population*, 4th ed. (Wiley-Blackwell), 296 pp.

17 For a good discussion of how the Demographic Transition unfolded differently in devel-oping countries than it did in Europe and North America, see the unparalleled book by J. E. Cohen, *How Many People Can the World Support?* (New York and London: W. W. Norton, 1995), 532 pp.

18 The Organisation for Economic Co-Operation and Development (OECD), a group of thirty developed and emerging-market countries with high global integration. Throughout this book I use *OECD* or *developed* to refer to this cohort rather than the term *first-world*. Today's OECD originated in the post–World War II Marshall Plan as the Organization of European Economic Cooperation, which later expanded to include non-European coun-tries. OECD members as of April 2010 were Australia, Austria, Belgium, Canada, Czech Republic, Denmark, Finland, France, Germany, Greece, Hungary, Iceland, Ireland, Italy, Japan, Korea, Luxembourg, Mexico, the Netherlands, New Zealand, Norway, Poland, Portugal, Slovak Republic, Spain, Sweden, Switzerland, Turkey, the United Kingdom, and the United States.

19 83%, computed from Human Influence Index (HII) grids, NASA Socioeconomic Data and Applications Center (SEDAC), http://sedac.ciesin.columbia.edu/wildareas/ (accessed October 8, 2008).

20 The following historical data on U.S. energy consumption taken from Appendix F, EIA (Energy Information Administration) *Annual Energy Review 2001*, U.S Department of Energy, http://tonto.eia.doe.gov/FTPROOT/multifuel/038401.pdf (accessed October 9, 2008).

21 The following numbers are calculated from British thermal unit (Btu) data. One Btu is the amount of energy required to raise the temperature of one pound of water by one degree Fahrenheit. One barrel of crude oil = 5,800,000 Btu, one short ton of coal = 20,754,000 Btu, one cubic foot of natural gas = 1,031 Btu, one cord wood=20,000,000 Btu.

22 Coal increased from 6,841 to 22,580 trillion Btu/year. Appendix F, EIA *Annual Energy Review,* 2001.

23 Oil increased from 229 to 38,404 trillion Btu/year. Ibid.

24 Wood-fuel increased from 2,015 to 2,257 trillion Btu/year. Ibid.

25 Jared Diamond, "What's Your Consumption Factor?" *The New York Times,* January 2, 2008.

26 For a brief introduction to globalization see Manfred Steger's *Globalization: A Very Short Introduction* (Oxford: Oxford University Press, 2003). See also *Global Transformations* by David Held et al., eds. (Palo Alto: Stanford University Press, 1999); *Runaway World* by Anthony Giddens (New York: Routledge, 2000); *Why Globalization Works* by Martin Wolf (New Haven: Yale University Press, 2004); *Globalization and the Race for Resources* by Steven Bunker and Paul Ciccantell (Baltimore: The Johns Hopkins University Press, 2005); *Hegemony: The New Shape of Global Power* by John A. Agnew (Philadelphia: Temple University Press, 2005); *In Defense of Globalization* by Jagdish Bhagwati (Oxford: Oxford University Press, 2007); *The Power of Place: Geography, Destiny, and Globalization's Rough Landscape* by Harm de Blij (USA: Oxford University Press, 2008); *Social Economy of the Metropolis: Cognitive-Cultural Capitalism and the Global Resurgence of Cities* by Allen J. Scott (Oxford: Oxford University Press, 2009); and *Globalization and Sovereignty* by John A. Agnew (Lanham, Md., and Plymouth, UK: Rowman & Littlefield Publishers, Inc., 2009).

27 T. L. Friedman, *The World Is Flat* (Gordonsville, Va.: Farrar, Straus & Giroux, 2005).

28 From "Store Openings," http://franchisor.ikea.com/ (accessed November 13, 2009).

29 P. 38, Steger, *Globalization: A Very Short Introduction* (Oxford: Oxford University Press, 2003).

30 For more on how the United States exported its business model to the world, see J. A. Agnew, *Hegemony: The New Shape of Global Power* (Philadelphia: Temple University Press, 2005).

31 The Washington Consensus is attributed to John Williamson of the Peterson Institute for International Economics, a think tank in Washington, D.C. (www.iie.com). Its policies have now been adopted by (or forced onto, depending on one's point of view) many developing countries. Neoliberals praise these reforms, citing new markets and jobs for struggling people. Critics point to two-dollar-a-day wages while multinational corporations grow rich. The Washington Consensus and similar policies remain highly controversial. If you have any antiglobalization friends, mention it to them sometime and watch their mouths foam.

32 "Expanding trade and investment has been one of the highest priorities of my administration. . . . When I took office, America had free trade agreements in force with only three nations. Today, we have agreements in force with fourteen." From November 22, 2008, speech in Lima, Peru, by outgoing U.S. president George W. Bush to the Asia-Pacific Economic Cooperation forum, his final summit gathering as president. See transcript http://www.whitehouse.gov/news/releases/2008/11/20081122-7.html, Office of the Press Secretary (accessed November 23, 2008). See also "At Summit, Bush Touts Free-Trade Record," www.cnn.com, November 22, 2008; and "Bush Wraps Up Asia Economic Meeting," *The New York Times,* November 23, 2008.

33 Some economists speculated the 2008–09 global financial crisis might tilt the world back toward tariffs and protectionism. This notion was rebuffed at a September 2009 G-20 summit in Pittsburgh, billed as a sort of "Bretton Woods II," which was toothless on banking regulations but strongly reaffirmed a common goal of continued free trade expansion in the developing world.

34 The most important greenhouse gas is water vapor, but unlike carbon dioxide its residence time in the atmosphere is extremely short. Without the greenhouse effect, global

temperatures would average about 0°F (–18°C) versus 59°F (15°C) today. Some details of this section drawn from Tim Hall's chapter on climate drivers, in G. Schmidt and J. Wolfe, *Climate Change: Picturing the Science* (New York: W. W. Norton & Co., 2009), 320 pp. See also R. Henson, *The Rough Guide to Climate Change* (London: Penguin Books Ltd., 2008), 374 pp. Both books provide very accessible introductions to the physics of climate and climate change.

35 The analogy to a closed car or glass greenhouse is imperfect because air circulation is not trapped in a moving atmosphere, but it's close enough for our purposes here.

36 Svante Arrhenius, "On the Influence of Carbonic Acid in the Air upon the Temperature of the Ground," *Philosophical Magazine and Journal of Science,* 5th Series 41 (April 1896): 237–276.

37 For more about Arrhenius and other early research on the greenhouse effect, see R. Henson, *The Rough Guide to Climate Change* (London: Penguin Books Ltd., 2008).

38 From global weather station data, the average hundred-year linear trend from 1906 to 2005 is +0.74°C (with error bars, between +0.56°C and +0.92°C). From air bubbles trapped in ice cores, we know atmospheric CO_2 concentrations averaged ~280 ppm in the pre-industrial era (before ~1750 A.D.) versus ~387 ppm in 2009. The first continuous direct sampling of CO_2 concentration was begun by Charles "Dave" Keeling at Mauna Loa Observatory in 1958 and continued by his son Ralph Keeling. Carbon dioxide levels have risen consistently every year from ~315 ppm in 1958 to ~387 ppm in 2009. For the latest data, see http://scrippsco2.ucsd.edu/. The 2007 IPCC SRES B1, A1T, B2, A1B, A2, and A1FI illustrative marker scenarios are about 600, 700, 800, 850, 1,250, and 1,550 ppm, by century's end respectively, with different scenarios reflecting different assumptions about controlling carbon emissions. Such numbers are two to five times preindustrial levels. *IPCC AR4 Synthesis Report,* Table 3.1. (Full reference *IPCC Fourth Assessment Report [AR4], Climate Change 2007: Synthesis Report,* Contribution of Working Groups I, II, and III to the Fourth Assessment Report of the Intergovernmental Panel on Climate Change, Core Writing Team, R. K. Pachauri, A. Reisinger (eds.), IPCC, Geneva, Switzerland: 104 pp.) available at http://www.ipcc.ch/pdf/assessment-report/ar4/syr/ar4_syr.pdf.

39 J. O'Neill, S. Lawson, "Things Are Heating Up: Economic Issues and Opportunities from Global Warming," *CEO Confidential,* Issue 2007-01, Goldman Sachs, February 8, 2007; J. Lash, F. Wellington, "Competitive Advantage on a Warming Planet," *Harvard Business Review,* March 2007.

40 USCAP Press Release, "Joint Statement of the United States Climate Action Partnership," January 19, 2007, www.us-cap.org/media/release_USCAPStatement011907.pdf (accessed November 20, 2008).

41 From www.us-cap.org/about/index.asp (accessed November 23, 2008). The Web page later showed the withdrawal of several members.

42 Atmospheric CO_2 variations have both natural cycles—which fall and rise with ice ages and warm interglacial periods—and anthropogenic sources, which are also substantial but rise much faster. Our current anthropogenic boost is perched on top of an already large natural interglacial peak, thus taking the atmosphere to levels not seen since the Miocene. Over the past 800,000 years of multiple ice age/warm interglacial cycles, including the current interglacial of the past ~12,000 years, preindustrial atmospheric CO_2 levels cycled within a range of ~172 (ice age) to 300 (interglacial) parts per million by volume (ppmv). Human activity has now boosted that to ~385 ppmv and we are projected to reach at least 450 ppmv and perhaps as much as 1,550 ppmv by the end of this century. See ice-core record, D. Lüthi et al., "High-Resolution Carbon Dioxide Concentration Record 650,000–800,000 Years before Present," *Nature* 453 (2008): 379–382, DOI:10.1038/nature06949; also Urs Siegenthaler et al., "Stable Carbon Cycle–Climate Relationship during the Late Pleistocene," *Science* 310, no. 131 (November 2005), DOI:10.1126/science.1120130, and others.

43 Much older Miocene PCO_2 now estimated from boron/calcium ratios in ocean core foraminifera, A. K. Tripati, C. D. Roberts, R. A. Eagle, "Coupling of CO_2 and Ice Sheet Stability over Major Climate Transitions of the Last 20 Million Years," *Science* 326, no. 5958 (December 4, 2009): 1394–1397, DOI:10.1126/science.1178296.

44 These events reconstructed from the victim's interview on Fox News ("Black Friday Tragedy," January 23, 2009); *Newsday* ("Trampled Wal-Mart Worker Had Helped Pregnant Woman," January 24, 2009); and materials provided by the Nassau County Police Department, courtesy Detective Anthony Repalone, January 8, 2009.

45 Population growth, commerce, and trade are not, of course, the only factors driving urban economic growth. For the past ten to twenty years, foreign direct investment has been at least as important. Effective governance and infrastructure are also critical. We will come to these later in the chapter. For more on how the level of urbanization is not always "coupled" to economic growth, see D. E. Bloom, D. Canning, G. Fink, "Urbanization and the Wealth of Nations," *Science* 319 (2008).

46 Even slum cities in our poorest countries usually offer better economic opportunities than do surrounding rural areas, although the job sector is informal and quality of life low. Global employment in services now averages 40% of total employment, versus 39% in agriculture. In developed countries and the European Union, service-sector jobs capture a whopping 73% of all employment. In contrast, they capture just 28% in sub-Saharan Africa. P. 330 and Table 11.2, P. Knox et al., *The Geography of the World Economy*, 5th ed. (London: Hodder Education, 2008), 464 pp.

47 Governments around the world are doing their part to help encourage all this. A new survey of 245 of the world's fastest-growing cities found them building transportation systems, designating "special economic zones," and streamlining their banking and financial systems. *State of the World's Cities 2008/2009*, United Nations Human Settlements Programme (UN-HABITAT) (UK and USA: Earthscan, 2008).

48 *World Urbanization Prospects: The 2007 Revision*, United Nations, Department of Economic and Social Affairs, Population Division, 2008.

49 *State of the World's Cities 2008/2009*, UN-HABITAT, 2008.

50 Press Conference, United Nations Department of Public Information, News and Media Division, New York, February 26, 2008.

51 UN-HABITAT Press Release, SOWC/08/PR2, 2008.

52 Table I.7, *World Urbanization Prospects: The 2007 Revision*, United Nations, Department of Economic and Social Affairs, Population Division, 2008.

53 66.2% urban in 2050 versus 40.8% urban in 2007; whereas Europe was 72.2% urban in 2007 and is projected to be 76.2% urban in 2050. Table I.5, *World Urbanization Prospects: The 2007 Revision*, United Nations, Department of Economic and Social Affairs, Population Division, 2008.

54 The 40% figure is relative to the year 2007. UN model projections for 2050 (medium variant) are population of the world 9.191 billion, Africa 1.998 billion, China 1.409 billion, India 1.658 billion, Europe 0.664 billion, South America 0.516 billion, North America 0.445 billion. These and most other population projections from *World Population Prospects: The 2006 Revision Population Database*, United Nations, Department of Economic and Social Affairs, Population Division, viewed January 30, 2009.

55 UN-HABITAT, 2008.

56 Hong Kong is ranked first. This index was created by the Heritage Foundation and *Wall Street Journal* and ranks the world's countries using ten descriptors ranging from free trade to corruption. Singapore received 87 out of 100 possible points in 2009; the United States received 80 points out of 100, ranking it sixth behind Hong Kong, Singapore, Australia, Ireland, and New Zealand. Nigeria received only 55 points, ranking it #117 out of 179 countries evaluated. Data from www.heritage.org/index, viewed January 28, 2009.

57 Government of Singapore Investment Corporation and Temasek Holdings, V. Shih, "Tools of Survival: Sovereign Wealth Funds in Singapore and China," *Geopolitics* 14, no. 2 (2009): 328–344; also http://www.temasekholdings.com.sg/media_centre_faq.htm (accessed November 16, 2009).

58 Mass transit is so efficient and appealing in Singapore that it has far fewer cars per capita than other comparable cities. Only 5% of Singapore's energy consumption goes into transportation, unlike Berlin (35%), London (26%), New York (36%), Tokyo (38%), Bologna (28%), Mexico City (53%), or Buenos Aires (49%). Figure 3.4.3, and 3.4.4 UN-HABITAT, 2008, p. 160.

59 Allen J. Scott, *Technopolis: High-Technology Industry and Regional Development in Southern California* (Berkeley: University of California Press, 1994), 322 pp.

60 H. Ghesquiere, *Singapore's Success: Engineering Economic Growth* (Singapore: Thomson Learning, 2007).

61 M. Gandy, "Planning, Anti-planning, and the Infrastructure Crisis Facing Metropolitan Lagos," *Urban Studies* 43, no. 2 (2006): 371–396.

62 E. Alemika, I. Chukwuma, "Criminal Victimization and Fear of Crime in Lagos Metropolis, Nigeria," CLEEN Foundation Monograph Series, no. 1, 2005.

63 J. Harnishfeger, "The Bakassi Boys: Fighting Crime in Nigeria," *Journal of Modern African Studies* 41, no. 1 (2003): 23–49.

64 "The State of Human Rights in Nigeria, 2005–2006," National Human Rights Commission, Nigeria, 2006, http://web.ng.undp.org/publications/governance/STATE_OF_HUMAN_RIGHTS_REPORT_IN_NIGERIA.pdf (accessed March 31, 2010). Note: The events described in this document were not independently verified.

65 P. 1, *Global Trends 2025: A Transformed World* (Washington, D.C.: U.S. National Intelligence Council, 2008), 99 pp.

66 "Dreaming with BRICs: The Path to 2050," Global Economics Paper no 99, Goldman Sachs (2003), 24 pp. Other, more recent model studies yield comparable results.

67 E.g., from the global accounting giant PricewaterhouseCoopers, "The World in 2050: How Big Will the Major Emerging Market Economies Get and How Can the OECD Compete?" J. Hawksworth, Head of Macroeconomics, PWC (2006), 46 pp.; and the Japan Center for Economic Research, "Long-term Forecast of Global Economy and Population 2006–2050: Demographic change and the Asian Economy," JCER (March 2007), 51 pp., and others.

68 These data are from the above econometric model study of the BRICs by Goldman Sachs. All figures in inflation-adjusted 2003 U.S. dollars, for years 2003 and 2050, Appendix II, Global Paper no. 99, Goldman Sachs (2003). Rather than simply extrapolating current growth rates, the model prescribes a set of clear assumptions capturing how growth and development work. Some of these—like continued financial and institutional stability, openness to trade, and education, for example—could certainly change with the choices of future political leaders. The extent to which the 2008–09 global economic collapse might delay these particular projections is unclear, but as of April 2010 these developing economies were recovering sharply (see next).

69 From 2007 to 2009, GDP grew 2.17%, 8.76%, 6.35% annually for Brazil, India, and China, respectively, and shrank -1.6%, -2.08%, and -3.07% in the U.S., Germany, and Japan. The revised Carnegie 2050 GDP projections are also from this study. U. Dadush and B. Stancil, "The G20 in 2050," *International Economic Bulletin*, November 2009, http://www.carnegieendowment.org/publications/index.cfm?fa=view&id=24195 (accessed November 26, 2009).

70 "Brazil Takes Off," *The Economist* 339, no. 8657 (November 12, 2009): 15.

71 Dadush and Stancil (2009).

72 The Goldman Sachs study projects Russia's per capita income to rise to around USD $50,000 by 2050 (all figures in inflation-adjusted 2003 U.S. dollars).

73   India's per capita income in 2010 was less than USD $1,000; it is projected to rise to around USD $17,000 by 2050 (all figures in inflation-adjusted 2003 U.S. dollars).

74   P. 99, *Global Trends 2025: A Transformed World* (Washington, D.C.: U.S. National Intelligence Council, 2008), 99 pp.

75   William A. V. Clark, *The California Cauldron: Immigration and the Fortunes of Local Communities* (New York: The Guilford Press, 1998), 224 pp.

76   See note 15.

77   For a good example of how population momentum is playing out in Asia, see S. B. Westley, "A Snapshot of Populations in Asia," *Asia-Pacific Population & Policy* 59 (2002).

78   0.55% per year in 2050, a global population doubling time of about 130 years, versus 1.02% in 2007, a doubling time of about 70 years. Data projections from *World Population Prospects: The 2006 Revision Population Database,* United Nations Population Division, viewed January 29, 2009.

79   L. Hayflick, "The Future of Ageing," *Nature* 408 (2000): 267–269.

80   One-half the population is older than the median age, and one-half is younger. All age data from *World Population Prospects: The 2006 Revision Population Database*, United Nations Population Division, viewed January 29, 2009.

81   Ibid.

82   In our least-developed countries this is also exacerbated by low life expectancy owing to poor health care, poor nutrition, and violence.

83   For example in Germany, a "rationing" of geriatric health care services is envisioned by 2050. R. Osterkamp, "Bevölkerungsentwicklung in Deutschland bis 2050 Demografische und ökonomische Konsequenzen für die Alterschirurgie," *Der Chirurg* 76, no. 1 (2005).

84   There is also the "youth dependency ratio," defined as the number of individuals aged zero to fourteen divided by the number of individuals aged fifteen to sixty-four, and the "total dependency ratio," defined as the sum of the youth dependency ratio and the elderly dependency ratio. The basic assumption behind these numeric ranges is that children under fifteen are in school and adults over sixty-four stop working, so both age groups are dependent, either upon working-age family members or upon state entitlement programs.

85   R. Hutchens, K. L. Papps, "Developments in Phased Retirement," in R. L. Clark, O. S. Mitchell, eds., *Reinventing the Development Paradigm* (New York: Oxford University Press, 2005).

86   E. Calvo, K. Haverstick, S. A. Sass, "Gradual Retirement, Sense of Control, and Retirees' Happiness," *Research on Aging* 31, no. 1 (2009).

87   "Japan's Pensioners Embark on 'Grey Crime' Wave," *The Independent,* April 13, 2006; "Report: More Elderly Japanese Turn to Petty Crime," CNN Asia, December 24, 2008.

88   See note 79.

89   P. 22, *Global Trends 2025: A Transformed World* (Washington, D.C.: U.S. National Intelligence Council, 2008), 99 pp.

90   "The People Crunch," *The Economist* 390, no. 8614 (January 13, 2009).

91   Direct material imports of 3.2 metric tons of fossil fuel, between 8 and 9 tons of renewable raw materials including water, and between 11 and 15 tons of ores and minerals. These estimates were calculated at the country level, but Sweden is 85% urban. V. Palm, K. Jonsson, "Materials Flow Accounting in Sweden Material Use for National Consumption and for Export," *Journal of Industrial Ecology* 7, no. 1, (2003): 81–92.

92   This materials accounting was monitored in 2004 for one full year. S. Niza, L. Rosado, "Methodological Advances in Urban Material Flow Accounting: The Lisbon Case Study," presented at ConAccount 2008, *Urban Metabolism, Measuring the Ecological City*, Prague, September 11–12, 2008.

93   For unknown reasons this link between urban growth and natural resource supply has been historically ignored in urbanization research. Of particular importance to China's

cities are cement, steel, aluminum, and coal. L. Shen, S. Cheng, A. J. Gunson, H. Wan, "Urbanization, Sustainability and the Utilization of Energy and Mineral Resources in China," *Cities* 22, no. 4 (2005): 287–302.

94   Both factors have contributed heavily to the export economies of newly industrializing countries. Often heavy industries have expanded even faster than consumer manufacturing. Such countries are exporting not only T-shirts and computer components but also steel, machinery, and chemicals.

95   Malthus' book was, in fact, hugely influential on the young Charles Darwin, helping him to arrive at his theory of Natural Selection some six decades later. The full title of the first edition, which Malthus published anonymously in 1798, was *An Essay on the Principle of Population as it Affects the Future Improvement of Society, with Remarks on the Speculations of Mr. Godwin, M. Condorcet, and other Writers* (London: printed for J. Johnson, in St. Paul's Church-Yard). Later versions appeared under his own name. This landmark book is still in print and remains controversial to this day.

96   Ehrlich wrote *The Population Bomb* (New York: Ballantine Books, 1968), discussed in Chapter 1, and a number of other books. The late Julian Simon rebuts Ehrlich in *The Ultimate Resource* (Princeton: Princeton University Press, 1981) and others, arguing that the only limit to human growth is human ingenuity.

97   This expansion of Malthus' ideas beyond issues of food production began in the 1800s, including by British economist David Ricardo, who discussed mineral deposits, and W. Stanley Jevons, who, in 1865, predicted that limits to coal reserves would ultimately halt the country's economic growth. Within a century Jevon's predictions of "peak coal" proved correct.

98   Data sources for the World Reserves table are the *BP Statistical Review of World Energy June 2008*, 45 pp., www.bp.com/statisticalreview (accessed February 12, 2009) (oil, gas, coal through 2007); and *World Metals & Minerals Review 2005* (London: British Geological Survey and Metal Bulletin, 2005), 312 pp. (through 2003). Natural gas is converted to LNG (1 metric ton liquefied natural gas = 48,700 cubic feet). "Titanium" is TiO_2. Platinum group includes platinum, palladium, rhodium, iridium, osmium, and ruthenium. Assumed human population is 6,830,000,000 (2010 estimate, United Nations).

99   A single cubic kilometer of average crustal rock contains 200,000,000 metric tons of aluminum, 100,000,000 metric tons of iron, 800,000 metric tons of zinc, and 200,000 metric tons of copper, so mineral exhaustion in the molecular sense is meaningless. D. W. Brooks, P. W. Andrews, "Mineral Resources, Economic Growth, and World Population," *Science* 185 (1974): 13–10.

100   For more on this discussion of mineral exhaustion and the perils of a fixed-stock approach to resource assessment, see John E. Tilton, *On Borrowed Time? Assessing the Threat of Mineral Depletion* (Washington, D.C.: RFF Press, 2002), 160 pp.

101   Matthew R. Simmons, *Twilight in the Desert: The Coming Saudi Oil Shock and the World Economy* (Hoboken, N.J.: John Wiley & Sons, 2005), 428 pp.

102   A very detailed analysis comes from the National Institute for Materials Science in Tsukuba, Japan. The authors use the Goldman Sachs BRICs and G6 economic projections discussed in Chapter 2 to project future demand for twenty-two metals. K. Halada, M. Shimada, K. Ijima, "Forecasting of the Consumption of Metals up to 2050," *Materials Transactions* 49, no. 3 (2008): 402–410.

103   J. B. Legarth, "Sustainable Metal Resource Management—the Need for Industrial Development: Efficiency Improvement Demands on Metal Resource Management to Enable a Sustainable Supply until 2050," *Journal of Cleaner Production* 4, no. 2 (1996): 97–104; see also C. M. Backman, "Global Supply and Demand of Metals in the Future," *Journal of Toxicology and Environmental Health, Part A,* 71 (2008): 1244–1254.

104   Unconventional oil is much more difficult to extract and includes materials that are often excavated, like oil shales and tar sands, and high-viscosity oils.

105 Based on their analysis of eight hundred oil fields, including all fifty-four "supergiants" containing five billion or more barrels, the International Energy Agency estimates the world average production-weighted decline rate is currently about 6.7% for fields that have passed their production peak, rising to 8.6% decline by 2030. *World Energy Outlook 2008*, OECD/IEA, 578 pp.

106 U.S. Crude Oil Field Production data, U.S. Energy Information Administration, http://tonto.eia.doe.gov/dnav/pet/hist/LeafHandler.ashx?n=pet&s=mcrfpus1&f=a (accessed March 31, 2010).

107 This paragraph drawn from remarks by James Schlesinger, p. 31, summary of the National Academies Summit on America's Energy Future, Washington, D.C., 2008.

108 This is not to suggest that these areas aren't or won't be developed. Turkmenistan, one of the last and most recent countries in the Caspian Sea region to be opened to foreign hydrocarbon development, had no fewer than fifteen petroleum companies seeking to launch activities in 2009, including China National Oil Corporation, Gazprom, Lukoil-ConocoPhillips, Midland Consortium, and Schlumberger, an oil field services company. *Turkmenistan's Crude Awakening: Oil, Gas and Environment in the South Caspian* (Alexandria, Va.: Crude Accountability, 2009), 87 pp.

109 Drawn from remarks by former U.S. secretaries of energy James Schlesinger and Samuel Bodman to the National Academies Summit on America's Energy Future, Washington, D.C., 2008.

110 This model projection by the International Energy Agency was revised downward from earlier forecasts to account for the 2008 global economic slowdown. It assumes that oil prices will average $100 per barrel during 2008–2015, then steadily rise to $120 by 2030. *World Energy Outlook 2008*, OECD/IEA (2008), 578 pp.

111 D. Goodstein, *Out of Gas: The End of the Age of Oil* (New York: W. W. Norton & Company, 2005), 148 pp.; M. Klare, *Resource Wars: The New Landscape of Global Conflict* (New York: Holt Paperbacks, 2002), 304 pp.; and *Rising Powers, Shrinking Planet: The New Geopolitics of Energy,* reprint ed. (New York: Holt Paperbacks, 2009), 352 pp.; M. Simmons, *Twilight in the Desert: The Coming Saudi Oil Shock and the World Economy* (Somerset, N.J.: John Wiley & Sons, 2005), 428 pp.

112 On average, postpeak oil field decline rates are 3.4% for supergiant fields, 6.5% for giant fields, and 10.4% for large fields, *World Energy Outlook 2008*, OECD/IEA (2008), 578 pp.

113 A successful Al Qaeda attack on the Abqaia facilities would have shocked world oil markets, as it handles two-thirds of the Saudi Arabian oil supply. National Academies Summit on America's Energy Future, Washington, D.C., 2008, p. 9.

114 There are major obstacles to a rapid transition to hydrogen fuel-cell cars, as will be described shortly.

115 Specifically from ozone and particulates. M. Jerrett et al., "Long-Term Ozone Exposure and Mortality," *New England Journal of Medicine* 360 (2009): 1085–1095.

116 Only if the electricity supplying the grid comes from clean, renewable sources does the plug-in automotive fleet become pollution- and carbon-free. But depending on the efficiency of the coal- or gas-fired power plant, and how many miles the electricity travels over high-voltage lines, the net balance of this trade-off still generally comes down on the side of plug-in electrics. Also, it is more feasible to recapture pollution and greenhouse gases from hundreds of power station smokestacks than from millions of car tailpipes, particularly with regard to carbon capture and storage (CCS) schemes.

117 Hydrogen is highly reactive and thus quickly combines with other elements, for example with oxygen to make water (H_2O).

118 Nearly all electric utility power is made using some outside source of energy to turn a mechanically rotating turbine, to spin a tightly wound coil of copper wire inside of a fixed magnetic field. This produces a flow of electrons in the copper wire that we call

electricity. Windmills, hydroelectric dams, coal-fired power plants, and nuclear power plants all use variants of this basic idea to make electricity, the main difference between them being the source of energy used to spin the turbine. For example, heat generated by burning coal or from a controlled nuclear reaction can be used to boil water, producing pressurized steam, which passes over a turbine. Building a dam across a river creates an artificial waterfall, allowing the weight of water to fall upon turbines, and so on.

119　In hydrolysis, electricity is used to split water molecules into pure hydrogen and oxygen. It is a common way to obtain pure hydrogen.

120　In terms of radiative physics, tropospheric water vapor is an even more potent greenhouse gas than carbon dioxide. However, owing to its short residence time in the atmosphere—on average just eleven days—it does not linger long before returning to the Earth's surface. In contrast, carbon dioxide can persist in the atmosphere for centuries, so its concentration steadily accumulates over time.

121　*Energy Technology Perspectives—Scenarios and Strategies to 2050* (OECD/International Energy Agency, 2006), 483 pp.

122　Ethanol is more corrosive than gasoline, so engines running on 100% ethanol require specially resistant plastic and rubber components and hardened valve seats. It also has lower energy content than gasoline, so can yield lower mileage results relative to gasoline. However, owing to its high octane of 115, ethanol can be used as an octane enhancer in gasoline instead of groundwater-polluting MTBE. R. E. Sims et al., "Energy Crops: Current Status and Future Prospects," *Global Change Biology* 12 (2006): 2054–2076.

123　Drawn from remarks by José Goldemberg, National Academies Summit on America's Energy Future, Washington, D.C., March 2008.

124　This forecast is not an extrapolation but is based on the number of ethanol plants licensed and under construction in Brazil, National Academies Summit on America's Energy Future, Washington, D.C., March 2008.

125　José Goldemberg, Suani Teixeira Coelho, Patricia Guardabassi, *Sugarcane's Energy: Twelve Studies on Brazilian Sugarcane Agribusiness and Its Sustainability, Energy Policy* 36, no. 6 (June 2008): 2086–2097. Multiple files available for free download from UNICA (Brazilian Sugarcane Industry Association) at http://english.unica.com.br/multimedia/publicacao/; also personal interview with Dr. Matthew C. Nisbitt, Columbus, Ohio, April 18, 2008.

126　Fig. 7.3, summary from National Academies Summit on America's Energy Future, Washington, D.C., March 2008.

127　"Brazil Ethanol Sales Pass Petrol," *Sydney Morning Herald,* December 31, 2008.

128　M. E. Himmel et al., "Biomass Recalcitrance: Engineering Plants and Enzymes for Biofuels Production," *Science* 315 (2007): 804–807.

129　Ethanol studies are all over the map in terms of net greenhouse gas (GHG) benefits or penalties, hinging notably on whether or not "coproducts" are included in the accounting. When these factors are considered, the GHG benefits of corn ethanol over petroleum become negligible, about a 13% reduction when the benefits of coproducts are included. But ethanol produced from cellulosic material (switchgrass) reduces both GHGs and petroleum inputs substantially. A. E. Farrell et al, "Ethanol Can Contribute to Energy and Environmental Goals, *Science* 311 (2006): 506–508.

130　Drawn from remarks by José Goldemberg, National Academies Summit on America's Energy Future, Washington, D.C., March 2008.

131　C. Gautier, *Oil, Water, and Climate: An Introduction* (New York: Cambridge University Press, 2008), 366 pp.

132　"Food Crisis Renews Haiti's Agony," *Time,* April 9, 2008; "Looters Running Wild in Haiti's Food Riots," *San Francisco Chronicle,* April 10, 2008; "Hunger, Strikes, Riots: The Food Crisis Bites," *The Guardian,* April 13, 2008; D. Loyn, "World Wakes Up to Food Challenge," BBC News, October 15, 2008.

133 Provided that areas currently used for grazing are converted to agriculture, especially in South America and the Caribbean, and sub-Saharan Africa. E. M. W. Smeets et al., "A Bottom-Up Assessment and Review of Global Bio-energy Potentials to 2050," *Progress in Energy and Combustion Science* 33 (2007): 56–106.

134 A. E. Farrell et al., "Ethanol Can Contribute to Energy and Environmental Goals," *Science* 311 (2006): 506–508.

135 For example, advanced conversion technologies like enzymatic hydrolysis, and new yeasts and microorganisms to convert five-carbon sugars. *Energy Technology Perspectives—Scenarios and Strategies to 2050*, International Energy Agency (2006), 483 pp.

136 The ecological footprint is a measure of environmental impact converted to units of land area. Holden and Høyer calculate ecological footprints of four different energy regimes and found that hydropower reduces ecological footprint by -75%, natural gas by -45% to -75% (highest for fuel cells), and oil by -15% to -30%, but cellulosic (wood) biofuel by 0% to +50%. E. Holden and K. G. Høyer, "The Ecological Footprints of Fuels," *Transportation Research Part D* 10 (2005): 395–403.

137 G. Fischer, L. Schrattenholzer, "Global Bioenergy Potentials through 2050," *Biomass and Bioenergy* 20 (2001): 151–159; and *Energy Technology Perspectives 2008: Scenarios and Strategies to 2050*, OECD/International Energy Agency (2008), 643 pp.

138 Up to 26% liquid biofuels by 2050. Ibid.

139 Table 9.1, "Nuclear Generating Units, 1955–2007," U.S. Energy Information Administration, http://www.eia.doe.gov/emeu/aer/nuclear.html (accessed March 11, 2009).

140 A. Petryna, *Life Exposed: Biological Citizens after Chernobyl* (Princeton: Princeton University Press, 2002), 264 pp.

141 The recovery workers now suffer a cancer rate several percent higher than normal, with up to four thousand additional people dying (over the expected one hundred thousand) by 2004. By 2002 about four thousand children had contracted thyroid cancer from drinking radioiodine-contaminated milk in the first months after the accident. The Chernobyl Forum: 2003–2005, "Chernobyl's Legacy: Health, Environmental and Socio-Economic Impacts," 2nd rev. ed. (Vienna: IAEA Division of Public Information, April 2006). Available from http://www.iaea.org/Publications/Booklets/Chernobyl/chernobyl.pdf. The Chernobyl Forum is an initiative of the IAEA, in cooperation with the WHO, UNDP, FAO, UNEP, UN-OCHA, UNSCEAR, the World Bank, and the governments of Belarus, the Russian Federation, and Ukraine. The mortality figures in this report are decried by some as being too low, but this comprehensive UN-led effort does represent a conservative assessment of the disaster.

142 M. L. Wald, "After 30 Slow Years, U.S. Nuclear Industry Set to Build Plants Again," *International Herald Tribune*, October 24, 2008; "EDF Nuclear Contamination," *The Economist*, November 21, 2009, 65–66; "Obama offers loan guarantees for first new nuclear power reactors in three decades," *USA Today*, February 16, 2010; S. Chu, "America's New Nuclear Option: Small modular reactors will expand the ways we use atomic power," *The Wall Street Journal*, March 23, 2010. A record 62% of Americans surveyed in a March 2010 Gallup poll favored the use of nuclear power, the highest since Gallup began polling on the issue in 1994. "Public support for nuclear power at new peak," *The Washington Post*, March 22, 2010.

143 The other being hydropower.

144 The white gas is water vapor, see note 120.

145 *Energy Technology Perspectives: Scenarios and Strategies to 2050* (OECD/International Energy Agency, 2008), 643 pp.

146 S. Fetter, "Energy 2050," *Bulletin of Atomic Scientists* (July/August 2000): 28–38.

147 Of particular promise are new "light water" reactors designed to be safer than today's nuclear plants, with core-damage probabilities lower than one in a million reactor-years. Ibid.

148   *Conventional* meaning "once-through" nuclear reactors of one thousand megawatt capacity each, with no spent-fuel recycling, thorium, or breeder reactors. *The Future of Nuclear Power: An Interdisciplinary MIT Study* (Cambridge: Massachusetts Institute of Technology, 2003), 170 pp.

149   Global electricity production from nuclear power was 2,771 TWh/yr in 2005, capturing 15% market share. By 2050, based on a range of global decision scenarios modeled by the International Energy Agency, it could fall as low as 3,884 TWh/yr and 8% market share ("Baseline 2050" scenario, with few new reactors built) or rise to as much as 15,877 TWh/yr and 38% market share ("BLUE HiNUC" scenario, with maximum expansion of nuclear power). Table 2.5, *Energy Technology Perspectives 2008: Scenarios and Strategies to 2050* (OECD/International Energy Agency, 2008), 643 pp.

150   Geothermal, ocean waves, and tidal energy are all carbon-free energy sources with high potential in certain places on Earth. However, none is foreseen as becoming more than a niche energy source by the year 2050.

151   Hydropower currently supplies about 2,922 TWh/yr, capturing 16% of the world electricity market. Based on a range of global decision scenarios modeled by the International Energy Agency, it will grow so slowly that it will actually lose market share, rising to between 4,590 TWh/yr and 9% market share ("Baseline 2050" scenario) to 5,505 TWh/yr and 13% market share by 2050 ("BLUE hiOil&Gas" scenario). Table 2.5, *Energy Technology Perspectives 2008: Scenarios and Strategies to 2050* (OECD/International Energy Agency, 2008), 643 pp.

152   C. Goodall, *Ten Technologies to Save the Planet* (London: Green Profile, 2008), 302 pp.

153   As of 2006, Germany, the United States, and Spain were leading the world in wind power with 22,247, 16,818, and 15,145 megawatts installed capacity, respectively. India and China had 8,000 and 6,050 megawatts, respectively. The United States is now installing more turbines per year than any other country. Table 10.1, *Energy Technology Perspectives 2008: Scenarios and Strategies to 2050* (OECD/International Energy Agency, 2008), 643 pp.

154   Technological advances, increased manufacturing capacity, and bigger turbines have helped to lower the cost of wind energy at least fourfold since the 1980s. Efficiency has steadily increased and the turbines themselves have become larger and taller, with mass-produced rotors growing from less than 20 meters in 1985 to >100 meters today, roughly the length of an American football field. While not yet price-competitive with coal or gas-fired power plants, wind-powered electricity is getting close.

155   Based on a range of global decision scenarios modeled by the International Energy Agency, global electricity production from wind power will rise from 111 TWh/yr and 1% market share in 2005 to at least 1,208 TWh/yr and 2% market share by 2050 ("Baseline 2050" scenario, with no new incentives), and could rise as high as 6,743 TWh/yr and 17% market share ("BLUE noCCS" scenario, with aggressive incentives and no established carbon sequestration technology). Table 2.5, *Energy Technology Perspectives 2008: Scenarios and Strategies to 2050* (OECD/International Energy Agency, 2008), 643 pp.

156   The Shockley-Queisser limit.

157   N. S. Lewis, "Toward Cost-Effective Solar Energy Use," *Science* 315 (2007): 798–801.

158   See note 118.

159   M. Lavelle, "Big Solar Project Planned for Arizona Desert," *U.S. News & World Report*, February 21, 2008.

160   For more information visit the Trans-Mediterranean Renewable Energy Cooperation (TREC) home page, www.desertec.org.

161   See D. J. C. Mackay, *Sustainable Energy without the Hot Air* (Cambridge, UK: UIT Cambridge, Ltd., 2009), 370 pp., available for free download at http://www.withouthotair .com. C. Goodall estimates the cost for undersea HVDC cable between Norway and the Netherlands, completed April 2008, at €1 million per kilometer. *Ten Technologies to Save the Planet* (London: Green Profile, 2008), 302 pp.

162 CSP plants, because they use the traditional turbine method for electricity generation, can also be designed to burn natural gas or coal during nights and cloudy days.

163 A newer concept, called compressed-air storage, is to pump air, rather than water, into a tank or sealed underground cavern.

164 www.google.org/recharge/index.html (accessed March 10, 2009).

165 Especially in thin-film photovoltaics and cheap catalysts, e.g., M. W. Kanan, D. G. Nocera, "In Situ Formation of an Oxygen-Evolving Catalyst in Neutral Water Containing Phosphate and Co^{2+}," *Science* 321 (2008): 1072–1075. According to the International Energy Agency the price of photovoltaic electricity in sunny climes could fall to $0.05 per kWh by 2050.

166 N. S. Lewis, "Toward Cost-Effective Solar Energy Use," *Science* 315 (2007): 798–801.

167 C. Goodall, *Ten Technologies to Save the Planet* (London: Green Profile, 2008), 302 pp.

168 Based on a range of global decision scenarios modeled by the International Energy Agency, global solar electricity production will rise from 3 TWh/yr (virtually zero market share) in 2005 to 167 TWh/yr (still virtually zero market share) in 2050 ("Baseline 2050" scenario, with no new incentives), to as high as 5,297 TWh/yr and 13% market share by 2050 ("BLUE noCCS" scenario, with aggressive incentives and no established carbon sequestration technology). Table 2.5, *Energy Technology Perspectives 2008: Scenarios and Strategies to 2050* (OECD/International Energy Agency, 2008), 643 pp.

169 Today, some 82% of the world's electricity is made from nonrenewable coal (40%), natural gas (20%), uranium (15%), and oil (7%). Hydropower and all other renewables combined provide just 18%. Depending on our choices, they could rise to capture as much as 64% market share by 2050 (in an extremely aggressive scenario) or drop slightly to 15%. The true outcome will likely lie somewhere in between these IEA model simulations, but under no imaginable scenario will we free ourselves from fossil hydrocarbon energy in the next forty years.

170 *Energy Technology Perspectives—Scenarios and Strategies to 2050* (OECD/International Energy Agency, 2006), 479 pp.; and Table 2.5, *Energy Technology Perspectives 2008: Scenarios and Strategies to 2050* (OECD/International Energy Agency, 2008), 643 pp.

171 "Explosive Growth: LNG Expands in Australia," *The Economist,* November 21, 2009, 66–67.

172 "BP Statistical Review of World Energy June 2009," 45 pp., www.bp.com/statisticalreview (accessed November 28, 2009).

173 More precisely, 150 times current annual production for hard coal, and over 200 times annual production for lignite. T. Thielemann, S. Schmidt, J. P. Gerling, "Lignite and Hard Coal: Energy Suppliers for World Needs until the Year 2100—An Outlook," *International Journal of Coal Geology* 72 (2007): 1–14.

174 Equivalent to five hundred 500-megawatt coal-fired power plants. J. Deutch, E. J. Moniz, I. Green et al, *The Future of Coal: Options for a Carbon-Constrained World* (Cambridge: Massachusetts Institute of Technology, 2007), 105 pp.

175 Fischer-Tropsch technology is one way to do this. Ibid.

176 L. C. Smith, G. A. Olyphant, "Within-Storm Variations in Runoff and Sediment Export from a Rapidly Eroding Coal-Refuse Deposit," *Earth Surface Processes and Landforms* 19 (1994): 369–375.

177 C. Gautier, *Oil, Water, and Climate: An Introduction* (New York: Cambridge University Press, 2008), 366 pp.

178 T. Thielemann, S. Schmidt, J. P. Gerling, "Lignite and Hard Coal: Energy Suppliers for World Needs until the Year 2100—An Outlook," *International Journal of Coal Geology* 72 (2007): 1–14.

179 J. Deutch, E. J. Moniz, I. Green et al, *The Future of Coal: Options for a Carbon-Constrained World* (Cambridge: Massachusetts Institute of Technology, 2007), 105 pp.

180 "Trouble in Store," *The Economist,* March 7, 2009, 74–75.

181   Iowa weather events reconstructed from personal interview with State Climatologist Harry Hillaker in Des Moines, July 16, 2008; also a written summary he prepared in December 2008; also press releases from the Iowa Department of Agriculture and Land Stewardship and the Federal Emergency Management Agency (FEMA).

182   "FEMA, Iowans Mark Six Month Anniversary of Historic Disaster," Federal Emergency Management Agency, Press Release Number 1763-222, November 26, 2008.

183   "Iowa Department of Agriculture and Land Stewardship Officials Brief Rebuild Iowa Commission on Damage to Conservation Practices from Flooding," press release, Iowa Department of Agriculture and Land Stewardship, July 31, 2008.

184   D. Heldt, "University of Iowa's New Flood Damage Estimate: $743 million," *The Gazette,* March 13, 2009.

185   *California Fire Siege 2007: An Overview*, California Department of Forestry and Fire Protection, 108 pp., http://www.fire.ca.gov/index.php (accessed March 22, 2009).

186   Executive Order S-06-08, signed June 4, 2008, by Arnold Schwarzenegger, governor of the State of California.

187   Proclamation, "State of Emergency—Water Shortage," issued February 27, 2009, by Arnold Schwarzenegger, governor of the State of California.

188   J. McKinley, "Severe Drought Adds to Hardships in California," *The New York Times,* February 22, 2009. The Central Valley has 4.7 million acres.

189   L. Copeland, "Drought Spreading in Southeast," *USA Today,* February 12, 2008; D. Chapman, "Water Fight May Ripple in Georgia," *The Atlanta Journal-Constitution,* August 24, 2008.

190   D. W. Stahle et al., "Early Twenty-first-Century Drought in Mexico," *Eos, Transactions, American Geophysical Union* 90, no. 11 (March 17, 2009).

191   Drought data from the University College London Global Drought Monitor, http://drought.mssl.ucl.ac.uk/drought.html (accessed March 25, 2009).

192   UN Food and Agricultural Organization Global Information and Early Warning System (FAO/GIEWS), Crop Prospects and Food Situation, no. 2, April 2008. Updates posted bimonthly at http://www.fao.org/giews/english/.

193   Severe drought hit 9.5 million hectares of winter wheat in Henan, Anhai, Shandong, Hebei, Shanxi, Shaanxi, and Gansu provinces. UN FAO/GIEWS Global Watch, January 4, 2009.

194   "1,500 Farmers Commit Mass Suicide in India," *Belfast Telegraph,* April 15, 2009.

195   Global flood inventory data downloaded from the Dartmouth Flood Observatory, www.dartmouth.edu/~floods/ (accessed March 25, 2009) indicate 4,553 fatalities and 17,487,312 people displaced between January 3 and November 4, 2008.

196   *Water for Food, Water for Life: A Comprehensive Assessment of Water Management in Agriculture* (London: Earthscan, and Colombo: International Water Management Institute, 2007), 665 pp.

197   I. A. Shiklomanov, "World Fresh Water Resources," in P. H. Gleick, ed., *Water in Crisis* (New York: Oxford University Press, 1993), 13–24. Note: It is necessary to cite all of I. A. Shiklomanov's initials because he also produced two famous geoscientist sons—Alexander Igor and Nikolai Igor—leading to three Shiklomanovs in overlapping fields, creating much confusion for everyone.

198   Average annual water withdrawal estimated at 3,800 km³. O. Taikan, S. Kanael, "Global Hydrological Cycles and World Water Resources," *Science* 313, no. 5790 (2006): 1068–1072. For definitions of withdrawal vs. consumption, see note 227.

199   Global water withdrawal is thought to be about 3,800 km³ per year and global artificial storage capacity is about 7,200 km³. Ibid. For definitions, see note 225.

200   Table 2, "Food and Water," *World Resources 2008 Data Tables* (Washington, D.C.: World Resources Institute, 2008).

201 Based on 2010 and 2050 population projections for Burkina Faso, Cape Verde, Chad, Gambia, Guinea-Bissau, Mali, Mauritania, Niger, and Senegal. United Nations, *World Population Prospects: The 2008 Revision*, http://esa.un.org/unpp/.

202 The Central Arizona Project.

203 R. G. Glennon, *Water Follies* (Washington, D.C.: Island Press, USA, 2002), 314 pp.

204 Note that in the United States, however, the trend over the last ~40 years has been declining total water consumption (not just per capita), owing to declining industrial use, as well as more efficient agricultural practices, appliances, low flush toilets, and higher density housing.

205 C. J. Vörösmarty, P. Green, J. Salisbury, R. B. Lammers, "Global Water Resources: Vulnerability from Climate Change and Population Growth," *Science* 289, no. 5477 (2000): 284–288. The study identifies "severe" water stress as areas where the ratio of human water withdrawal to available river discharge is 0.4 or higher. The described three maps are found in Figure 3 of this paper. They are slightly deceptive in places like the western United States, where the source areas of water (e.g., mountain snowpack) differ from where the water is used (e.g., Tucson, Los Angeles, etc).

206 E.g., "Impending global-scale changes in population and economic development," the authors conclude, "will dictate the future . . . to a much greater degree than will changes in mean climate." Ibid.

207 Piped, protected wells or springs, rainwater cisterns, or boreholes.

208 Ethiopians (22%), Somalians (29%), Afghanis and Papua New Guineans (39%), Cambodians (41%), Chadians (42%), Equatorial Guineans and Mozambicans (43%). Data Table 3, P. H. Gleick et al., *The World's Water 2008–2009* (Washington, D.C.: Island Press, 2009), 432 pp.

209 J. Bartram, K. Lewis, R. Lenton, A. Wright, "Focusing on Improved Water and Sanitation for Health," *The Lancet* 365, no. 9461 (2005): 810–812.

210 M. Barlow, *Blue Gold: The Fight to Stop the Corporate Theft of the World's Water* (New York: The New Press, 2003), 296 pp.; *Blue Covenant: The Global Water Crisis and the Coming Battle for the Right to Water* (New York: The New Press, 2007), 196 pp.

211 Mission statement of the World Water Council, www.world watercouncil.org (accessed April 5, 2009).

212 A good account of these battles is the award-winning documentary *Flow* (2008), www .flowthefilm.com.

213 P. 189, UN World Water Assessment Programme, *The United Nations World Water Development Report 3: Water in a Changing World* (Paris: UNESCO, and London: Earthscan, 2009), 318 pp.

214 Virtually all countries negotiate water-sharing agreements for transboundary rivers crossing their borders. For emerging ideas on how satellites could change the game, see D. E. Alsdorf et al., "Measuring Surface Water from Space," *Reviews of Geophysics* 45, no. 2, article no. RG2002 (2007); D. E. Alsdorf et al., "Measuring global oceans and terrestrial freshwater from space," *Eos, Transactions, American Geophysical Union* 88, no. 24 (2007): 253; F. Hossain, "Introduction to the Featured Series on Satellites and Transboundary Water: Emerging Ideas," *Journal of the American Water Resources Association* 45, no. 3 (2009): 551–552; S. Biancamaria et al., "Preliminary Characterization of SWOT Hydrology Error Budget and Global Capabilities," *IEEE JSTARS* 3, no. 1 (2010): 6–19.

215 The Surface Water Ocean Topography (SWOT) satellite will also measure oceans. It is a joint venture between the space agencies of the United States and France (NASA and CNES).For more, see http://swot.jpl.nasa.gov/index.cfm.

216 E.g., global topography data from SRTM (http://srtm.csi.cgiar.org/) and ASTER (http://asterweb.jpl.nasa.gov/gdem.asp); global image data from Landsat (http://www.landcover.org/index.shtml); and many others.

217   D. Ignatius, "The Climate-Change Precipice," *The Washington Post*, March 2, 2007; F. Al-Obaid, "Water Scarcity and Resource War," *Kuwait Times*, March 9, 2008; H. A. Amery, "Water Wars in the Middle East: A Looming Threat, *The Geographical Journal* 168, no. 4 (2002): 313–23; N. L. Poff et al., "River Flows and Water Wars: Emerging Science for Environmental Decision Making," *Frontiers in Ecology and the Environment 1*, no. 6 (2003): 298–306; and others.

218   P. 19, UN World Water Assessment Programme, *The United Nations World Water Development Report 3: Water in a Changing World* (Paris: UNESCO, and London: Earthscan, 2009), 318 pp.

219   P. 163, M. Klare, *Resource Wars: The New Landscape of Global Conflict* (New York: Holt Paperbacks, 2002), 304 pp.

220   Ibid., p. 139.

221   Between 1948 and 1999 there were 1,831 interactions between countries over water resources, ranging from verbal exchanges to written agreements to military activity. Of these, 67% were cooperative, 28% conflictive, and 5% neutral or insignificant. There were no formal declarations of war made specifically over water. W. Barnaby, "Do Nations Go to War over Water?" *Nature* 458 (2009): 282–283; other material drawn from S. Yoffe et al., *Journal of the American Water Resources Association* 39 (2003): 1109–1126; A. T. Wolf, "Shared Waters: Conflict and Cooperation," *Annual Review of Environment and Resources* 32 (2007): 241–69.

222   See http://biblio.pacinst.org/conflict/ and http://worldwater.org/conflictchronology.pdf and http://www.transboundarywaters.orst.edu/.

223   J. I. Uitto, A. T. Wolf, "Water Wars? Geographical Perspectives: Introduction," *The Geographical Journal* 168, no. 4 (2002): 289–292; T. Jarvis et al., "International Borders, Ground Water Flow, and Hydroschizophrenia," *Ground Water* 43, no. 5 (2005): 764–770.

224   W. Barnaby, "Do Nations Go to War over Water?" *Nature* 458 (2009): 282–283.

225   Water "withdrawal" refers to the gross amount of water extracted from any source in the natural environment for human purposes. Water "consumption" refers to that part of water withdrawn that is evaporated, transpired, incorporated into products or crops, consumed by humans or livestock, or otherwise removed from the immediate water environment. Global "blue water" withdrawals from rivers, reservoirs, lakes, and aquifers are estimated at 3,830 cubic kilometers, of which 2,664 cubic kilometers are used for agriculture. Pp. 67–69, *Water for Food, Water for Life: A Comprehensive Assessment of Water Management in Agriculture* (London: Earthscan, and Colombo: International Water Management Institute, 2007), 665 pp.

226   The term *virtual water* was coined by J. A. Allan in the early 1990s, e.g., "Policy Responses to the Closure of Water Resources," in *Water Policy: Allocation and Management in Practice*, P. Howsam, R. Carter, eds. (London: Chapman and Hall, 1996).

227   The global transfer of virtual water embedded within commodities is estimated at 1,625 billion cubic meters per year, about 40% of total human water consumption. A. K. Chapagain, A. Y. Hoekstra, "The Global Component of Freshwater Demand and Supply: An Assessment of Virtual Water Flows between Nations as a Result of Trade in Agricultural and Industrial Products," *Water International* 33, no. 1 (2008): 19–32. See also pp. 35 and 98, UN World Water Assessment Programme, *The United Nations World Water Development Report 3: Water in a Changing World* (Paris: UNESCO, and London: Earthscan, 2009), 318 pp.

228   R. G. Glennon, *Water Follies: Groundwater Pumping and the Fate of America's Fresh Waters* (Washington, D.C.: Island Press, 2002), 314 pp. Windmills and other early technology could lift water from a maximum depth of only seventy to eighty feet, but the centrifugal pump, powered by diesel, natural gas, or electricity, could lift water from depths as great as three thousand feet.

229 Figure 7.6, UN World Water Assessment Programme, *The United Nations World Water Development Report 3: Water in a Changing World* (Paris: UNESCO, and London: Earthscan, 2009), 318 pp.

230 U.S. Geological Survey, "Estimated Use of Water in the United States in 2000," USGS Circular 1268, February 2005.

231 Other materials can also make good aquifers, for example gravel or highly fractured bedrock.

232 See M. Rodell, I. Velicogna and J. S. Famiglietti, "Satellite-based Estimates of Groundwater Depletion in India," *Nature* 460 (2009): 999-1002, DOI:10.1038/nature08238; and V. M. Tiwari, J. Wahr, and S. Swenson, "Dwindling Groundwater Resources in Northern India, from Satellite Gravity Observations," *Geophysical Research Letters* 36 (2009), L18401, DOI:10.1029/2009GL039401.

233 Also known as the High Plains Aquifer, the Ogallala underlies parts of Kansas, Nebraska, Texas, Oklahoma, Colorado, New Mexico, Wyoming, and South Dakota. Other material in this section drawn from V. L. McGuire, "Changes in Water Levels and Storage in the High Plains Aquifer, Predevelopment to 2005," U.S. Geological Society Fact Sheet 2007-3029, May 2007.

234 Human drawdown averages around one foot per year, but natural replenishment is less than an inch per year. Telephone interview with Kevin Mulligan, April 21, 2009.

235 "Useful lifetime" is projected time left until the saturated aquifer thickness falls to just thirty feet. When the aquifer is thinner than thirty feet, conventional wells start sucking air, owing to a thirty-foot cone of depression that forms in the water table around the borehole. The described GIS data and useful lifetime maps for the Ogallala are found at http://www.gis.ttu.edu/OgallalaAquiferMaps/.

236 LEPA drip irrigation systems create a smaller cone of depression, allowing water to be sucked from the last thirty feet of remaining aquifer saturated thickness. Therefore a switch to LEPA can prolong the usable aquifer lifetime another ten to twenty years, but cannot stop the outcome.

237 Notably the Netherlands, France, Germany, and Austria. P. H. Gleick, "Water and Energy," *Annual Review of Energy and the Environment* 19 (1994): 267-299. This is not to say all of the water used is irrevocably lost; most power plants return most of the heated water back to the originating river or lake. See note 225 for withdrawal vs. consumption.

238 This is the legal maximum in the European Union, but recommended "guideline" temperatures are lower, around 12-15 degrees Celsius in the EU and Canada. Ibid.

239 See also his book on wind power. M. Pasqualetti, P. Gipe, R. Righter, *Wind Power in View: Energy Landscapes in a Crowded World* (San Diego: Academic Press, 2002), 248 pp.

240 The reason for this is the very large water losses that evaporate from the open reservoirs behind hydroelectric dams.

241 For example, see P. W. Gerbens-Leenes, A. Y. Hoekstra, T. H. van der Meer, "The Water Footprint of Energy from Biomass: A Quantitative Assessment and Consequences of an Increasing Share of Bio-energy in Energy Supply," *Ecological Economics* 68 (2009): 1052-1060.

242 Telephone interview with M. Pasqualetti, April 14, 2009.

243 T. R. Curlee, M. J. Sale, "Water and Energy Security," *Proceedings,* Universities Council on Water Resources, 2003.

244 For climate model simulations of Hadley Cell expansion, see J. Lu, G. A. Vecchi, T. Reichler, "Expansion of the Hadley Cell under Global Warming," *Geophysical Research Letters* 34 (2007): L06805; for direct observations from satellites, see Q. Fu, C. M. Johanson, J. M. Wallace, T. Reichler, "Enhanced Mid-latitude Tropospheric Warming in Satellite Measurements," *Science* 312, no. 5777 (2006): 1179.

245 P. C. D. Milly, K. A. Dunne, A. V. Vecchia, "Global Pattern of Trends in Streamflow and Water Availability in a Changing Climate," *Nature* 438 (2005): 347-350.

246   G. M. MacDonald et al., "Southern California and the Perfect Drought: Simultaneous Prolonged Drought in Southern California and the Sacramento and Colorado River Systems," *Quaternary International* 188 (2008): 11–23.

247   The medieval warming was triggered by increased solar output combined with low levels of volcanic sulfur dioxide in the stratosphere, whereas today the driver is greenhouse gas forcing. The comparison between the medieval warm period and today is imperfect because the former saw temperatures rise most in summer, whereas greenhouse gas forcing causes maximum warming in winter and spring. Still, the medieval warm period is the best "real world" climate analog scientists have for examining possible biophysical responses to projected greenhouse warming. For more, see G. M. MacDonald et al., "Climate Warming and Twenty-first Century Drought in Southwestern North America," *EOS, Transactions, AGU* 89 no. 2 (2008). For more on the Pacific Decadal Oscillation, see G. M. MacDonald and R. A. Case, "Variations in the Pacific Decadal Oscillation over the Past Millennium," *Geophysical Research Letters* 32, article no. L08703 (2005), DOI:10.1029/2005GL022478.

248   R. Seager et al., "Model Projections of an Imminent Transition to a More Arid Climate in Southwestern North America," *Science* 316 (2007): 1181–1184.

249   P. C. D. Milly, J. Betancourt, M. Falkenmark, R. M. Hirsch, Z. W. Kundzewicz, D. P. Lettenmaier, R. J. Stouffer, "Stationarity Is Dead: Whither Water Management?" *Science* 319 (2008): 573–574.

250   The confusion arises from the fact that the "hundred-year flood," "five-hundred-year flood," etc., are simply statistical probabilities expressed as a flood height. This leads the common misperception that a hundred-year flood happens only once every hundred years, a five-hundred-year flood happens only once every five hundred years, and so on. In fact, the probability is 1/100 and 1/500 in any given year. The likelihood of enjoying a hundred consecutive years without suffering at least one hundred-year flood is just $(99/100)^{100} = 37\%$.

251   For example, it now appears likely that climate change will increase risk uncertainty with crop yields. B. A. McCarl, X. Villavicencio, X. Wu, "Climate Change and Future Analysis: Is Stationarity Dying?" *American Journal of Agricultural Economics* 90, no. 5 (2008): 1241–1247.

252   P. C. D. Milly, J. Betancourt, M. Falkenmark, R. M. Hirsch, Z. W. Kundzewicz, D. P. Lettenmaier, R. J. Stouffer, "Stationarity Is Dead: Whither Water Management?" *Science* 319 (2008): 573–574.

253   D. P. Lettenmaier, "Have We Dropped the Ball on Water Resources Research?" *Journal of Water Resources Planning and Management* 134, no. 6 (2008): 491–492.

254   The company, State Farm Florida, sent cancellation notices to nearly a fifth of its 714,000 customers after failing to win a 47.1% rate hike from state regulators. In the same year Florida's Office of Insurance Regulation projected that 102 of the 200 largest Florida insurance carriers were running net underwriting losses. "State Farm Cancels Thousands in Florida," February 23, 2010, http://www.msnbc.msn.com/id/35220269/ns/business-personal_finance/.

255   P. W. Mote et al., *Bulletin of the American Meteorological Society* 86, no. 1 (2005): 39–49.

256   T. P. Barnett et al., "Human-Induced Changes in the Hydrology of the Western United States," *Science* 319 (2008): 1080–1083.

257   J. Watts, "China Plans 59 Reservoirs to Collect Meltwater from Its Shrinking Glaciers," *The Guardian,* March 2, 2009; "Secretary Salazar, Joined by Gov. Schwarzenegger, to Announce Economic Recovery Investments in Nation's Water Infrastructure," U.S. Bureau of Reclamation Press Release, April 14, 2009; "California to Get $260 Million in U.S. Funds for Water," Reuters, April 15, 2009.

258   Melting glacier ice and the thermal expansion of ocean water as it warms are the two most important contributors to sea-level rise. Thermal expansion of ocean water is a relatively

sluggish process that is still responding to warming of past decades and will continue in response to more warming in the pipeline. To date, roughly 80% of the heat from climate warming has been absorbed by oceans. A very recent post-IPCC study estimates that over the period 1900–2008 thermal expansion caused 0.4 ± 0.2 mm/yr of sea-level rise, small glaciers and ice caps 0.96 ± 0.44 mm/yr, the Greenland Ice Sheet 0.3 ± 0.33 mm/yr, the Antarctic Ice Sheet 0.14 ± 0.26 mm/yr, and terrestrial runoff 0.17 ± 0.1 mm/yr. C. Shum, C. Kuo, "Observation and Geophysical Causes of Present-day Sea Level Rise," in *Climate Change and Food Security in South Asia*, ed., R. Lal, M. Sivakumar, S. M. A. Faiz, A. H. M. Mustafizur Rahman, K. R. Islam (Springer Verlaag, Holland: in press). Construction of twentieth-century impoundments may have trapped back ~30 mm sea level equivalent in total, an average of -0.55 mm/yr. B. F. Chao, Y. H. Wu, and Y. S. Li, "Impact of artificial reservoir water impoundment on global sea level," *Science* 320 (2008): 212–214. However, the trapping effect of human impoundments has since slowed or even reversed. D. P. Lettenmaier, P. C. D. Milly, "Land Waters and Sea Level," *Nature Geoscience* 2 (2009): 452–454, DOI:10.1038/ngeo567.

259 S. Rahmstorf et al., Response to Comments on "A Semi-Empirical Approach to Projecting Future Sea-Level Rise," *Science* 317, 1866d (2007). (See erratum for updated sea-level rise rates.)

260 M. Heberger, H. Cooley, P. Herrera, P. H. Gleick, E. Moore, "The Impacts of Sea-Level Rise on the California Coast," Final Paper, California Climate Change Center, CEC-500-2009-024-F (2009), 115 pp., available at http://pacinst.org/reports/sea_level_rise/report.pdf.

261 The 2007 *IPCC AR4* "consensus estimate" of 0.18 to 0.6 meters by 2100 may be too low. Other estimates suggest a possible range of 0.8–2.0 meters (W. T. Pfeffer et al., "Kinematic Constraints on Glacier Contributions to 21st-Century Sea-Level Rise," *Science* 321, no. 5894 2008: 1340–1343) and 0.5–1.4 meters (S. Rahmstorf, "A Semi-Empirical Approach to Projecting Future Sea-Level Rise," *Science* 315, no. 5810 [2007]: 368–370, DOI:10.1126/science.1135456.)

262 The main reason for this is that hurricanes and typhoons are fueled by sea surface temperatures. The Fourth Assessment of the Intergovernmental Panel on Climate Change estimates their intensity is "likely" to increase, meaning a >66% statistical probability. *IPCC AR4* (2007).

263 Calculated from Table 2 of R. J. Nicholls et al., "Ranking Port Cities with High Exposure and Vulnerability to Climate Extremes: Exposure Estimates," *OECD Environment Working Papers*, no. 1 (OECD Publishing, 2008), 62 pp., DOI:10.1787/011766488208. See also J. P. Ericson et al., "Effective Sea-Level Rise and Deltas: Causes of Change and Human Dimension Implications," *Global and Planetary Change* 50 (2006): 63–82.

264 Monetary amounts are in international 2001 U.S. dollars using purchasing power parities. Ibid.

265 Short for "Water Global Assessment and Prognosis." See Center for Environmental Systems Research, http://www.usf.uni-kassel.de/cesr/.

266 The climate-change component of this particular simulation is from the HadCM3 circulation model assuming a B2 SRES scenario. For more on other, nonclimatic assumptions, see Alcamo, M. Flörke, and M. Marker, "Future Long-term Changes in Global Water Resources Driven by Socio-economic and Climatic Changes," *Hydrological Sciences* 52, no. 2 (2007): 247–275.

267 P. Alpert et al., "First Super-High-Resolution Modeling Study that the Ancient 'Fertile Crescent' Will Disappear in This Century and Comparison to Regional Climate Models," *Geophysical Research Abstracts* 10, EGU2008-A-02811 (2008); A. Kitoh et al., "First Super-High-Resolution Model Projection that the Ancient 'Fertile Crescent' Will Disappear in This Century," *Hydrological Research Letters* 2 (2008): 1–4.

268 T. H. Brikowski, "Doomed Reservoirs in Kansas, USA? Climate Change and Groundwater Mining on the Great Plains Lead to Unsustainable Surface Water Storage," *Journal*

of Hydrology 354 (2008): 90–101; S. K. Gupta and R. D. Deshpande, "Water for India in 2050: First-Order Assessment of Available Options," *Current Science* 86, no. 9 (2004): 1216–1224.

269　Global climate models almost unanimously project that human-induced climate change will reduce runoff in the Colorado River region by 10%–30%. T. P. Barnett., D. W. Pierce, "Sustainable Water Deliveries from the Colorado River in a Changing Climate," *Proceedings of the National Academy of Sciences* 106, no. 18 (2009), DOI:10.1073/pnas.0812762106. See also T. P. Barnett D. W. Pierce, "When Will Lake Mead Go Dry?" *Water Resources Research* 44 (2008), W03201.

270　This is not necessarily so dire as it sounds. Water rights are about withdrawals, not consumptive use, so some share of the withdrawn water is recycled and returned to the river system, allowing it to be reused again downstream.

271　J. L. Powell, *Dead Pool: Lake Powell, Global Warming, and the Future of Water in the West* (London: University of California Press, 2008), 283 pp.

272　The 2003 pact, called the Quantification Settlement Agreement, also requires the Imperial Irrigation District to sell up to 100,000 acre-feet to the cities of the Coachella Valley. California's total Colorado River allocation is 4.4 million acre-feet per year. The Metropolitan Water District of Southern California serves twenty-six cities. Press releases of the Imperial Irrigation District, November 10, 2003, and April 30, 2009 (www.iid.com); also M. Gardner, "Water Plan to Let MWD Buy Salton Sea Source," *Union-Tribune*, signonsandiego.com, April 6, 2009.

273　Unlike water vapor, which is quickly recycled, other greenhouse gases tend to linger longer in the atmosphere, especially CO_2, which can persist for centuries. S. Solomon et al., "Irreversible Climate Change Due to Carbon Dioxide Emissions," *PNAS* 106, no. 6 (2009): 1704–1709. About half will disappear quite quickly and some 15% will stick around even longer, but on balance carbon dioxide persists in the atmosphere for a very long time.

274　More precisely, volcanic eruptions release sulfur dioxide gas (SO_2), which oxidizes to sulphate aerosols (SO_4). If aerosols penetrate the stratosphere, they can circulate globally for several years, creating brilliant sunsets and blocking sunlight to create a temporary climate cooling.

275　Some of these mechanisms can persist for several decades, especially long-lived ocean circulation phenomena like the Pacific Decadal Oscillation, e.g., G. M. MacDonald and R. A. Case, "Variations in the Pacific Decadal Oscillation over the Past Millennium," *Geophysical Research Letters* 32, article no. L08703, DOI:10.1029/2005GL022478 (2005).

276　By averaging model simulations over a twenty-year period (2046–2064), this map smooths out most of the short-term variability described earlier, thus revealing the strength of the underlying greenhouse effect. Yet even after this smoothing process, we still find a geographically uneven pattern of warming. For map source see next endnote.

277　*IPCC AR4*, Figure 10.8, Chapter 10, p. 766 (Full citation: G. A. Meehl et al., Chapter 10, "Global Climate Projections," in S. Solomon, D. Qin, M. Manning, Z. Chen, M. Marquis, K. B. Averyt, M. Tignor, H. L. Miller, eds., *Climate Change 2007: The Physical Science Basis. Contribution of Working Group I to the Fourth Assessment Report of the Intergovernmental Panel on Climate Change* (Cambridge, UK, and New York: Cambridge University Press, 2007). See Chapter 1 for more on the IPCC Assessment Reports.

278　These outcomes are called SRES scenarios, of which three are shown here (i.e., each row is a different SRES scenario). There are many economic, social, and political choices contained within different SRES scenarios, but the differences are not important for our purposes here. SRES refers to the IPCC *Special Report on Emissions Scenarios.* They are grouped into four families (A1, A2, B1, and B2) exploring alternative development pathways, covering a wide range of demographic, economic, and technological driving forces and resultant greenhouse gas emissions. B1 describes a convergent, globalized world with a rapid transition toward a service and information economy. The A1 family assumes rapid economic growth, a global population that peaks around 2050, and rapidly advancing energy technology, with A1B assuming a balance between fossil and nonfossil energy. A2 describes a nonglobalized world with high population growth, slow economic development, and slow technological change. For more, see N. Nakicenovic, R. Swart, eds., *Special Report on Emissions Scenarios: A Special Report of Working Group III of the Intergovernmental Panel on Climate Change* (Cambridge, UK: Cambridge University Press, 2000), 570 pp.

279 The three SRES scenarios shown, which I have renamed for clarity, are B1, A1B, and A2, respectively. There are a number of other scenarios but these three illustrate a representative cross-section from the IPCC AR4 Assessment.

280 P.217, R. Henson, *The Rough Guide to Climate Change* (London: Penguin Books Ltd., 2008).

281 These are discussed further in Chapter 9.

282 Projected temperature increases average about 50% higher over land than over oceans. The stubborn bull's-eye marks where warm, north-flowing waters of the Meridional Overturning Current (MOC)—also known as the North Atlantic Deep Water Formation (NADW)—cool and sink. Weakened MOC overturning is expected to counter the climate warming effect locally in this area. There are other physical reasons why the warming effect is amplified in the high northern latitudes, including low evaporation rate, a thinner atmosphere, and reduced albedo (reflectivity) over land. But the most important reason by far is the disappearance of sea ice over the Arctic Ocean, changing it from a high-albedo surface that reflects incoming sunlight back out to space to an open ocean that absorbs it.

283 E.g., Figure 10.12, *IPCC AR4*, Chapter 10, p. 769. The models also concur pretty well in the Mediterranean region, southern South America, and the western United States, where precipitation is projected to decrease. They concur well around the equator, over the southern oceans around Antarctica, and throughout the northern high latitudes, where it is projected to increase. Except for Canada's western prairies, precipitation is projected to rise significantly across the northern territories and oceans of all eight NORC countries.

284 Among other things the Clausius-Clapeyron relation, i.e., a warmer atmosphere holds more water vapor.

285 The 2050 projections are from P. C. D. Milly et al., "Global Pattern of Trends in Streamflow and Water Availability in a Changing Climate," *Nature* 438 (2005): 347–350. That the projected northern runoff increases surpass all bounds of natural climate variability is shown by Hulme et al., "Relative Impacts of Human-Induced Climate Change and Natural Climate Variability," *Nature* 397, no. 6721 (1999): 688–691. The twentieth-century river discharge increases appeared first and most strongly in Russia, B. J. Peterson et al., "Increasing River Discharge to the Arctic Ocean," *Science* 298, no. 5601 (2003): 2171–2173; J. W. McClelland et al., "A Pan-Arctic Evaluation of Changes in River Discharge during the Latter Half of the Twentieth Century," *Geophysical Research Letters* 33, no. 6 (2006): L06715. In Canada, runoff experienced late-century declines in total runoff to Hudson's Bay but increases in the Northwest Territories. S. J. Déry, "Characteristics and Trends of River Discharge into Hudson, James, and Ungava Bays, 1964–2000," *Journal of Climate* 18, no. 14 (2005): 2540–2557; J. M. St. Jacques, D. J. Sauchyn, "Increasing Winter Baseflow and Mean Annual Streamflow from Possible Permafrost Thawing in the Northwest Territories, Canada," *Geophysical Research Letters* 36 (2009): L01401. An excellent recent synopsis is A. K. Rennermalm, E. F. Wood, T. J. Troy, "Observed Changes of Pan-Arctic Cold-Season Minimum Monthly River Discharge," *Climate Dynamics*, DOI: 10.888/1748-9326/4/2/024011.

286 L. C. Smith et al., "Rising Minimum Daily Flows in Northern Eurasian Rivers: A Growing Influence of Groundwater in the High-Latitude Hydrologic Cycle," *Journal of Geophysical Research* 112, G4, (2007): G04S47.

287 Ice caps are large glacier masses on land. Unlike Antarctica, a continent buried beneath mile-thick glaciers and surrounded by oceans, the Arctic is an ocean surrounded by continents. It is thinly covered with just one to two meters of seasonally frozen ocean water called "sea ice."

288 The Fall Meeting of the American Geophysical Union, which convenes each December in San Francisco, California.

289 The Arctic Ocean freezes over completely in winter but partially opens in summer. The annual sea-ice minimum occurs in September.

290 By September 2009 sea-ice cover was nearing recovery to its old trajectory of linear decline. However, the extreme reductions of 2007–2009 were a major excursion from the long-term trend and clearly demonstrate the surprising rapidity with which the Arctic's summer sea-ice cover can disappear.

291 Unlike land-based glaciers, the formation or melting of sea ice does not significantly raise sea level because the volume of buoyant ice is compensated by the volume of water displaced (Archimedes' Principle). A slight exception (about 4%) to this does arise because sea ice is fresher than the ocean water it is displacing (thus taking up slightly more volume than the equivalent mass of sea water).

292 This albedo feedback works in the opposite direction, too, by amplifying global cooling trends. If global climate cools, then Arctic sea ice expands, reflecting more sunlight, thus causing more local cooling and more sea-ice formation, and so on.

293 Sea ice does form around the edge of the Antarctic continent, but its areal extent is much less than in the Arctic Ocean and it does not survive the summer. Other reasons for the warming contrast between the Arctic and Antarctica include the strong circumpolar vortex around the southern oceans, which divorce Antarctica somewhat from the global atmospheric circulation, and the cold high elevations of interior Antarctica, where air temperatures will never reach the melting point, unlike the Arctic Ocean, which is at sea level.

294 The sea-ice albedo feedback is the most important factor causing the global climate warming signal to be amplified in the northern high latitudes, but there are also others. Reduced albedo over land (from less snow), a thinner atmosphere, and low evaporation in cold Arctic air are some of the other positive warming feedbacks operating in the region. The transition to a new summertime ice-free state is likely to happen rapidly once the ice pack thins to a vulnerable state. M. C. Serreze, M. M. Holland, J. Stroeve, "Perspectives on the Arctic's Shrinking Sea-Ice Cover," *Science* 315, no. 5815 (2007): 1533–1536. Not all northern albedo feedbacks are positive—for example, more forest fires, an expected consequence of rising temperatures, actually raise albedo over the long term. E. A. Lyons, Y. Jin, J. T. Randerson, "Changes in Surface Albedo after Fire in Boreal Forest Ecosystems of Interior Alaska Assessed Using MODIS Satellite Observations," *Journal of Geophysical Research* 113: (2008) G02012.

295 Based on projections of the NCAR CCSM3 climate model. You can view these results in D. M. Lawrence, A. G. Slater, R. A. Tomas, M. M. Holland, and C. Deser, "Accelerated Arctic Land Warming and Permafrost Degradation during Rapid Sea Ice Loss," *Geophysical Research Letters* 35, no. 11, (2008): L11506, DOI:10.1029/2008GL033985.

296 Hill and Gaddy use the term *Siberian Curse* to argue that Soviet planners shortchanged their country economically by seeking to develop its cold hinterlands. I am co-opting the term here to more broadly include biological factors as well. F. Hill and C. Gaddy, *The Siberian Curse* (Washington, D.C.: Brookings Institution Press, 2003), 303 pp.

297 This summary drawn from Chapter 2, "Arctic Climate: Past and Present," of the *Arctic Climate Impact Assessment (ACIA)* (Cambridge, UK: Cambridge University Press, 2005),

1,042 pp.; and *Working Group II Report*, Chapter 15, "Polar Regions," of the *IPCC AR4* (2007). See also S. J. Déry, R. D. Brown, "Recent Northern Hemisphere Snow Cover Extent Trends and Implications for the Snow-Albedo Feedback," *Geophysical Research Letters* 34, no. 22 (2007): L22504. Much of the observed warming is not caused by greenhouse forcing directly, but instead to atmospheric circulation changes, suggesting that the Arctic is just in the early stages of the human-induced greenhouse gas signature. M. C. Serreze, J. A. Francis, "The Arctic Amplification Debate," *Climatic Change* 76 (2006): 241–264.

298 For example, a +8% increase in peak greenness north of 65° N latitude from 1982 to 1990; a +17% increase in northern Alaska from 1981 to 2001. R. Myneni et al., "Increased Plant Growth in the Northern Latitudes from 1982 to 1991," *Nature* 386 (1997): 698–702; G. J. Jia, H. E. Epstein, D. A. Walker, "Greening of Arctic Alaska, 1981–2001," *Geophysical Research Letters* 30, no. 20 (2003): 2067; also M. Sturm, C. Racine, K. Tape, "Climate Change: Increasing Shrub Abundance in the Arctic," *Nature* 411 (2001): 546–547; I. Gamach, S. Payette, "Height Growth Response of Tree Line Black Spruce to Recent Climate Warming across the Forest-Tundra of Eastern Canada," *Journal of Ecology* 92 (2004): 835–845.

299 Arctic-wide average net primary productivity is forecast to rise from 2.8 to 4.9 Pg C/year by the 2080s under the "optimistic" IPCC B2 scenario, Table 7.13, ACIA (2005).

300 This paragraph and others drawn from personal interviews and anecdotes collected 2006/2007 throughout Canada, Alaska, and Finland, including Fort Chipewyan, Fort McMurray, Cumberland House, Whitehorse, High Level, Hay River, Yellowknife, Churchill, Fairbanks, and Barrow. Also G. Beaugrand et al., "Reorganization of North Atlantic Marine Copepod Biodiversity and Climate," *Science* 296 (2002): 1692–1694; A. L. Perry et al., "Climate Change and Distribution Shifts in Marine Fishes," *Science* 308 (2005): 1912–1915; N. S. Morozov, "Changes in the Timing of Migration and Winter Records of the Common Buzzard *(Buteo buteo)* in the Central Part of European Russia: The Effect of Global Warming?" *Zoologichesky Zhurnal* 86, no. 11 (2007): 1336–1355; G. Jansson, A. Pehrson, "The Recent Expansion of the Brown Hare *(Lepus europaeus)* in Sweden with Possible Implications to the Mountain Hare *(L. timidus),*" *European Journal of Wildlife Research* 53 (2007): 125–130; N. H. Ogden, "Climate Change and the Potential for Range Expansion of the Lyme Disease Vector *Ixodes scapularis* in Canada," *International Journal for Parasitology* 36, no. 1 (2006): 63–70; S. Sharma et al., "Will Northern Fish Populations Be in Hot Water Because of Climate Change?" *Global Change Biology* 13 (2007): 2052–2064; S. Jarema et al., "Variation in Abundance across a Species' Range Predicts Climate Change Responses in the Range Interior Will Exceed Those at the Edge: A Case Study with North American Beaver," *Global Change Biology* 15 (2009): 508–522.

301 Cartoons and children's books that show penguins and polar bears coexisting together perpetuate a widespread myth about their geographic distribution. Polar bears are found only in the far northern hemisphere. Penguins are found only in the southern hemisphere. Unlike the Arctic, with its bears, foxes, and humans, there are no land-based predators in Antarctica. This is why penguins and elephant seals are fearless of humans whereas ringed seals are not.

302 These events happened in 2004. S. C. Amstrup et al., "Recent Observations of Intraspecific Predation and Cannibalism among Polar Bears in the Southern Beaufort Sea," *Polar Biology* 29 (2006): 997–1002. Increasing polar bear interaction with human settlements is described by I. Stirling, Parkinson, "Possible Effects of Climate Warming on Selected Populations of Polar Bears *(Ursus maritimus)* in the Canadian Arctic," *Arctic* 59, no. 3 (2006): 261–275; also E. V. Regehr et al., "Effects of Earlier Sea Ice Breakup on Survival and Population Size of Polar Bears in Western Hudson Bay," *Journal of Wildlife Management* 71 (2007): 2673–2683. For more on projected future declines in

polar bear sea-ice habitat, see G. M. Durner et al., "Predicting 21st-Century Polar Bear Habitat Distribution from Global Climate Models," *Ecological Monographs* 79, no. 1 (2009): 25–58.

303   S. C. Amstrup et al., *Forecasting the Range-wide Status of Polar Bears at Selected Times in the 21st Century: Administrative Report to Support U.S. Fish and Wildlife Service Polar Bear Listing Decision* (Reston, Va.: U.S. Department of the Interior/U.S. Geological Survey, 2007), 126 pp.

304   C. D. Thomas et al., "Extinction Risk from Climate Change," *Nature* 427 (2004): 145–148. The *IPCC AR4* similarly estimates a 20%–30% species extinction for a global temperature rise of 1.5°–2.5°C.

305   For example, since the early twentieth century the western United States has suffered a 73% loss in the coverage area of alpine tundra. H. F. Diaz et al., "Disappearing 'Alpine Tundra' Koppen Climatic Type in the Western United States," *Geophysical Research Letters* 34, no. 18 (2007): L18707. Under the high-end A2 emissions scenario, 12%–39% and 10%–48% of the Earth's terrestrial surface is projected to experience novel and disappearing climates by 2100 A.D.; corresponding projections for the low-end B1 scenario are 4%–20% and 4%–20%. J. W. Williams et al., "Projected Distributions of Novel and Disappearing Climates by 2100 A.D.," *Proceedings of the National Academy of Sciences* 104, no. 14 (2007): 5738–5742.

306   Note that I said least disturbed, not undisturbed. The myth of a pristine North is exposed in Chapter 7.

307   More precisely, up to 44% of all species of vascular plants and 35% of all species in four vertebrate groups. N. Myers et al., "Biodiversity Hotspots for Conservation Priorities," *Nature* 403 (2000): 853–858, DOI:10.1038/35002501. Seven million is a conservative estimate and refers to eukaryotes, meaning species generally recognized as plants or animals but excluding things like bacteria.

308   Owing to increased forest disturbance from insect pests and wildfires, e.g., Gillett et al., "Detecting the Effect of Climate Change on Canadian Forest Fires," *Geophysical Research Letters* 31 (2004): L18211; E. S. Kasischke, M. R. Turetsky, "Recent Changes in the Fire Regime across the North American Boreal Region—Spatial and Temporal Patterns of Burning across Canada and Alaska," *Geophysical Research Letters* 33 (2006): L09703.

309   From personal interviews with Ron Brower of Barrow, Alaska, August 9, 2006; Mayor E. Sheutiapik of Iqualuit, Nunavut, August 5, 2007; Mayor E. Kavo and J. Meeko of Sanikiluaq, Nunavut, August 7, 2007.

310   Personal interview with Ron Brower, Barrow, Alaska, August 9, 2006.

311   Drawn from J. Painter, "Greenland Sees Bright Side of Warming," BBC News, September 14, 2007; C. Woodward, "Global Warming Is a Boon for Farmers and Fishermen but a Hardship for Ice-Dependent Inuit," *Christian Science Monitor,* October 1, 2007; and "Greenlandic Super Potatoes," *The Copenhagen Post,* May 18, 2009.

312   Workshop on Conservation of Crop Genetic Resources in the Face of Climate Change, Bellagio, Italy, September 3–6, 2007.

313   More specifically South Asia wheat, Southeast Asia rice, and southern Africa corn. The editors of *Science* must have also been impressed, as the research appeared there five months later. D. B. Lobell, M. B. Burke et al., "Prioritizing Climate Change Adaptation Needs for Food Security in 2030," *Science* 319 (2008): 607–610.

314   W. Schlenker, D. B. Lobell, "Robust negative impacts of climate change on African agriculture," *Environmental Research Letters* 5 (2009), DOI:10.1088/1748-9326/5/1/014010.

315   D. S. Battisti, R. L. Naylor, "Historical Warnings of Future Food Insecurity with Unprecedented Seasonal Heat," *Science* 323 (2009): 240–244.

316   The experiment assumed a doubling of atmospheric CO_2. R. M. Adams et al., "Global Climate Change and U.S. Agriculture," *Nature* 345 (1990): 219–224.

317 E.g., J. E. Olesen, M. Bindi, "Consequences of Climate Change for European Agricultural Productivity, Land Use and Policy," *European Journal of Agronomy* 16 (2002): 239–262. G. Maracchi, O. Sirotenko, and M. Bindi, "Impacts of Present and Future Climate Variability on Agriculture and Forestry in the Temperate Regions: Europe," *Climatic Change* 70 (2005): 117–135; N. Dronin, A. Kirilenko, "Climate Change and Food Stress in Russia: What If the Market Transforms as It Did during the Past Century?" *Climatic Change* 86 (2008): 123–150.

318 There's more to it than just temperature and rain. A key issue is the so-called CO_2 fertilization effect. Plants like CO_2, so having more of it in the air tends to increase crop yields. Most agro-climate models build in a hefty benefit for this, based on early greenhouse experiments using enclosed chambers. This enables the models to offset a large share of the damages of summer heat and drought, owing to the anticipated fertilizing benefit from elevated CO_2 levels. However, more realistic experiments staged outdoors, using blowers over actual farm fields, show a much lower fertilization benefit. This suggests that the models may be seriously underestimating the negative impacts of climate change to world food production. S. P. Long et al., "Food for Thought: Lower-than-Expected Crop Yield Stimulation with Rising CO_2 Concentrations," *Science* 312 (2006): 1918–1921.

319 For example, crop declines from a doubling of extreme weather events by the 2020s. J. Alcamo et al., "A New Assessment of Climate Change Impacts on Food Production Shortfalls and Water Availability in Russia," *Global Environmental Change* 17 (2007): 429–444.

320 For example, Russia's West Siberian, East Siberian, Northwestern, Northern, and Far East regions are all forecast to experience increased cereal and potato productivity by the 2020s, but its Central, Central Chernozem, North Caucasian, Volga-Vyatka, and Volga regions are projected to decline. A. P. Kirilenko et al., "Modeling the Impact of Climate Changes on Agriculture in Russia," *Doklady Earth Sciences* 397, no. 5 (2004): 682–685 (translated from Russian).

321 T. Parfitt, "Russia's Polar Hero," *Science* 324, no. 5933 (2009): 1382–1384. See also "Artur Chilingarov: Russia's Arctic Explorer," *The Moscow News,* July 17, 2008.

322 Tom Casey, a U.S. State Department spokesman, said, "I'm not sure whether they put a metal flag, a rubber flag, or a bedsheet on the ocean floor. Either way, it doesn't have any legal standing." "Russian Subs Seek Glory at North Pole," *USA Today,* August 2, 2007. See also "Russia Plants Flag on North Pole Seabed," *The Guardian* UK; "Russia Plants Flag under N Pole," BBC News; "Russia Plants Underwater Flag at North Pole," *The New York Times*; "Russia to Claim Energy Wealth beneath Arctic Ocean," *Pravda*; and many others (all August 2, 2007).

323 ArcticNet is a Canadian government-funded research consortium that coordinates big projects in the Arctic, including the CCGS *Amundsen* expedition, http://www.arcticnet .ulaval.ca/.

324 The 2007–09 International Polar Year (IPY, www.ipy.org) was an international science program focused on the Arctic and Antarctic that lasted from March 2007 to March 2009. More than two hundred projects, sixty countries, and thousands of scientists participated in IPY. It was actually the fourth such Polar Year, following earlier ones in 1882–83, 1932–33, and 1957–58.

325 2007 was the astonishing record year in which nearly 40% of the Arctic's late-summer Arctic sea disappeared. See Chapter 5.

326 "A Mad Scramble for the Shrinking Arctic," *The New York Times,* September 10, 2008.

327 In 2008 a test shipment of this very pure ore was delivered to Europe from the Baffinland Mine in Mary River. P. 77, *Arctic Marine Shipping Assessment 2009 Report,* Arctic Council, April 2009, 190 pp.

328   "Circum-Arctic Resource Appraisal: Estimates of Undiscovered Oil and Gas North of the Arctic Circle," digital data and USGS Fact Sheet 2008–3049, 2008; D. L. Gautier et al., "Assessment of Undiscovered Oil and Gas in the Arctic," *Science* 324 (2009): 1175–1179.

329   S. G. Borgerson, "Arctic Meltdown: The Economic and Security Implications of Global Warming," *Foreign Affairs*, March/April 2008.

330   S. G. Borgerson, "The Great Game Moves North," *Foreign Affairs,* March 25, 2009. See also T. Halpin, "Russia Warns of War within a Decade over Arctic Oil and Gas Riches," *The Times,* May 14, 2009; A. Doyle, "Arctic Nations Say No Cold War; Military Stirs," Reuters, June 21, 2009.

331   M. Galeotti, "Cold Calling—Competition Heats Up for Arctic Resources," *Jane's Intelligence Review,* September 23, 2008.

332   R. Huebert, "In the Grip of Climate Change: The Circumpolar Dimension," Session Paper no. 1, 2030 NORTH National Planning Conference, Ottawa, June 1–4, 2009.

333   Canada asserts that the "Northwest Passage" (it actually contains several possible routes) constitutes a domestic waterway, whereas the United States, Russia, and European Union maintain it is an international strait. At present the tacit policy between the United States and Canada is to agree to disagree on this issue.

334   Russia's aircraft approached but did not enter Canadian airspace. B. Smith-Windsor, "The Perils of Sexing Up Arctic Security," *Toronto Star,* June 26, 2009. See also "Two Russian Bombers Fly over Icelandic Airspace," http://www.icenews.is/index.php/2009/08/10/two-russian-bombers-fly-over-icelandic-airspace/;*IceNews,* August 10, 2009; and others.

335   Much of this paragraph and the next are drawn from the work of Rob Huebert at the University of Calgary, "In the Grip of Climate Change: The Circumpolar Dimension," Session Paper no. 1, 2030 NORTH National Planning Conference, Ottawa, June 1–4, 2009; and the School of Public Policy, University of Calgary, "United States Arctic Policy: The Reluctant Arctic Power," *SPP Briefing Papers* 2, no.2 (May 2009), 27 pp.

336   R. Huebert, "United States Arctic Policy: The Reluctant Arctic Power," *SPP Briefing Papers* 2, no. 2 (May 2009), 27 pp.

337   Captain L. W. Brigham, Ph.D., personal communication, June 2, 2009.

338   Reportedly a 2009 "ice exercise" using attack submarines. R. Huebert, "In the Grip of Climate Change: The Circumpolar Dimension," Session Paper no. 1, 2030 NORTH National Planning Conference, Ottawa, June 1–4, 2009, p. 18.

339   This 2009 directive lists four developments as justification for a change in U.S. Arctic policy, namely "(1) Altered national policies on homeland security and defense; (2) The effects of climate change and increasing human activity in the Arctic region; (3) The establishment and ongoing work of the Arctic Council; and (4) A growing awareness that the Arctic region is both fragile and rich in resources." United States White House, Office of the Press Secretary, National Security Presidential Directive/NSPD 66, Homeland Security Presidential Directive/HSPD 25, Washington, D.C., January 9, 2009, http://media.adn.com/smedia/2009/01/12/15/2008arctic.dir.rel.source.prod_affiliate.7.pdf.

340   Personal interview with R. Huebert, Ottawa, June 3, 2009.

341   M. Gorbachev, "The Speech in Murmansk at the Ceremonial Meeting on the Occasion of the Presentation of the Order of Lenin and the Gold Star Medal to the City of Murmansk," October 1, 1987 (Novosti Press Agency: Moscow, 1987), http://www.barentsinfo.fi/docs/Gorbachev_speech.pdf; see also K. Åtland, Mikhail Gorbachev, "The Murmansk Initiative, and the Desecuritization of Interstate Relations in the Arctic," *Cooperation and Conflict* 43, no. 3 (2008): 289–311, DOI:10.1177/0010836708092838.

342   This assistance was often done at the grassroots level. For example, by securing research funding to do fieldwork in Siberia, I was able to hire Russian scientists and locals for logistics support and scientific collaboration during this very difficult time.

343 The Arctic Environmental Protection Strategy, or AEPS, signed June 14, 1991, in Rovaniemi. AEPS is a nonbinding multilateral agreement signed by Canada, Denmark, Finland, Iceland, Norway, Sweden, Union of Soviet Socialist Republics, and the United States, with participation by the Inuit Circumpolar Conference, Nordic Sámi Council, USSR Association of Small Peoples of the North, Federal Republic of Germany, Poland, United Kingdom, United Nations Economic Commission for Europe, United Nations Environment Program, and the International Arctic Science Committee. See http://arcticcouncil .org/filearchive/arctic_environment.pdf.

344 The Arctic Council is an intergovernmental forum established in 1996 "to provide a means for promoting cooperation, coordination, and interaction among the Arctic States, with the involvement of the Arctic Indigenous communities and other Arctic inhabitants on common Arctic issues, in particular issues of sustainable development and environmental protection" (http://arctic-council.org). Its "member states" are the eight Arctic countries Canada, the United States, Denmark/Greenland/Faroe Islands, Iceland, Norway, Sweden, Finland, and Russia; other categories of membership include six "permanent participant" aboriginal groups; and non-Arctic observer states like the United Kingtom, Spain, China, Italy, Poland, and South Korea. The Arctic Council focuses on environmental protection and sustainable development issues; it is strictly forbidden to engage issues of security or territory. Nonetheless it is the premier "Arctic" polity as of 2010.

345 By the turn of the millennium, even before the shock wave of 9/11, things had started to tighten up. People were beginning to consider the prospect of new economic opportunities for oil and gas exploration, shipping, and fisheries made possible by the reduction of summer Arctic sea ice. Under the Putin administration, Russia began funding her own scientists again, while also rolling up the welcome mat for western scientists. I and two graduate students—informed we were no longer allowed to do fieldwork even if escorted by Russian colleagues—packed up and left.

346 *ACIA, Arctic Climate Impact Assessment* (Cambridge, UK: Cambridge University Press, 2005), 1042 pp. Available for free download at http://www.acia.uaf.edu.

347 *AMSA, Arctic Marine Shipping Assessment 2009 Report,* Arctic Council, 190 pp., April 2009.

348 These things are specifically barred from the Arctic Council's mandate. The United States would not have supported its creation otherwise. This is perhaps unsurprising, as few, if any, superpowers will cede discussion of military matters to an intergovernmental forum. At high policy levels, U.S. support for the Arctic Council has always been reluctant, unlike lower policy levels, and among scientists, where U.S. support is strong.

349 J. Broadus, R. Vartanov, *Environmental Security: Shared U.S. and Russian Perspectives* (Woods Hole, Mass.: Woods Hole Oceanographic Institute, 2002), 60–61.

350 The Canada-U.S. dispute derives from differing interpretations of an 1825 treaty between Great Britain and Russia. However, Norway and Russia announced resolution of their decades-old dispute in April 2010, W. Gibbs, "Russia and Norway Reach Accord on Barents Sea," *The New York Times,* April 27, 2010; "Norway, Russia Strike Deal to Divide Arctic Undersea Territory," *The Moscow Times,* April 27, 2010; "Thaw in the Arctic, *Financial Times,* April 29, 2010.

351 UN Commission for the Limits of the Continental Shelf (CLCS). The extension is for the seafloor only, called an "Extended Continental Shelf," or ECS, extending the standard EEZ up to 350 nautical miles. It does not include control over pelagic fishing as does the standard EEZ.

352 Sweden and Finland do not have coasts fronting the Arctic Ocean. The United States is unable to file an Article 76 claim until it ratifies UNCLOS. However, the United States is behaving as if it has, and has been carrying out the scientific investigations needed to make an UNCLOS Article 76 claim. The United States has also assisted other countries, especially Canada, in the collection of scientific data for their claims.

353   Resolution of Norway's Article 76 claim was not perfect. The CLCS found that both Russia and Norway have legitimate cases for their overlapping claims in one area of the Barents Sea. The two countries had to reach their own agreement to resolve it. "UN Backs Norway Claim to Arctic Seabed Extension," *Ottawa Citizen,* April 15, 2009. They did so in April 2010; see note 350.

354   The so-called "Ilulissat Declaration" was released May 28, 2008. Denmark invited Canada, Norway, Russia, and the United States to Ilulissat, Greenland, to craft this statement of these countries' solidarity and commitment to existing legal frameworks, i.e., UNCLOS. It is widely perceived as a message to other entities, like the European Union, which had been issuing its own documents with proposals for shared Arctic Ocean governance, to stay out. Even the other Arctic countries of Sweden, Finland, and Iceland, and aboriginal organizations, were excluded from the meeting in Ilulissat. See http://www.oceanlaw.org/downloads/arctic/Ilulissat_Declaration.pdf.

355   D. L. Gautier et al., "Assessment of Undiscovered Oil and Gas in the Arctic," *Science* 324 (2009): 1175–1179.

356   The current boundary between Canada and Denmark runs down the center of Lomonosov Ridge, thus both countries have the possibility of proving it is a geological extension of their continental shelves.

357   The Northern Sea Route offers a 35%–60% distance savings between Europe and the Far East. To go from Yokohama to Rotterdam via the Arctic Ocean would take just 6,500 nautical miles, versus 11,200 through the Suez Canal.

358   "Multiyear ice" (MYI) is sea ice that survives through at least one summer, and can grow considerably thicker and harder than "first-year ice" (FYI), normally only one to two meters thick. FYI is easier for icebreakers and fortified ships to pass through than MYI.

359   Russia's newest nuclear icebreaker, the world's largest, is named *50 Years of Victory.* A. Revkin, "A Push to Increase Icebreakers in the Arctic," *The New York Times,* August 16, 2008.

360   *AMSA 2009,* Table 5.2, p. 79.

361   *AMSA 2009,* p. 72. The "six-thousand" figure includes vessels traveling on the North Pacific's Great Circle Route between Asia and North America through the Aleutian Island chain, which the United States defines as being within the "Arctic."

362   Adapted from maps 5.5 and 5.6, AMSA 2009, p. 85.

363   Personal interview with J. Marshall, vice-president, Northern Transportation Co. Ltd., Hay River, NWT, July 6, 2007. For more about this long-running company, now aboriginal-owned, see http://www.ntcl.com/.

364   Personal interview with ConocoPhillips Russia president Don Wallette, January 22, 2007, Tromsø.

365   Because ice is fresh but ocean water salty, pockets of highly saline brine develop within sea ice as it first begins to freeze. As the ice grows over multiple winters, the brine pockets drain and the ice thickens, increasing its strength and hardness.

366   Sea ice, including first-year ice, is always dangerous, and will always be a limiting factor in the Arctic Ocean.

367   Ships must have fortified hulls, powerful engines, and other technical requirements to operate safely in sea ice. A ship's polar class designates the allowable conditions it can handle (summer or year-round operation, first-year or multiyear ice, etc.). The design requirements for a given polar class are set by the International Maritime Organization (IMO), and the International Association of Classification Societies (IACS) defines a range of categories. The higher the polar class, the more expensive the ship is to build.

368   World fleets typically travel at fifteen to twenty or more knots. A Russian icebreaker can break ice at speeds as high as twelve to fifteen knots, but risks of damage are higher. Six- to ten knots are more typical in ice. Personal communication with Captain Lawson Brigham, November 25, 2009.

369 Canada and Russia maintain that these passages are domestic waters under their control; the United States and others maintain they are international straits and thus freely available to use without declaration or permission. These and other nontrivial impediments to transnational shipping in the Arctic are described in *AMSA 2009*.

370 I suppose someday there might be more of them—perhaps by 2100 or 2150, if globalization hasn't collapsed into a pile of fiefdoms—together with booming new Arctic port cities. The geography of distance, along with further sea-ice reductions in store, is just too compelling. But this won't happen by 2050, the time frame of this book's thought experiment.

371 Some 1.2 million passengers took cruise ships to the region in 2004; three years later the number had more than doubled. By 2008 some 375 cruise-ship port calls were scheduled for Greenland's ports and harbors alone (*AMSA 2009*, p. 79).

372 From personal interviews with Mike Spence, mayor of Churchill, June 28, 2007, and L. Fetterly, general manager, Hudson Bay Port Co. (owned by OmniTRAX), June 30, 2007. Apparently there is a powerful lobby for keeping Canada's grain running east-west on its longer, nationalized rail link to Thunder Bay, rather than on the shorter, privately held north-south line to Churchill.

373 Permafrost is also commonly studded with massive lenses of ice, which occupy less volume and may drain away entirely if it melts. This sets the stage for some highly irregular ground settling if the permafrost starts to thaw. Trees lean drunkenly and fall over. Oddly shaped sinkholes called "thermokarst" appear and fill with water, and other odd phenomena.

374 Borehole temperatures in permafrost are generally warming everywhere around the northern latitudes, but to varying degrees as a function of depth and location. In Alaska it has warmed as much as +3°C since the 1980s, but a more typical range is 0.5°–2°C. For a summary of observed permafrost temperature changes, see Table 6.8 and associated discussion on pp. 210–213, *ACIA* (2005).

375 The presence of permafrost helps to hold water near the land surface. L. C. Smith, Y. Sheng, G. M. MacDonald, L. D. Hinzman, "Disappearing Arctic Lakes," *Science* 308 (2005): 1429.

376 North of 45° N latitude, the single most important determinant of northern lake abundance is glaciation history, followed by the presence or absence of permafrost. On average, glaciated landscapes contain about four times as many lakes as nonglaciated landscapes; permafrost roughly doubles lake numbers. From GIS analysis of northern hemisphere lake distribution, I estimate that in a "permafrost-free" world, the number of known, mapped lakes north of 45° N latitude would be reduced from roughly 192,000 to 103,000 (-46%) and their total inundation area reduced from about 560,000 to 325,000 km^2 (-42%). However, that is an extreme scenario. More realistic for 2050 is an overall reduction of known lakes to 155,000 (-15%) and 476,000 km^2 (-15%), respectively. These numbers are underestimates because the true number of Arctic lakes (i.e., unmapped) is in the millions. L. C. Smith, Y. Sheng, G. M. MacDonald, "A First Pan-Arctic Assessment of the Influence of Glaciation, Permafrost, Topography and Peatlands on Northern Lake Distribution," *Permafrost and Periglacial Processes* 18 (2007): 201–208, DOI:10.1002/ppp.581.

377 So-called "continuous" permafrost will decline even more, by 19%–53%. 2050 forecasts from the CGCM2, ECHAM4/OPYC3, GFDL-R30, HadCM3, and CSM climate models, *ACIA* (2005), Table 6.9. Seasonal thaw depth refers to the depth of the active layer at the ground surface, which thaws out in summer and refreezes in winter. Typical active layer depths are ten to one hundred centimeters.

378 The percentage of dangerous buildings in large villages and cities ranges from 22% in Tiksi to 80% in Vorkuta, including 55% in Magadan, 60% in Chita, 35% in Dudinka, 10% in Noril'sk, 50% in Pevek, 50% in Amderma, and 35% in Dikson. On the Baikal-Amur Mainline railroad 10%–16% of the subgrade in permafrost was deformed by permafrost in the early 1990s, rising to 46% by 1998. *ACIA* (2005), pp. 935–936.

379   This map is assembled from several types of data. The permafrost load-bearing capacity model (gray tones) is very new and will comprise the Ph.D. dissertation of D. Streletskiy, University of Delaware. Permafrost is warmed by rising air temperatures and/or deeper winter snowpack (deeper snow insulates the ground). In general, warmer permafrost means lower load-bearing capacity, but other factors like geology, ice content, and thermal properties are also important. These processes have recently been incorporated into Streletskiy's semiempirical model, driven here by NCAR CCSM3 projections of surface temperature and snow depth averaged over fifteen-year periods, 2000–2014 and 2045–2059, assuming an SRES A1B emissions scenario. The map shows the projected changes occurring between those two time intervals. "Severe loss" is strength loss of >50%, "moderate" is 25–50%, and "mild" is under 25%. The hatched markings refer to increased travel cost from reduced winter road suitability, work done at UCLA by my graduate student Scott Stephenson. Winter roads may only be used for transport where climate provides suitable conditions for their construction and use. Winter road suitability is strongly correlated with freezing index, which is a function of temperature. Land area was classified as suitable for winter road use where mean temperature was 0°C or lower and snow depth exceeded 20 cm. Rivers and lakes were classified as suitable if they received at least 23 cm of freeze depth. Suitability losses were cumulated from November to March. Again, NCAR CCSM3 projections of surface temperature were averaged over fifteen-year periods 2000–2014 and 2045–2059 assuming a SRES A1B emissions scenario, with the map showing the projected change in areal extent of suitability occurring between those two time intervals. Note that this map does not require that winter roads are currently being used in these areas, but instead measures the climatic suitability for their potential use.

380   Personal interview with D. Augur, assistant deputy minister, NWT Department of Transportation, Yellowknife, July 9, 2007. On average, permanent roads cost $0.5–$1.0 M/km to build, whereas winter roads average $1,300 M/km.

381   The Tibbitt–Contwoyto is jammed with heavy trucks during its brief operating season. In 2007 it absorbed eleven thousand loaded trips in just seventy-two days. D. Hayley and S. Proskin, "Managing the Safety of Ice Covers Used for Transportation in an Environment of Climate Warming," 4th Canadian Conference on Geohazards, May 20–24, 2008, Québec City, Canada.

382   Geologically speaking, a kimberlite pipe. Diamonds form under extreme pressure deep in the Earth's crust but can sometimes be found in kimberlite pipes, narrow chimneys of igneous rock that can reach the surface. In the NWT kimberlites are often found under lakes because they are softer than the surrounding granitic rocks, thus becoming eroded depressions that fill with water.

383   Personal interview with Tom Hoefer, manager of external and internal affairs, Diavik Diamond Mines, Inc., Yellowknife, NWT, July 9, 2007.

384   Personal interview with Divisional Forester Jeremy Beal, Tolko Industries Ltd., High Level, Alberta, June 4, 2007.

385   Compared with other types of road, properly constructed and used winter roads have surprisingly low impact on the environment, especially over lakes and wetlands. See S. Guyer, B. Keating, "The Impact of Ice Roads and Ice Pads on Tundra Ecosystems," National Petroleum Reserve–Alaska, U.S. Bureau of Land Management, BLM-Alaska Open File Report 98 (April 2005), 57 pp.

386   L. D. Hinzman et al., "Evidence and Implications of Recent Climate Change in Northern Alaska and Other Arctic Regions," *Climatic Change* 72 (2005): 251–298.

387   One of the ways to mitigate the climate-warming effect is to deploy sweepers to clear snow from the planned roadway, reducing its insulating effect on the ground.

388   A $270 million proposal is pending to build a port road from Bathurst Inlet, which would help the diamond mines to offset decline of the Tibbitt–Contwoyto ice road as

well as enabling other mining activity in the area. G. Quenneville, "Bathurst Inlet Project Reconsidered," Northern News Services, June 15, 2009.

389 Obviously, in terms of sheer numbers, most of the U.S. increase will be in southern states. However, the United States as a whole is still a NORC country and the +15 million figure for its northern states is probably conservative. Alaska today has fewer than a million people, for example, but is one of the fastest-growing U.S. states, projected to grow nearly 40% by 2030. In contrast, New York is projected to grow less than 3%. U.S. Census Bureau, Population Division, Interim State Population Projections, 2005, www.census.gov/population/www/projections/projectionsagesex.html. Table data are from United Nations Population Division: The 2008 Revision Population Database (medium variant), http://esa.un.org/u/npp (accessed July 26, 2009).

390 This calculation is from GIS analysis for land area of the northern quarter of the planet, i.e., between 45° and 90° N latitude. About twenty-one million square kilometers is underlain by some form of permafrost, and eighteen million were glaciated in the last ice age, leaving a smoothed landscape (except in mountain belts) that is relatively easy to get around on. Adding in all the coastal and low-lying areas (here assumed simply as land elevations three hundred meters a.s.l. or less), because they are warmer and more accessible than high-elevation terrain, yields about twenty-seven million square kilometers, of which thirteen million is currently in some stage of permafrost. Subtracting the permafrost areas leaves roughly fourteen million square kilometers of ostensibly livable land.

391 Unlike North America and northern Europe, Eurasia was not extensively ice covered during the last ice age. Most of modern-day Russia has been occupied by humans for at least the past forty to forty-five thousand years and perhaps longer. Even in the high Arctic, new archaeological discoveries at Mamontovaya Kurya and the Yana River indicate human activities thirty to forty thousand years old. See Pavel Pavlov et al., *Nature* 413 (September 6, 2001): 64–67, and Richard Stone, *Science* 303 (January 2, 2004): 33.

392 They are loaded with ancient gene haplogroup U, especially U5B1B1, the so-called "Sámi motif," dating back fifty-five thousand years to the Iberian Peninsula, from where they migrated north at the end of the ice age. T. Lappalainen et al., "Migration Waves to the Baltic Sea Region," *Annals of Human Genetics* 72 (2008): 337–348.

393 Country-averaged population densities for Canada, China, and India are 3,141, and 369 persons per square kilometer, respectively, equivalent to 82.4, 1.75, and 0.67 acres of land per person.

394 This has to do with the generally clockwise rotation of gyres in the northern hemisphere oceans, transporting southern ocean water north along the western edges and northern ocean water south along the eastern edges of the Atlantic and Pacific Basins. Thermohaline ocean circulation is also vitally important, as we shall see shortly. Finally, prevailing wind directions are westerly for much of the northern hemisphere meaning advection of warm ocean air over the land moves generally from west to east rather than east to west.

395 In the northern hemisphere. Ibid.

396 For more on how physical geography can influence human settlement, see Harm de Blij, *The Power of Place: Geography, Destiny, and Globalization's Rough Landscape*, (USA: Oxford University Press, (2008), 304 pp.

397 This was done under the U.S. Lend-Lease program to supply massive amounts of military material to its allies during the war. P. 42, K. S. Coates, W. R. Morrison, *The Alaska Highway in World War II* (Norman and London: University of Oklahoma Press, 1992), 309 pp.

398 All told, the United States poured at least $4 billion (in 2009 dollars) into the projects. U.S. expenditures from 1942 through 1945 were roughly $41 million for airfields, $20 million for the initial temporary highway, $133–$144 million for the Canol Road and pipeline, $131 million for the finished highway; no data for the Haines Road. K. S. Coates, W. R. Morrison, *The Alaska Highway in World War II* (Norman and London: University of Oklahoma Press, 1992), 309 pp.

399 When a Japanese invasion became unlikely, the U.S. soldiers and contractors were recalled from northwestern Canada and the newly built infrastructure soon turned over as promised. Other northern bases were retained for decades, including a large military presence at Keflavík, not turned over to Iceland until 2006. Sondre Stromfjord (now Kangerlussuaq) was turned over to Greenland in 1992. Thule Air Base is still operated by the United States.

400 A. Applebaum, *GULAG: A History* (London: Penguin Books 2003), 610 pp. Highly recommended.

401 The acronym *GULAG* or *Gulag* comes from *Glavnoe upravlenie legerei*, meaning Main Camp Administration. Work camps had long antecedents in tsarist Russia and were implemented by Lenin almost immediately after the Russian Revolution. But Stalin's expansion of the camp system in 1929 took it to a new level of scale and economic significance. For more, see A. I. Solzhenitsyn, *The Gulag Archipelago 1918–1956* (New York: Harper Collins, 1974), 660 pp., and A. Applebaum, *GULAG: A History* (London: Penguin Books, 2003), 610 pp. See also F. Hill and C. Gaddy, *The Siberian Curse* (Washington, D.C.: Brookings Institution Press, 2003).

402 F. Hill and C. Gaddy, Ibid.

403 Ph.D. dissertation of T. Mikhailova, "Essays on Russian Economic Geography: Measuring Spatial Inefficiency," Pennsylvania State University, Department of Economics, 2004. See also F. Hill and C. Gaddy, Ibid.

404 Geological evolution and other material for this section drawn from June 5, 2009, personal interview with John D. Grace of Earth Science Associates, Long Beach, California, and his superb book *Russian Oil Supply: Performance and Prospects* (New York: Oxford University Press, 2005), 288 pp.

405 A primary reason for this is economic "discounting" of up-front capital, in which money is valued higher today than tomorrow. The anticipated future profits for a proposed project are weighed against the alternative profits that could be generated by placing the project's up-front cost into some other interest-bearing investment today. If the second number is larger, it makes no financial sense to proceed. Massive projects with longtime horizons to profitability, like building a freeway system or developing West Siberia, are thus unattractive to private capital. The key parameter in these calculations is the "discount rate," i.e., the interest rate. The steeper the discount rate (the higher the interest rate offered by alternative investments), the sooner a project must be completed to make sense. Economic discounting is extremely important in energy development: Whether a proposed oil or gas field will take five years or seven before production can make the difference between its making economic sense or not.

406 I led a three-year National Science Foundation project to study peatland carbon dynamics in the West Siberian Lowland from 1998 to 2000. Its purpose was to drill cores across the region and involved dozens of Russian and American scientists and graduate students, including Olga Borisova, Konstantine Kremenetski, and Andrei Velichko at the Russian Academy of Sciences and David Beilman, Karen Frey, Glen MacDonald, and Yongwei Sheng at UCLA. For publications and results, see http://lena.sscnet.ucla.edu.

407 The Federal Security Service of the Russian Federation (FSB) is the successor to the Soviet KGB and Russia's main domestic security agency. Upon arrival, foreign visitors to West Siberian cities must register/interview with local FSB officers and surrender passports at hotels. Some towns are completely closed to foreigners.

408 Including a CAD$1.2 billion bid for the rights to explore an offshore area of 611,000 hectares, p. 77, *AMSA 2009.*

409 "Circum-Arctic Resource Appraisal: Estimates of Undiscovered Oil and Gas North of the Arctic Circle," digital data and USGS Fact Sheet 2008-3049 (2008); D. L. Gautier et al., "Assessment of Undiscovered Oil and Gas in the Arctic," *Science* 324 (2009): 1175–1179.

410 More specifically, the other promising geological provinces for oil are the Canning-Mackenzie (6.4 BBO), North Barents Basin (5.3 BBO), Yenisei-Khatanga (5.3 BBO), Northwest Greenland Rifted Margin (4.9 BBO), the South Danmarkshavn Basin (4.4 BBO), and the North Danmarkshavn Salt Basin (3.3 BBO). Other promising geological provinces for natural gas are South Barents Basin (184 TCF), North Barents Basin (117 TCF), and again the Alaska Platform (122 TCF). P. 1178, D. L. Gautier et al., Ibid.

411 Interview with Alexei Varlomov, deputy minister for natural resources of the Russian Federation, Tromsø, January 22, 2007.

412 In 2008 Russia produced 602.7 billion cubic meters of natural gas and had 43.3 trillion more in proved reserves, both greater than any other country. Russia produced an average of 9,886,000 barrels of oil per day, second only to Saudi Arabia (10,846,000 barrels per day). *BP Statistical Review of World Energy June 2009*, available at www.bp.com/statisticalreview.

413 See Chapter 3.

414 J. D. Grace, *Russian Oil Supply: Performance and Prospects* (New York: Oxford University Press, 2005), 288 pp.

415 At peak production West Siberia's giant Urengoi, Yambur, and Medvezhye gas fields produced almost 500 billion cubic meters of natural gas per year; by 2030 production will decline to 130 billion cubic meters per year. E. N. Andreyeva, V. A. Kryukov, "The Russian Model—Merging Profit and Sustainability," pp. 240–287 in A. Mikkelsen and O. Lenghelle, eds., *Arctic Oil and Gas* (New York: Routledge, 2008), 390 pp.

416 Gazprom commenced laying pipeline across the floor of Baydaratskaya Bay in 2009, hoping to open the Bovanenkovo gas field for European markets by 2011. July 24, 2009, "Yamal Pipeline Laying Proceeds," www.barentsobserver.com.

417 Some producers skip the upgrading step to produce lower-grade bitumen. The described process is used by Syncrude, Canada's largest tar sands producer. B. M. Testa, "Tar on Tap," *Mechanical Engineering* (December 2008): 30–34.

418 In 2008 a flock of about five hundred mallard ducks died after landing in a Syncrude tailing pond. "Hundreds of Ducks Die after Landing in Oil Sands in Canada," Fox News, May 8, 2008. See also E. A. Johnson, K. Miyanishi, "Creating New Landscapes and Ecosystems: The Alberta Oil Sands," *Annals, New York Academy of Sciences* 1134 (2008): 120–145; and M. J. Pasqualetti, "The Alberta Oil Sands from Both Sides of the Border," *The Geographical Review* 99, no. 20 (2009): 248–267.

419 T. M. Pavelsky, L. C. Smith, "Remote Sensing of Hydrologic Recharge in the Peace-Athabasca Delta, Canada," *Geophysical Research Letters* 35 (2008):L08403, DOI:10.1029/2008GL033268.

420 Oil sands operators self-report that a total of 65 square kilometers have been reclaimed in some form, or about 12% of the total disturbed area. According to the nonprofit Pembina Institute, only 1 square kilometer has been fully restored and certified by the government of Alberta. Regardless of this discrepancy both numbers are small compared with the 530 square kilometers disturbed.

421 E. A. Johnson, K. Miyanishi, "Creating New Landscapes and Ecosystems: The Alberta Oil Sands," *Annals, New York Academy of Sciences* 1134 (2008): 120–145.

422 The Mackenzie Gas Project has been proposed since the early 1970s but was previously suspended pending settlement of aboriginal land claims. This obstacle is now settled and the project pending as is described in Chapter 8.

423 Under the Kyoto Protocol, Canada pledged to reduce carbon emissions to -6% below 1990 levels by 2008–2012. Instead by 2009 her emissions grew +27% and will rise again in 2010 if Alberta tar sands development intensifies. "Canada's northern goal," in *The World in 2010*, special supplement to *The Economist* (2009): 53–54. Syncrude and Suncor, two of the largest tar sands operators, are the third- and sixth-largest emitters of greenhouse gases

in Canada. M. J. Pasqualetti, "The Alberta Oil Sands from Both Sides of the Border," *The Geographical Review* 99, no. 20 (2009): 248–267.

424   The most promising current underground extraction technology is steam-assisted gravity drainage, in which pressurized steam is forced underground in long horizontal injection wells to heat the bitumen. After about six months of heating the bitumen begins to flow and can be pumped from a second, parallel recovery well to the surface.

425   From Alberta Energy, the total area leased for in situ (underground) development as of May 19, 2009, is 79,298 square kilometers. J. Grant, S. Dyer, D. Woynillowicz, "Clearing the Air on Oil Sands Myths" (Drayton Valley, Alberta: The Pembina Institute, June 2009), 32 pp., www.pembina.org. Future projections from B. Söderbergh et al., "A Crash Programme Scenario for the Canadian Oil Sands Industry," *Energy Policy* 35, no. 3 (2007): 1931–1947. As of 2009, oil production from Alaska's North Slope averaged about seven hundred thousand barrels per day.

426   Government of Canada, Policy Research Initiative, "The Emergence of Cross-Border Regions between Canada and the United States, Final Report" (November 2008), 78 pp., www.policyresearch.gc.ca. See also D. K. Alper, "The Idea of Cascadia: Emergent Regionalisms in the Pacific Northwest–Western Canada," *Journal of Borderland Studies* 11, no. 2 (1996): 1–22; S. E. Clarke, "Regional and Transnational Discourse: The Politics of Ideas and Economic Development in Cascadia," *International Journal of Economic Development* 2, no. 3 (2000): 360–378; H. Nicol, "Resiliency or Change? The Contemporary Canada–U.S. Border," *Geopolitics* 10 (2005): 767–790; V. Conrad, H. N. Nicol, *Beyond Walls: Re-inventing the Canada-United States Borderlands* (Aldershot, Hampshire, and Burlington, Vt.: Ashgate, 2008), 360 pp.

427   See www.atlantica.org.

428   This discovery of common sociocultural values within cross-border superregions is based on survey data, Government of Canada, Policy Research Initiative, "The Emergence of Cross-Border Regions between Canada and the United States," Final Report (November 2008), 78 pp, www.policyresearch.gc.ca.

429   The U.S. State Department recently quelled any hint of a U.S. claim to a half-dozen islands off Russia's Arctic coast, even though Americans were involved with the discovery and exploration of some of them. "Status of Wrangel and Other Arctic Islands," U.S. Department of State, Bureau of European and Eurasian Affairs, Washington, D.C., May 20, 2003. While Canadian politicians like to fret about protecting Canada's vast northern territories from the United States and Russia, there is little evidence that either country has designs on them. Indeed, the United States provides tacit military backing of Canadian sovereignty there. For more on the relative success of U.S.-Canada relations, see K. S. Coates et al., *Arctic Front: Defending Canada in the Far North* (Toronto: Thomas Allen Publishers, 2008), 261 pp. However, while the likelihood of conflict between Arctic nation-states is low, there is ongoing domestic tension from aboriginal groups over land title, as is discussed in Chapter 8.

430   Another area of increasing cross-border economic ties is between Russia and the U.S., with Chukotka Autonomous Okrug in the Russian Far East increasingly importing fuel and other supplies from Alaska. J. Newell, *The Russian Far East* (Simi Valley, Calif.: Daniel & Daniel Publishers, Inc., 2004), 466 pp.

431   This table was constructed using data from the following sources: 2009 Index of Economic Freedom, Heritage Foundation and Wall Street Journal (179 countries, www.heritage.org); 2008 Economic Freedom of the World Index (141 countries, http://www.freetheworld .com/2008/EconomicFreedomoftheWorld2008.pdf); 2009 KOF Index of Globalization (208 countries, http://globalization.kof.ethz.ch/); 2009 Global Peace Index (144 countries, http://www.visionofhumanity.org/gpi/results/rankings.php); 2008 Economist Intelligence Unit Democracy Index (167 countries, http://graphics.eiu.com/PDF/Democracy%20 Index%202008.pdf); 2009 Freedom in the World Country Rankings (193 countries,

http://www.freedomhouse.org). To allow comparison between these indices, numeric index data were converted to percentile country rank. Taking an average of these percentile rankings provides the composite score in the right-most column of the table.

432 Each index has its own agenda, which is why I prefer to look at all of them. Jeffrey Sachs, for example, questions the contention in *Index of Economic Freedom* that trade liberalization necessarily leads to GDP growth, citing examples, like China, which have very strong economic growth despite low scores on the index. J. Sachs, *The End of Poverty: Economic Possibilities for Our Time* (New York: Penguin Group, 2005), 416 pp.

433 Most oil and gas outfits operating in the northern high latitudes are private multinational companies, except in the Russian Federation, where the industry is increasingly returning to state control.

434 The 2010 Economist Intelligence Unit assessed 140 countries in their global livability index. The four NORC cities making the top ten were Vancouver, Toronto, Calgary, and Helsinki; the others were Vienna, Melbourne, Sydney, Perth, Adelaide, and Auckland. The world's lowest-ranked cities were Dakar, Colombo, Kathmandu, Douala, Karachi, Lagos, Port Moresby, Algiers, Dhaka, and Harare. EIU Press Release, "Winter Olympics Host, Vancouver, Ranked World's Most Liveable City," February 10, 2010, http://www.eiuresources.com/mediadir/default.asp?PR=2010021001 (accessed February 16, 2010).

435 Indeed, without immigration the populations and labor forces of most European countries will shrink. Germany, for example, now has a total fertility rate of just 1.3 and is in population decline. Western Europe has a total fertility rate of 1.6, which, combined with a growing elderly population, suggests that the European Union must admit 1.1 million immigrants per year just to maintain its current labor force. P. 129, K. B. Newbold, *Six Billion Plus: World Population in the 21st Century* (Lanham, Md.: Rowman & Littlefield Publishers, Inc., 2007), 245 pp.

436 As of 2009 Russia's total fertility rate was just 1.4 births per woman; the replacement rate is 2.1. Russia's crude death rate was 16.2 per 1,000 people versus a crude birth rate of 10 per 1,000 people. *The Economist, Pocket World in Figures* (London: Profile Books, 2009), 256 pp.

437 I. Saveliev, "The Transition from Immigration Restriction to the Importation of Labor: Recent Migration Patterns and Chinese Migrants in Russia," *Forum of International Development Studies* 35 (2007): 21–35.

438 G. Kozhevnikova, "Radical Nationalism in Russia in 2008, and Efforts to Counteract It," *Sova Center Reports and Analyses* (April 15, 2009), http://xeno.sova-center.ru/.

439 More precisely, in 2008 the United States granted 1,107,126 people legal permanent resident status, and 1,046,539 were naturalized. There were 175 million visitors, of whom 90% were short-term, e.g., tourists and business travelers, and 10% (3.7 million) were longer-term temporary residents like specialty workers, students, and nurses. Between 2005 and 2008 U.S. border apprehensions ranged from 723,840 to 1,189,031 people per year. Drawn from the following reports by the U.S. Department of Homeland Security, Office of Immigration Statistics: R. Monger, N. Rytina, "U.S. Legal Permanent Residents: 2008," Annual Flow Report, March 2009; J. Lee, N. Rytina, "Naturalizations in the United States: 2008," Annual Flow Report, March 2009; R. Monger, M. Barr, "Nonimmigrant Admissions to the United States: 2008," Annual Flow Report, April 2009; N. Rytina, J. Simanski, "Apprehensions by the U.S. Border Patrol: 2005–2008," Fact Sheet, June 2009; J. Napolitano et al., *2008 Yearbook of Immigration Statistics*, August 2009.

440 Canada admitted 247,243 legal permanent residents in 2008, of whom 149,072 were in the "Economic Class" (skilled workers), 65,567 were in the "Family Class" (reunification), and 32,602 were "Refugees" or "Other" classes. "Facts and Figures 2008—Immigration Overview: Permanent and Temporary residents," Citizenship and Immigration Canada Web site, www.cic.gc.ca/english/resources/statistics/facts2008/index.asp (accessed August 22, 2009).

441   See pp. 121–128, K. B. Newbold, *Six Billion Plus: World Population in the 21st Century* (Lanham, Md.: Rowman & Littlefield Publishers, Inc., 2007), 245 pp.

442   Through their memberships in the European Free Trade Association (EFTA) and the Schengen Agreement, Iceland and Norway have essentially opened their labor markets to the EU.

443   As of 2005 the percent foreign born in the United States and Germany was 12.3% and 12.5%, respectively. Canada had the most with 19.3%. Data from Table 1, J.-C. Dumont, G. Lemaître, "Counting Immigrants and Expatriates in OECD Countries: A New Perspective," *OECD Social, Employment and Migration Working Papers,* no. 25 (2005), 41 pp. See http://www.oecd.org/dataoecd/34/59/35043046.pdf.

444   Unusual warm spells in winter cause snow to partly melt, then refreeze, encasing the snowpack in ice. Starvation can result for herbivores unable to break through. Rain-on-snow events are particularly deadly; in October 2003 a particularly severe rainstorm killed approximately twenty thousand musk oxen, one-fourth of the herd, in Banks Island, Canada. J. Putkonen et al., "Rain on Snow: Little Understood Killer in the North," *Eos, Transactions, American Geophysical Union* 90, no. 26 (2009): 221–222.

445   In 2007–08 crude birth rates in Nunavut averaged 25.2 per 1,000 versus 11.1 for all of Canada and 10.6 for Ontario. Total fertility rate (TFR) averaged 2.84 children per woman versus 1.59 TFR for all of Canada. Median age was 23.1 years in Nunavut versus 39.5 years for Canada. Source: Statistics Canada, www.40.statcan.gc.ca/l01/cst01/demo04b-eng .htm and www.statcan.gc.ca/pub/84f0210x/2006000/5201672-eng.htm (accessed August 28, 2009).

446   Personal interview with Iqaluit mayor Elisapee Sheutiapik, on the CCGS *Amundsen* ice-breaker, August 5, 2007. For a strategic plan of Iqaluit's deepwater port ambitions, see www.city.iqaluit.nu.ca/i18n/english/pdf/portproject.pdf.

447   Canada is comprised of provinces and territories. There are currently three territories: the Northwest Territories (NWT), Yukon, and Nunavut. Territories are politically autonomous but less powerful than provinces, which are constitutionally enshrined.

448   The Russian Federation recognizes almost 200 "nationalities," of which 130 (~20 million people, or 14% of Russia's population) are likely aboriginal. However, only 45 groups (~250,000 people) are officially recognized as such ("indigenous numerically small peoples of the north"), or about 0.2% of Russia's total population. See B. Donahoe et al., "Size and Place in the Construction of Indigeneity in the Russian Federation," *Current Anthropology* 49, no. 6 (2008): 993–1009.

449   North American aboriginal population data from the U.S. Census Bureau and Statistics Canada. For the Nordic countries, which do not collect ethnicity data during census, estimates are from *UN World Directory of Minorities and Indigenous Peoples,* available at http://www.minorityrights.org/directory.

450   As of the 2000 U.S. Census the aboriginal population of Alaska was 85,698 out of 550,043 (15.6%): U.S. Census Brief C2KBR/01-15, "The American Indian and Alaska Native Population: 2000," February 2002, http://www.census.gov/prod/2002pubs/c2kbr01-15 .pdf (accessed August 30, 2009). The Sámi population of Sweden averages about 11% (5,900/53,772) across Kiruna, Gällivare, Jokkmokk, and Arvidsjaur municipalities: Minority Rights Group International, *World Directory of Minorities and Indigenous Peoples—Sweden: Sámi,* 2008, http://www.unhcr.org/refworld/docid/49749ca35.html; in Finland about 40% (7,500/18,990) across Utsjoki, Inari, Enontekiö, and Sodankylä: Minority Rights Group International, *World Directory of Minorities and Indigenous Peoples—Finland: Sámi,* 2008, http://www.unhcr.org/refworld/ docid/49749d2319.html; in Norway's Finnmark County about 34% (25,000/73,000): Minority Rights Group International, *World Directory of Minorities and Indigenous Peoples—Norway: Sámi,* 2008, http://www.unhcr.org/refworld/docid/49749cd45.html. Denmark/Greenland and Sakha Yakut data from the *Arctic Human Development Report*

(Akureyri: Stefansson Arctic Institute, 2004), 242 pp. But in the Russian North, aboriginals officially number only about 250,000 and thus comprise just 0.2% of the total population: Government of the Russian Federation, "Yedinyy perechen' korennykh malochislennykh narodov Rossiyskoy Federatsii (Unified list of indigenous numerically small peoples of the Russian Federation)," Confirmed by Decree 255 of the Russian Government, March 24, 2000.

451 American Indians and Alaska Natives, currently numbering 4.9 million, are expected to rise to 8.6 million by 2050. U.S. Census Bureau, Press Release CB08-123, "An Older and More Diverse Nation by Midcentury," August 14, 2008, http://www.census.gov/Press-Release/www/releases/archives/population/012496.html (accessed August 29, 2009). Canada's 2006 census recorded 1,172,790 people as North American Indian (First Nations), Inuit, or Métis (mixed race), versus 976,305 in 2001 and 799,010 in 1996. Statistics Canada, Press Release, Aboriginal Peoples in Canada in 2006: Inuit, Métis and First Nations, 2006 Census, January 15, 2008, http://www.statcan.gc.ca/daily-quotidien/080115/dq080115a-eng.htm (accessed August 30, 2009).

452 Tlingit Nation had even recorded a protest with Russia on this issue, T. Penikett, *Reconciliation: First Nations Treaty Making in British Columbia* (Vancouver: Douglas & McIntyre Ltd., 2006), 303 pp.

453 For a history of the circumstances and politics leading to the landmark ANCSA bill, see W. R. Borneman, *Alaska: Saga of a Bold Land* (New York: HarperCollins Perennial, 2004), 608 pp.

454 After ANCSA the U.S. federal government owned nearly 60% of the land in Alaska, the state 28%, and the regional corporations 12%. All other private lands combined totaled less than 2%.

455 Subsurface mineral rights are retained by the regional corporations, but village corporations can obtain surface rights, e.g., water and timber. Alaska's twelve regional corporations are Ahtna, Inc.; The Aleut Corporation; Arctic Slope Regional Corporation; Bering Straits Native Corporation; Bristol Bay Native Corporation; Calista Corporation; Chugach Alaska Corporation; Cook Inlet Region, Inc.; Doyon Ltd.; Koniag, Inc.; NANA Regional Corporation, Inc.; and Sealaska Corporation. A thirteenth, aptly called The 13th Regional Corporation, received cash only for Alaska aboriginals residing outside the state.

456 These included political organizations by Inuit, Yukon Indians, Métis, Cree, and other groups. F. Abele, "Northern Development: Past, Present and Future," in N. F. Abele et al., eds., *Northern Exposure: Peoples, Powers and Prospects in Canada's North* (Montreal: McGill-Queen's University Press, 2009), 605 pp.

457 Support for the Mackenzie Gas Project is not yet unanimous, as the Dehcho claim isn't done and they currently don't support the pipeline. Also, within the NWT the Akaitcho and Northwest Territories Métis claims are not yet settled.

458 "Imperial Says Earliest Startup Date for Mackenzie Gas Project in 2018," *Oilweek,* March 15, 2010, www.oilweek.com/news.asp?ID=27306 (accessed April 4, 2010).

459 The amount and details of resource royalty returns vary greatly between settlements. In general, ANCSA lands retain all mineral and subsurface rights on granted land, but receive no royalties from surrounding public land. Canadian land claims agreements retain only a portion of subsurface revenues from their actual owned holdings, but also receive royalties for extraction from surrounding public lands, which are also under land claims management. Thus, the geographic reach of the Canadian settlements extends across public as well as aboriginal-owned land, whereas in Alaska it does not.

460 Parts of this discussion drawn from personal interview with land claims attorney John Donihee, Ottawa, June 3, 2009.

461 At least twenty-two comprehensive land claims agreements have entered effect in Canada. Most recent are the Nunavik Inuit Land Claims Agreement and Tsawwassen First Nation Final Agreement Act beginning 2008 and 2009, respectively. Earlier ones are

the James Bay and Northern Québec Agreement (1975), Northeastern Québec Agreement (1978), Inuvialuit Final Agreement (1984), Gwich'in Agreement (1992), Sahtu Dene and Métis Agreement (1994), Nunavut Land Claims Agreement (1995), Nisga'a Final Agreement (2000), Tlicho Agreement (2005), Labrador Inuit Land Claims Agreement (2005), Nunavik Inuit Land Claims Agreement (2008), the Council for Yukon Indians Umbrella Final Agreement (1993), and corresponding self-government agreements: Vuntut Gwich'in First Nation (1995), First Nation of Nacho Nyak Dun (1995), Teslin Tlingit Council (1995), Champagne and Aishihik First Nations (1995), Little Salmon/Carmacks First Nation (1997), Selkirk First Nation (1997), Tr'ondek Hwech'in First Nation (1998), Ta'an Kwach'an Council (2002), Kluane First Nation (2004), Kwanlin Dun First Nation (2005), Carcross/Tagish First Nation (2005). Indian and Northern Affairs Canada, www .ainc-inac.gc.ca/al/ldc/ccl/pubs/gbn/gbn-eng.asp (accessed September 3, 2009).

462 A final wave of LCAs will be in British Columbia, the only British colony in North America that refused to extinguish aboriginal title through treaties. BC tribes are now actively negotiating modern land claims treaties. T. Penikett, *Reconciliation: First Nations Treaty Making in British Columbia* (Vancouver: Douglas & McIntyre Ltd., 2006), 303 pp.; personal interview with former Yukon premier T. Penikett, Ottawa, June 2, 2009. Also, Indian and Northern Affairs Canada is still negotiating claims agreements with the Dehcho, Akaitcho, and Northwest Territories Métis Nation in NWT, plus two Denesuline overlaps in the southernmost NWT and southern Nunavut. Claims are also being negotiated, or are entering negotiations, in Québec, Labrador, the Maritime Provinces, and Eastern Ontario; personal communication with D. Perrin, Indian and Northern Affairs Canada, November 24, 2009.

463 Greenland's highest elected body prior to the introduction of Home Rule in 1979 was the *Landsråd*, roughly translated as "Provincial Council." J. Brøsted and H. V. Gulløv, "Recent Trends and Issues in the Political Development of Greenland," *Actes du XLII Congrés International des Américanistes,* Paris (September 1976): 76–84.

464 Home Rule was introduced on May 1, 1979. In 1982 Greenland voters passed another referendum to withdraw from the European Community. Certain areas, such as foreign affairs and justice, are still managed by Danish authorities, but the Danish government must consult Greenland on all matters relevant to it. The chief connection between the two countries today is economic, as Greenland depends on heavy subsidies from Denmark for solvency. In 2008 Greenland voters overwhelmingly passed another referendum moving Greenland toward full independence from Denmark.

465 As noted in the preceding note, full independence for Greenland, which some speculate could be declared in 2021, the 300th anniversary of Danish colonial rule, will require weaning from generous Danish subsidies averaging $11,000 annually for every Greenlander. The most likely mechanism for this weaning is revenue from oil and gas development, which is being actively encouraged by the Greenland government. So far, thirteen exploration licenses have been issued to companies like ExxonMobil, and another round of licensing will take place in 2010. "Greenland, the New Bonanza," in *The World in 2010*, special supplement to *The Economist* (2009): 54.

466 Canada's Constitution Act of 1982.

467 The Dene of the Northwest Territories and the southern Yukon were signatories of Treaty 8 or Treaty 11, but these treaties were never fully implemented. Personal communication, D. Perrin, Indian and Northern Affairs Canada, November 24, 2009.

468 To make this map, multiple data sources from the Alaska Bureau of Land Management, the Alaska Department of Natural Resources, the U.S. National Atlas, Natural Resources Canada, and Indian and Northern Affairs Canada were combined in a Geographic Information System (GIS) as follows: (1) Alaska land claim data were extracted from the Alaska Bureau of Land Management's Spatial Data Management System. Land claims are represented by Native Patent or Interim Conveyance zones and Native Selected zones, data

accessed from http://sdms.ak.blm.gov/isdms/imf.jsp?site=sdms (2) Alaska Native Claims Settlement Act (ANCSA) Corporation boundaries were downloaded from the Alaska Department of Natural Resources Geospatial Data Extractor. Boundaries were created from the Bureau of Land Management's "Alaska Land Status Map" dated June 1987, data accessed from http://www.asgdc.state.ak.us/. (3) Indian lands of the United States were downloaded from the National Atlas and show areas recognized by the Federal Government as territory in which American Indian tribes have primary governmental authority, administered by the Bureau of Indian Affairs, data accessed from http://nationalatlas.gov/mld/indlanp.html. (4) Indian lands in Canada were downloaded from Natural Resources Canada's GeoBase. These include surrendered lands or a reserve, as defined in the Indian Act, and Sechelt lands, as defined in the Sechelt Indian Band Self-Government Act, data accessed from http://www.geobase.ca/geobase/en/data/admin/alta/description.html (5) Canada land claims were extracted from the 'Comprehensive Land Claims Map' from Indian and Northern Affairs Canada, updated through late 2009 at http://www.ainc-inac.gc.ca/al/ldc/ccl/pubs/gbn/gbn-eng.asp.

469 After the bloody Pontiac Uprising in which nine British forts were captured, King George III issued the Royal Proclamation of 1763, which declared that Indians should "not be molested or disturbed" and only the Crown, not private citizens, was allowed to purchase land from them. To this day it is credited as a first legal acknowledgment of aboriginal land claims in Canada. Also, British Columbia refused to extinguish aboriginal title, as per note 462.

470 A second type of modern agreement, called "Specific Claims," exists in Canada to redress past grievances of aboriginal groups who did sign historic treaties. Many aboriginal groups have pursued, or are pursuing, Specific Claims. However these are typically cash settlements and do not relate to land title.

471 From GIS analysis of aforementioned spatial data I estimate 284,247 km^2 of Indian reservations in the conterminous United States and 4,358,247 km^2 covered by Canadian land claims agreements as of 2009.

472 As a rule, *Lapp* is now considered derogatory and should be avoided in favor of Sámi or Saami.

473 Personal interviews with Aili Keskitalo, president, Norwegian Sámi Parliament (Tromsø, January 23, 2007); Nellie Couroyea, chair/CEO, Inavialuit Regional Corporation and former NWT premier (Tromsø, January 23, 2007); Lars-Emil Johansen minister of foreign affairs and former prime minister (Greenland, May 24, 2007); Mike Spence, mayor of Churchill (Manitoba, June 28, 2007); Elisapee Sheutiapik, mayor of Iqaluit (Nunavut, August 5, 2007); Eli Kavik, mayor of Sanikiluaq (Nunavut, August 7, 2007); Richard Glenn, vice-president, Arctic Slope Regional Corporation (Barrow, Alaska, August 22, 2008); Tony Penikett, former Yukon premier (Ottawa, June 1, 2009); Mary Simon, president, ITK (Inuit Tapiriit Kanatami, Canada's national Inuit organization, Ottawa, June 2, 2009); Ed Schultz, executive director, Council of Yukon First Nations (Ottawa, June 4, 2009).

474 The United Nations Permanent Forum on Indigenous Issues (UNPFII) produced the Declaration on the Rights of Indigenous Peoples, the "most comprehensive statement of the rights of indigenous peoples ever developed, giving prominence to collective rights to a degree unprecedented in international human rights law," adopted by the General Assembly September 13, 2007, http://www.un.org/esa/socdev/unpfii/ (accessed September 6, 2009). All five Nordic countries voted in favor of this declaration. Australia, the United States, and Canada voted against it; Russia was one of eleven countries abstaining.

475 Norway's Finnmark Act of 2005 transferred 96% of Finnmark County's land ownership to a council called the Finnmark Commission, comprised of representatives from the Sámi Parliament as well as the local and central governments. Minority Rights Group International, *World Directory of Minorities and Indigenous Peoples—Norway: Overview, 2007,* http://www.unhcr.org/refworld/docid/4954cdff23.html (accessed September 10, 2009).

476   According to Aili Keskitalo, president, Norwegian Sámi Parliament, personal interview, Tromsø, January 23, 2007.

477   J. Madslien, "Russia's Sami Fight for Their Lives," BBC News, December 21, 2006, http://news.bbc.co.uk/2/hi/business/6171701.stm.

478   M. M. Balzer, "The Tension between Might and Rights: Siberians and Energy Developers in Post-Socialist Binds," *Europe-Asia Studies* 58, no.4 (2006): 567–588. See also A. Reid, *The Shaman's Coat: A Native History of Siberia* (New York: Walker & Company, 2002), 226 pp.

479   However, outright land ownership is a backburner issue in Russia. Most Russians, including aboriginals, view private land ownership as nonessential and even inappropriate. Aboriginal people are more concerned with winning stewardship, protections from competing uses, and the ability to pass use of the land on to their descendants. G. Fondahl and G. Poelzer, "Aboriginal Land Rights in Russia at the Beginning of the Twenty-first Century," *Polar Record* 39, no. 209 (2003): 111–122.

480   A very small aboriginal group called the Yukagir people successfully fought for the adoption of a special law guaranteeing them self-governance in the two townships of Nelemnoe and Andrushkino, where much of their population (1,509 people in 2002) lives. P. 97, *Arctic Human Development Report* (Akureyri, Iceland: Stefansson Arctic Institute, 2004), 242 pp.

481   S. N. Kharyuchi, "Option (sic) letter by the delegates of the VI Congress of indigenous small-numbered peoples of the North, Siberia and the Far East of the Russian Federation" (open letter to President Dmitry Medvedev and Chairman Vladimir Vladimirovich Putin regarding the sale of twenty-year commercial salmon fishing leases in Kamchatka), May 12, 2009, RAIPON, http://www.raipon.org/RAIPON/News/tabid/523/mid/1560/newsid1560/3924/Option-letter-by-the-delegates-of-the-VI-Congress-of-indigenous-small-numbered-peoples-of-the-North-Siberia-and-the-Far-East-of-the-Russian-Federation/Default.aspx (accessed September 15, 2009). See also G. Fondahl, A. Sirina, "Rights and Risks: Evenki Concerns Regarding the Proposed Eastern Siberia–Pacific Ocean Pipeline," *Sibirica* 5, no. 2 (2006): 115–138.

482   On September 7, 1995, Aleksandr Pika and eight others disappeared after setting out from the town of Sireniki, Chukotka, by boat. Five days later the overturned boat and five bodies were found, with Pika's among the unrecovered. Quote is from p. 16, Aleksander Pika, ed., *Neotraditionalism in the Russian North* (Edmonton: Canadian Circumpolar Institute Press, and Seattle: University of Washington Press, 1999), 214 pp.

483   Russian Federal Law 82-F3, April 30, 1999, *O garantiyakh prav korennykh malochislennykh narodov Rossiyskoy Federatsii* ("On guarantees of the rights of the indigenous numerically small peoples of the Russian Federation"); Russian Federal Law 104-F3, July 20, 2000, *Ob obshchikh printsipakh organizatsii obshchin korennykh malochislennykh narodov Severa, Sibiri i Dal'nego Vostoka Rossiyskoy Federatsii* ("On general principles for organization of obshchinas of the indigenous numerically small peoples of the north, Siberia, and the Far East of the Russian Federation"); Russian Federal Law 104-F3, July 20, 2000, *O territoriyakh traditsionnogo prirodopol'-zovaniya korennykh malochislennykh narodov Severa, Sibiri i Dal'nego Vostoka Rossiyskoy Federatsii* ("On territories of traditional nature use of the indigenous numerically small peoples of the north, Siberia, and the Far East of the Russian Federation"). Translations by G. Fondahl and G. Poelzer, "Aboriginal Land Rights in Russia at the Beginning of the Twenty-first Century," *Polar Record* 39, no. 209 (2003): 111–122.

484   P. 50, *Arctic Human Development Report* (Akureyri: Stefansson Arctic Institute, 2004), 242 pp.

485   Unlike other NORC countries, Canada currently has no university in the far north, but there is growing pressure to found one. In general, the fights in North America and Greenland will move on from issues of property title and political governance to other

problems of education, public health, and the devolution of natural resource revenues, which are beyond the scope of this chapter.

486 E.g., conservation of mass and energy, gas laws, radiative transfer and cloud physics, fundamental geography like the positions and elevations of the continents and size and rotation rate of the planet, proper parameterizations for subgrid processes, and aerosols.

487 R. B. Alley, *The Two-Mile Time Machine: Ice Cores, Abrupt Climate Change, and Our Future* (Princeton: Princeton University Press, 2000), 229 pp.

488 K. C. Taylor et al., "The 'Flickering Switch' of Late Pleistocene Climate Change," *Nature* 361 (1993): 432–436, DOI:10.1038/361432a0; R. B. Alley et al., "Abrupt Increase in Greenland Snow Accumulation at the End of the Younger Dryas Event," *Nature* 362 (1993): 527–529, DOI:10.1038/362527a0.

489 B. L. Isacks et al., "Seismology and the New Global Tectonics," *Journal of Geophysical Research* 73, no. 18 (1968): 5855–5899.

490 The project ended up with some interesting results after all, thanks in part to Richard Alley. It found that a mathematical technique called wavelet analysis is useful for detecting hidden climate signals in river flow data. L. C. Smith, D. L. Turcotte, B. L. Isacks, "Streamflow Characterization and Feature Detection Using a Discrete Wavelet Transform," *Hydrological Processes* 12 (1998): 233–249.

491 Greenland Ice Sheet Project 2, drilled between 1989 and 1993 near the center of Greenland.

492 R. B. Alley et al., "Abrupt Increase in Greenland Snow Accumulation at the End of the Younger Dryas Event," *Nature* 362 (1993): 527–529, DOI:10.1038/362527a0.

493 The new CIA climate-change center will assess "the national security impact of phenomena such as desertification, rising sea levels, population shifts, and heightened competition for natural resources." CIA Press Release, "CIA Opens Center on Climate Change and National Security," September 25, 2009, www.cia.gov/news-information/press-releases-statements/center-on-climate-change-and-national-security.html (accessed November 26, 2009). See also J. M. Broder, "Climate Change Seen as Threat to U.S. Security," *The New York Times,* August 8, 2009.

494 M.B. Burke et al., "Warming Increases the Risk of Civil War in Africa," *PNAS* 106, no. 49 (2009): 20670–20674, www.pnas.org/cgi/DOI/10.1073/pnas.0907998106.

495 The most famous and dramatic reversal is the so-called "Younger Dryas" event, an abrupt return to nearly ice-age conditions that began suddenly about 12,700 years ago, then persisted nearly 1,300 years before resumption of warming. Its cause and also the cause of the 8.2 ka event is thought to be a shutdown in ocean thermohaline circulation owing to freshening of the North Atlantic, as will be described shortly. For reviews of the 8.2 ka event, see R. B. Alley, A. M. Ágústsdóttir, "The 8k Event: Cause and Consequences of a Major Holocene Abrupt Climate Change," *Quaternary Science Reviews* 24 (2005): 1123–1149; and E. R. Thomas et al., "The 8.2 ka Event from Greenland Ice Cores," *Quaternary Science Reviews* 26 (2007): 70–81.

496 Peter Schwartz, Doug Randall, "An Abrupt Climate Change Scenario and Its Implications for United States National Security" (October 2003), 22 pp., www.accc.gv.at/pdf/pentagon_climate_change.pdf (accessed September 27, 2009).

497 The flood path for the smaller 8.2 ka event was probably through the Hudson Strait. D. C. Barber et al., "Forcing of the Cold Event of 8,200 Years Ago by Catastrophic Drainage of Laurentide Lakes," *Nature* 400 (July 22, 1999): 344–348, DOI:10.1038/22504. It is also hypothesized that the Younger Dryas event was triggered by a flood draining ancient Lake Agassiz through the St. Lawrence Seaway, or possibly a longer route through the Mackenzie River and Arctic Ocean to the North Atlantic. L. Tarasov, W. R. Peltier, "Arctic Freshwater Forcing of the Younger Dryas Cold Reversal, *Nature* 435 (June 2, 2005): 662–665, DOI:10.1038/nature03617.

498 The story begins with W. S. Broecker, D. M. Peteet, D. Rind, "Does the Ocean-Atmosphere System Have More than One Stable Mode of Operation?" *Nature* 315 (1985): 21–26. A recent development is Z. Liu et al., "Transient Simulation of Last Deglaciation with a New Mechanism for Bølling-Allerød Warming," *Science* 325 (2009): 310–314.

499 A. K. Rennermalm et al., "Relative Sensitivity of the Atlantic Meridional Overturning Circulation to River Discharge into Hudson Bay and the Arctic Ocean," *Journal of Geophysical Research* 112 (2007), G04S48, DOI:10.1029/2006JG000330. The *IPCC AR4* (2007) gave >90% chance the thermohaline conveyor will remain functioning for the next century.

500 Even at the lowest carbon dioxide scenarios, with stabilization at 450 ppm, this critical threshold is eventually crossed in thirty-four out of thirty-five stabilization scenarios. J. M. Gregory et al., "Climatology: Threatened Loss of the Greenland Ice-Sheet," *Nature* 428 (April 8, 2004): 616, DOI:10.1038/428616a.

501 Table 1, G. A. Milne et al., "Identifying the Causes of Sea-Level Change," *Nature Geoscience* 2 (June 14, 2009): 471–478, DOI:10.1038/ngeo544. However, keep in mind the Earth had 70% more ice then than it does today, so a four-meters-per-century sea-level rise is not likely to be repeated.

502 Ibid., 496.

503 Ice sheets help to preserve their own existence by creating an elevated surface at high, cold altitudes and by reflecting back much of the sun's energy. If Greenland's ice sheet were removed, temperatures over its low, dark bedrock surface would be much warmer than today and the ice sheet unlikely to form again.

504 Especially Shanghai, Osaka–Kobe, Lagos, and Manila. Also affected will be Buenos Aires, Chennai, Dhaka, Guangzhou, Istanbul, Jakarta, Karachi, Kolkata, Los Angeles, Mumbai, New York, Rio de Janeiro, Shenzhen, and Tokyo.

505 Geological data suggests the WAIS collapsed 400,000 years ago, and perhaps even 14,500 years ago. P. U. Clark et al., "The Last Glacial Maximum," *Science* 325, no. 5941 (August 7, 2009): 710–714, DOI:10.1126/science.1172873. It is also clear the WAIS is currently losing mass, and there is evidence this has been happening for the past 15,000 years in response to rising sea levels initiated by deglaciation in the northern hemisphere. Thus, even limiting greenhouse warming may not lead to the desired stabilization of the ice sheet. J. Oerlemans, "Freezes, Floes, and the Future, *Nature* 462 (2009): 572–573, DOI:10.1038/462572a.

506 Sea levels are not the same everywhere but vary owing to water pile-up from currents, gravitational attraction, water temperature, crustal rebound, and other factors. The above-average sea-level rise along the U.S. coastline is shown by J. X. Mitrovica et al., "The Sea-Level Fingerprint of West Antarctic Collapse," *Science* 323, no. 5915 (February 6, 2009): 753, DOI:10.1126/science.1166510; and J. L. Bamber et al., "Reassessment of the Potential Sea-Level Rise from a Collapse of the West Antarctic Ice Sheet," *Science* 324, no. 5929 (May 15, 2009): 901–903, DOI:10.1126/science.1169335. The latter study also suggests a global average sea-level increase of 3.2 meters for a WAIS collapse, lower than the five-meter estimate by the IPCC AR4.

507 For more, see D. G. Vaughan, R. Arthern, "Why Is It Hard to Predict the Future of Ice Sheets?" *Science* 315, no. 5818 (2007): 1503–1504, DOI:10.1126/science.1141111; and R. B. Alley et al., "Understanding Glacier Flow in Changing Times," *Science* 322 (2008): 1061–1062.

508 S. A. Zimov et al., "Permafrost and the Global Carbon Budget," *Science* 312, no. 5780 (2006): 1612–1613, DOI:10.1126/science.1128908; E. A. G. Schuur et al., "Vulnerability of Permafrost Carbon to Climate Change: Implications for the Global Carbon Cycle," *Bioscience* 58, no. 8 (2008): 701–714; C. Tarnocai et al., "Soil Organic Carbon Pools in the Northern Circumpolar Permafrost Region," *Global Biogeochemical Cycles* 23, GB2023 (2009), DOI:10.1029/2008GB003327.

509 For more on the challenges surrounding this problem, see S. E. Trumbore, C. I. Czimczik, "An Uncertain Future for Soil Carbon," *Science* 321 (2008): 1455–1456.

510 By drilling cores to the bottom of peatlands and radiocarbon dating their age, we know that northern peatlands started spreading quickly about 11,700 years ago as the Younger Dryas cold period ended. This methane shows up in ice cores of Greenland and Antarctica. L. C. Smith et al., "Siberian Peatlands a Net Carbon Sink and Global Methane Source since the Early Holocene," *Science* 303 (2004): 353–356; and G. M. MacDonald et al., "Rapid Early Development of Circumarctic Peatlands and Atmospheric CH_4 and CO_2 Variations," *Science* 314 (2006): 285–288. Sweden study is E. Dorrepaal et al., "Carbon Respiration from Subsurface Peat Accelerated by Climate Warming in the Subarctic," *Nature* 460 (2009): 616–619, DOI:10.1038/nature08216. The two West Siberia studies are K. E. Frey and L. C. Smith, "Amplified Carbon Release from Vast West Siberian Peatlands by 2100," *Geophysical Research Letters* 32, L09401 (2005), DOI:10.1029/2004GL022025, 2005; and D. W. Beilman et al., "Carbon Accumulation in Peatlands of West Siberia over the Last 2000 Years," *Global Biogeochemical Cycles* 23, GB1012 (2009), DOI:10.1029/2007GB003112. Alaska study is E. A. G. Schuur et al., "The Effect of Permafrost Thaw on Old Carbon Release and Net Carbon Exchange from Tundra," *Nature* 459 (2009): 556–559, DOI:10.1038/nature08031.

511 In other words a large generation of parents born when fertility was still high. Population momentum also works in reverse—for example, elderly countries would continue to shrink even if fertility were increased, owing to a small generation of parents born when fertility was low.

512 As a percentage of GNP, over the period 1880–1913 national investment and national savings were more strongly correlated in the industrialized countries than they were in 1999, meaning that investment today relies more on domestic saving and less on foreign investment than it did in 1913. Pp. 89–90 and Figure 3.3, P. Knox et al., *The Geography of the World Economy,* 4th ed. (New York: Oxford University Press, 2003), 437 pp.

513 Just before World War I broke out, merchandise trade averaged 12% of gross national output for industrialized nations, a level not attained again until the 1970s. P. 32, M. B. Steger, *Globalization: A Very Short Introduction* (New York: Oxford University Press, 2003), 147 pp.

514 *Global Trends 2025: A Transformed World* (Washington, D.C.: U.S. National Intelligence Council, 2008), 99 pp.

515 "Green with Envy: The Tension between Free Trade and Capping Emissions," *The Economist,* November 21, 2009.

516 Nataliya Ryzhova and Grigory Ioffe document hyberbolic assertions ranging from ten to twelve million Chinese already inside Russia to predictions of forty million by the year 2020. Russian migration scholars estimate a current figure of only four hundred thousand Chinese. N. Ryzhova, G. Ioffe, "Trans-border Exchange between Russia and China: The Case of Blagoveshchensk and Heihe," *Eurasian Geography and Economics* 50, no. 3 (2009): 348–364, DOI:10.2747/1539-7216.50.3.348.

517 Ryzhova and Ioffe note thirty-four thousand Chinese labor migrants in Amur Oblast versus an official statistic of just 435. Ibid.

518 B. Lo, "The Long Sunset of Strategic Partnership: Russia's Evolving China Policy," *International Affairs* 80, no. 2 (2004): 295–309. This contested island was finally ceded to China in 1991.

519 W-J Kim, "Cooperation and Conflict among Provinces: The Three Northeastern Provinces of China, the Russian Far East, and Sinuiju, North Korea," *Issues & Studies* 44, no. 3 (September 2008): 205–227. "Development of Trade and Economic Collaboration between China and Primorye Discussed in Vladivostok," http://vladivostoktimes.ru/show/?id=48916&p= (accessed March 11, 2010).

520   In 2004 Turkey signed a deal to send water by supertanker to Israel. The program has since struggled off and on, but Israel has floated the idea of a water pipeline from Turkey. C. Recknagel, "Can 'Wet' Countries Export Water to 'Dry' Ones?" Radio Free Europe, March 21, 2009, www.rferl.org/Content/Can_Wet_Countries_Export_Water_To_Dry_Ones/1514322.html.

521   As of 2009 the eastern route is mostly done, the central route is anticipated for 2014, and the controversial western route through mountains slated for completion in 2050. S. Oster, "China Slows Water Project," *The Wall Street Journal*, December 31, 2008.

522   P. Annin, *The Great Lakes Water Wars* (Washington, D.C.: Island Press, 2006), 303 pp.

523   Québec premier Robert Bourassa and future prime minister John Turner. R. MacGregor, "A Visionary's Epiphany about Water," *The Globe and Mail*, October 5, 2009, www.theglobeandmail.com/news/national/a-visionarys-epiphany-about-water/article1311853/. See also pp. 60–63, P. Annin, *The Great Lakes Water Wars* (Washington, D.C.: Island Press, 2006), 303 pp.

524   Modeling studies suggest that the GRAND Canal project would delay spring ice-out on Hudson Bay as much as a month each year, causing colder, wetter conditions locally during the peak of the growing season, a change in coastal flora, the retreat of forests from the coast, and the growth of permafrost. W. R. Rouse, M-K Woo, J. S. Price, "Damming James Bay: 1. Potential Impacts on Coastal Climate and the Water Balance," *The Canadian Geographer* 36, no. 1 (1992): 2–7.

525   F. Pierre Gingras, "Northern Waters: A Realistic, Sustainable and Profitable Plan to Exploit Quebec's Blue Gold," Montreal Economic Institute, Economic Notes (special edition, July 2009), www.iedm.org/uploaded/pdf/juillet09_en.pdf.

526   P. Micklin, "'Project of the Century': The Siberian Water Transfer Scheme," paper prepared for Engineering Earth; the Impacts of Megaengineering Projects, University of Kentucky, July 21–24, 2008.

527   In 2004. "Luzhkov Wants to Reverse a River," *The Moscow Times*, December 10, 2002; N. N. Mikheyev, "*Voda bez granits* (Water without Limits)," *Melioratsiya i vodnoye khozyaystvo* 1 (2002):32–34; see also F. Pearce, "Russia Reviving Massive River Diversion Plan," *New Scientist*, February 9, 2009, www.newscientist.com/article/dn4637-russia-reviving-massive-river-diversion-plan.html?full=true; and P. Micklin, "The Aral Sea Crisis and Its Future: An Assessment in 2006," *Eurasian Geography and Economics* 47, no. 5 (2006): 546–567, DOI:10.2747/1538-7216.47.5.546.

528   The Ob', Yenisei, and Lena rivers dump significant amounts of freshwater into the Arctic Ocean, much of which freezes into sea ice, then eventually flushes out through Fram Strait or the Canadian Archipelago toward the North Atlantic, where it melts, freshening ocean surface waters and thus impeding deepwater sinking of the thermohaline circulation.

529   The European Space Agency's first CryoSat satellite cost about €140 million but was destroyed in a 2005 launch failure; a follow-up CryoSat-2 was built and launched successfully in April 2010. NASA launched its first ICESat in 2003 and is building two more ice-seeking satellites, ICESat-II and DESDynI, slated for launch around 2015; a total capital investment of USD $2 billion seems likely for these three satellite missions. For more background, see *Earth Science and Applications from Space: National Imperatives for the Next Decade and Beyond*, Committee on Earth Science and Applications from Space: A Community Assessment and Strategy for the Future (Washington, D.C.: National Research Council, 2007), ISBN: 978-0-309-10387-9, 456 pp.

530   This paragraph refers to details presented earlier in the book, including U.S. naval exercises off Alaska's North Slope, Norway's recent purchase of five Aegis-armed frigates and nearly fifty F-35 fighter jets, and Samsung's pursuit of a polar tanker vessel to transport liquefied natural gas from Arctic waters. The total amount received by the U.S. Minerals Management Service from energy companies for an Arctic offshore lease sale totaled USD $2.7 billion in 2008.

531 Arctic Council, *AMSA, Arctic Marine Shipping Assessment 2009*: 77–79.

532 From GIS analysis of global map data I calculate the following in square kilometers for the world's planetary surface area, land extent, ice-free land extent, and ice-free/permafrost-free land extent, respectively. World: 508,779,504 km², 147,263,072 km², 132,801,596 km² and 109,508,640 km², respectively. North of 45° N: 74,697,936 km², 40,364, 452 km², 38,212,960 km², and 17,100,072 km². North of Arctic Circle: 21,239,512 km², 7,930,424 km², 6,159,648 km², and 271,632 km². By all measures, we see the Arctic proper (between 66.55° and 90° N latitude) is truly a tiny place.

533 This particular geographic definition of the "Arctic," proposed in the 2004 Arctic Human Development Report, encompasses all of Alaska; Canada north of 60° N latitude together with northern Québec and Labrador; all of Greenland and the Faroe Islands; Iceland; the northernmost counties of Norway, Sweden, and Finland; and in Russia the Murmansk Oblast, the Nenets, Yamalo-Nenets, Taimyr, and Chukotka autonomous okrugs, Vorkuta City in the Komi Republic, Noril'sk and Igarka in Krasnoyarsky Kray, and parts of the Sakha Republic lying closest to the Arctic Circle. Pp. 17–18, *Arctic Human Development Report* (Akureyri, Iceland: Stefansson Arctic Institute, 2004), 242 pp.

534 USA North, defined as states touching or lying north of 45° N latitude. Alaska, Idaho, Maine, Michigan, Minnesota, Montana, New Hampshire, New York, North Dakota, South Dakota, Vermont, Washington, and Wisconsin all graze the 45th parallel and are contained within a NORC country as per the "North" definition in Chapter 1. Excluding New York State would lower the NORC totals to $5.944 trillion GDP, 31,837,087 km² land area, and 235,059,000 people.

535 The so-called "resource curse" refers to empirical evidence that states with abundant resource wealth perform less well than resource-poor ones, but there is little consensus about why this is. See M. L. Ross, "The Political Economy of the Resource Curse," *World Politics* 51 (1999): 297–322; C. N. Brunnschweiler, E. H. Bulte, "The Resource Curse Revisited and Revised: A Tale of Paradoxes and Red Herrings," *Journal of Environmental Economics and Management* 55, no. 3 (2008): 248–264.

536 The bulk of the Arctic economy is based on commodity exports. Public services comprise 20%–40% GDP, transportation accounts for some 5%–12%, with tourism and retail significant only in particular areas. In 2001 the total Arctic economy was U.S. $230 billion (in purchasing power parity), with *Arctic* defined as all of Alaska (USA); Yukon, NWT, Nunavut, Nunavik, and Labrador (Canada); Greenland and the Faroe Islands (Denmark); Iceland; Nordland, Troms, Finnmark, and Svalbard (Norway); Västerbotten and Norrbotten (Sweden); Oulu and Lapland (Finland); and the republics of Karelia, Komi, and Sakha; the oblasts of Arkhangelsk, Murmansk, Tyumen, Kamchatka, and Magadan; the autonomous okrugs of Nenets, Khanty-Mansii, Yamal-Nenets, Krasnoyarsk Krai, Taimyr, Evenk, Koryak, and Chukchi (Russian Federation). "Public services" includes public administration, health care, and education. "Economic Systems," pp. 59–84 of *Arctic Human Development Report* (Akureyri, Iceland: Stefansson Arctic Institute, 2004), 242 pp.

537 NTCL, founded in 1934 as Northern Waterways Limited, was purchased in 1985 by the Inuvialuit Development Corporation and Nunasi Corporation, making it a 100% private, aboriginal-owned company. For more, see www.ntcl.com/about-us/history-timeline .html.

CREDITS

Frontmatter maps by author.

Page 51. Musical lyrics from "Whoever You Are" by Tommy C. Jordan and Greg Kurstin © 1996 Nudo Music/Warner Bros. Records, Inc., reprinted by permission of Tommy C. Jordan and Hal Leonard Corporation (Whoever You Are, Words and Music by Greg Kurstin and Tommy Jordan, ©2004 EMI BLACKWOOD MUSIC, INC., TUCANO MUSIC, AND NUDO MUSIC, all Rights for TUCANO MUSIC Controlled and Administered by EMI BLACKWOOD MUSIC INC., All Rights Reserved, International Copyright Secured, Used by Permission).

Page 118: Maps by author using model data courtesy of Joseph Alcamo and Martina Flörke, Center for Environmental Systems Research, University of Kassel.

Page 126, 128: Climate model projections reprinted courtesy IPCC AR4 (see endnote 277 for full reference). Climate-change projection maps presented in Chapter Five were modified by permission of the IPCC, Climate Change 2007: The Physical Science Basis. Working Group I Contribution to the Fourth Assessment Report of the Intergovernmental Panel on Climate Change, Figure 10.8, Cambridge University Press. Please note that the modifications made to these maps ("optimistic," "moderate," "pessimistic") are for the purposes of this book only, and are not suggested or used by the IPCC.

Pages 158–159: Maps by author using 2006 shipping data from AMSA, 2009 (see endnote 362).

Page 166: Map by author.

Page 212: Map by author.

Page 250: "Abandonment of the Jeannette" reprinted from *Wonders of the Polar World,* National Publishing Co.: Philadelphia, Chicago, St. Louis, 1885. "The Last Polar Bear" used by permission from Freezingpictures/Dreamstime.com/GetStock.com.

For photo insert (numbers refer to photograph sequence):

1. Photo used by permission from James Martell; 2, 3. Photos by author; 4. Photo used by permission from John Rasmussen, Narsaq Foto; 5. Photo used by permission from Dr. Ivan Frolov, Arctic and Antarctic Research Institute, Saint Petersburg; 6. Photo used by permission from ITAR-TASS News Agency, Russian Federation, 7–11. Photos by author; 12. Photo used by permission from Dr. Vladimir Romanovsky, University of Alaska—Fairbanks; 13. Photo by author; 14. Photo used by permission from *Toronto Star*/GetStock.com; 15. Photo used by permission from Dr. Richard Forster, University of Utah; 16. Photo used by permission from David Dodge, The Pembina Institute (www.oilsandswatch.org); 7. Photo used by permission from Benjamin Jones, Alaska Science Center, U.S. Geological Survey, Anchorage. Backmatter author photo used by permission from Karen Frey, Clark Univertiy.

ACKNOWLEDGMENTS

This project came about thanks to the urging of two highly accomplished colleagues: Judith Carney and William A. V. Clark, of the UCLA Department of Geography. The conversation over a brief coffee with Clark, a hard-boiled statistician not known for mincing words, went like this:

> *Clark:* You need to apply for a Guggenheim and write a book.
> *Smith:* I'm a scientist. We don't write books.
> *Clark:* Nonsense. I know plenty that do. You need to apply for a Guggenheim and write a book.

Four years later the book is done thanks to their advice, a fellowship from the John Simon Guggenheim Foundation, and the support of three other especially encouraging colleagues—John Agnew, Jared Diamond, and David Rigby, also UCLA geographers.

I thank my agent, Russell Weinberger of Brockman, Inc., for taking on a first-time book author with no experience writing for a lay audience. His patience in fielding my many questions was surpassed only by the patience of my editors Stephen Morrow (Dutton, New York) and Duncan Clark (Profile Books, London).

I am indebted to my amazing wife, Abbie Tingstad, who I met researching the book. Aside from her graciousness upon discovering her new husband would embed at his desk for nearly two years, she was the book's foremost critic and sounding board. I thank my parents, Norman and Judith Smith, and brother, Daniel, for being so supportive of me throughout the project.

Maps and illustrations were drawn by the superb cartographer and artist Chase Langford. Vital research assistance was provided by UCLA students Vena Chu, Nora Hazzakzadeh, and Scott Stephenson.

The manuscript was substantially improved thanks to critical expert reviews of one or more chapters by John Agnew, Richard Alley, Doug Alsdorf, Lawson

Brigham, Marshall Burke, Richard Glenn, John Grace, Richard Forster, Dennis Lettenmaier, David Perrin, and Gavin Schmidt. Scott Lefavour and Gary Levy provided helpful feedback on the final chapter. Norman Smith and Abbie Tingstad read and commented on the manuscript in its entirety.

Many people sacrificed time from their busy lives to grant interviews. These include Trevor Amiot, Daniel Augur, David Barber, Jeremy Beal, Kathryn Boivin, Cathie Bolstad, Jason Box, Ron Brower, Guylaine Charbonneau, Nellie Couroyea, Joanne Delaronde, Lloyd Dick, John Donihee, Ken Drinkwater, Kamyar Enshayan, Lyle Fetterly, Patrick Frank, Beth Freeman, Melissa Gibbons, Richard Glenn, Michael Goodyear, John Grace, Robert Grandjambe, Jackie Grebmeier, James Hansen, Udloriak Hanson, David Henry, Tony Hill, Harry Hillaker, Tom Hoefer, Stella Hoksbergen, Robert Huebert, Richard Janowicz, Anne Jensen, Lars-Emil Johansen, Brenda Jones, Eli Kavik, Aili Keskitalo, Andrei Kortunov, Jason Langis, Brian and Susan Lendrum, Diana Liverman, Kim Ma, Lise Marchand, John Marshall, Stephanie Martin, Dan McKenney, Jim McLaughlin, Jobie Meeko, Josee Michaud, Ellen Mosley-Thompson, Kevin Mulligan, Dona Novecosky, Adrian Orr, Pentti and Ritva Peltokangas, Tony Penikett, Dorothy Peteet, Laurie Renauer, Andrew Revkin, John Richardson, Ed Schultz, Glenn Sheehan, Elisapee Sheutiapik, Mary Simon, Duane Smith, Rodney Smith, Guy Smith, Mike Spence, Sara Tabbert, Greg Thessen, Lonnie Thompson, Daniela Tommasini, Wayne Tuck, Paningoak' Vaengtoft, Sophie Vandenbergh, Alexei Varlamov, Aino Viker, Don Wallette, George Wandering Spirit, Sheila Watt-Cloutier, and Robert Zywotko.

Individuals who provided data, readings, photographs, advice, or other forms of assistance include Joseph Alcamo, Kim Barnes, Jason Box, Marsha Branigan, Lawson Brigham, David Dodge, Gebisa Ejita, Kamyar Enshayan, Martina Flörke, Gail Fondahl, Louis Fortier, Cary Fowler, Karen Frey, Ivan Frolov, Harry Gill, Maya Gold, Ken Hinkel, Larry Hinzman, Ben Jones, Tommy Jordan, Sergey Kirpotin, David Lawrence, Glen MacDonald, Ross MacDonald, James Martell, Philip Micklin, Tatiana Mikhailova, Kevin Mulligan, Tom Narins, Heather Nicol, Matthew Nisbitt, Samuel Niza, Trevor Paglen, Martin Pasqualetti, Tamlin Pavelsky, Fred Pearce, Dorothy Peteet, Tom Puleo, John Rasmussen, Åsa Rennermalm, Anthony Repalone, Bruce Robison, Vladimir Romanovsky, Michael Shermer, Nikolay Shiklomanov, C. K. Shum, Dimas Streletskiy, and Sara Wheeler.

Travel for this project was supported in part by the John Simon Guggenheim Foundation, with sabbatical release time granted by the University of California–Los Angeles. Several months of office space were kindly provided by the NASA Goddard Institute for Space Studies in New York, and a Bellagio Residency from the John D. Rockefeller Foundation.

INDEX

Note: Page numbers in *italics* indicate photographs and illustrations. Page numbers followed by the letter "n" indicate note reference.